AF580866

ANALOG DESIGN FOR CMOS VLSI SYSTEMS

ANALOG DESIGN FOR CMOS VLSI SYSTEMS

by

Franco Maloberti
Texas A & M University, U.S.A. and
University of Pavia, Italy

KLUWER ACADEMIC PUBLISHERS
BOSTON / DORDRECHT / LONDON

A C.I.P. Catalogue record for this book is available from the Library of Congress.

ISBN 0-7923-7550-5

Published by Kluwer Academic Publishers,
P.O. Box 17, 3300 AA Dordrecht, The Netherlands.

Sold and distributed in North, Central and South America
by Kluwer Academic Publishers,
101 Philip Drive, Norwell, MA 02061, U.S.A.

In all other countries, sold and distributed
by Kluwer Academic Publishers,
P.O. Box 322, 3300 AH Dordrecht, The Netherlands.

Printed on acid-free paper

Printed in the Netherlands.

CONTENTS

Chapter 2

Resistors, Capacitors, Switches 59

Chapter 3

Basic Building Blocks 99

Chapter 4

Current and Voltage Sources 155

Chapter 5

CMOS Operational Amplifiers 217

Chapter 6

PREFACE

The purpose of this book is to describe the design techniques of analog integrated circuits and to teach the reader how to properly design CMOS operational amplifiers and comparators for mixed analog-digital integrated systems. Analog circuits have become an increasingly critical factor for the systems design. The huge amount of transistors made available by sub-micron technologies allows us to integrate entire systems on chip (SoC). Therefore, analog sections, digital parts, and possibly sensors must use the same technology, the same supply voltage, the same silicon substrate and so on. Moreover, the increased speed of operation and the augmented resolution of digital processors require performing analog functions: such as speed, dynamic range, power supply and noise rejection. Without a deep understanding of the operation and the limits of basic analog circuits, the designer can not properly design the analog processing functions required by modern systems. Thus, this book gives a fairly detailed study of CMOS circuit configurations, learning performances and limits.

This book evolved from a set of lecture notes written in 1986 for an in-house training short course presented in a semiconductor company. Later the material formed the basis of a graduate course on analog CMOS integrated circuit offered since 1998 at the Pavia University, Italy. The initial set of lecture notes progressed in time following the technology evolution. The reference technology migrated from a 2 μm CMOS to the 0.25 μm CMOS used in this book. The circuit techniques moved from design targets like 5 V and tens of MHz bandwidth to the present 1.8 V and hundreds of MHz bandwidth. However, such an amazing change did not modify much of the basic philosophy: to follow a bottom-up approach. The teaching starts from basic physics elements, discusses the features of the MOS transistor, studies the passive components, and considers circuit design of basic blocks, current and voltage reference. In this way the reader acquires all the elements necessary to properly design op-amps and comparators.

The book contains six chapters. Chapter 1 provides those physical, technological and device modelling issues necessary to properly comprehend the behaviour of MOS transistors and their modelling. Chapter 1 starts with a resume of the basic principles of solid state physics and discusses the properties of the basic materials used in microelectronics. This enables the modelling of MOS transistors, both at a simple level, and at a more complex level for computer simulation. Finally, the chapter studies noise performances and discusses layout techniques.

Chapter 2 examines the basic properties of integrated resistors, capacitors and analog switches. The features of integrated components are quite different from the ones of discrete elements. Therefore, it is essential that the reader knows limits and performances well.

Chapter 3 studies simple gain stages, differential pairs, differential to single ended convertors, output stages, etc. The approach conforms the hierarchical view of the teaching pattern: the reader must become familiar with and understand the features and performances of simple cells before studying more complex cells.

The basic building blocks studied in Chapter 3 and other analog functions require, as essential elements for their operation, current generators and voltage biases. Chapter 4 covers the basic architecture of current and voltage sources. This enables the reader to know how to design these "auxiliary" blocks appropriately, to recognize their functional limits, and to estimate costs and benefits for the best design decision.

Chapter 5 deals with operation amplifiers (usually referred to as op-amps). The function and operation of op-amps should be well known to the reader. For this reason, the chapter deals with those circuit implementations that are specifically used in CMOS integrated VLSI systems. Namely, the chapter studies a special category of op-amp: the operational transconductance amplifiers (OTAs). An OTA achieves a large gain exploiting its large output resistance. The reader will learn that when used inside an integrated architecture an op-amp drives capacitive loads. This makes the request of having a low output impedance of little importance.

Chapter 6 studies CMOS comparators. A comparator is together with the op-amp the basic block used in analog signal processors. Ideally, it generates an output logic signal as response to an analog input. Since a real circuit does not achieve the ideal function it is essential to know how the limitations affect the performance of systems where comparators are used.

An important feature of the book comes from a number of examples on computer simulations. Three different hypothetical technologies are envisioned: 0.8 μm, 0.35 μm and 0.25 μm CMOS. The Appendix provides the Spice models for a normal Spice, a Tanner Spice and a Cadence Spice. The reader can repeat the examples and solve the many problems dealing with simulation analysis to acquire experience on an important phase of the design process. However, it is essential to keep in mind that computer simulations must be driven by the designer and not vice versa. The designer must know exactly what the simulation is doing and must predict with sufficient confidence the expected results. The confidence and the capability to assess results will, hopefully, result from the study of this book.

Another characteristic of the book is the use of numbers. The performances of a circuit depends on equations (and the reader must be familiar with them). Often, equations are merely guidelines for the designer. The approximation used can lead to modest accuracy in the results. Therefore, it is important to assess performances with numbers. They can come form the use of design equations and computer simulations. The accordance or the discrepancy of numbers permit the reader to acquire design experience and the set of "rule of thumbs" that characterize an expert designer. To facilitate this process the book includes a number of insets with tip and warnings on what various chapters discuss.

I am grateful to many students that helped (and encouraged) me along the way to accomplish this book. I can mention only a few of them: P. Malcovati, D. Gardino, A, Centuori, G. Cangani, and, more recently, A. Váldez-García, J. Koh, D. Aksin, J. Wang, and P. Estrada Gutiérrez,. Last, but not least, I want to express my deep gratitude to my wife Pina for the immense moral and practical support during the past thirty five years. Without her understanding and help this work would not have been possible.

Franco Maloberti

Chapter 1

THE MOS TRANSISTOR

In this chapter we shall deal with those physical and device modelling issues that constitute the foundations of analogue integrated circuit design. We shall start with a resumé of the basic principles of solid state physics. Following this, we shall discuss the most important properties of the basic materials used in microelectronics. The fabrication steps of a typical CMOS process will also be considered. Subsequently, we shall analyse the modelling of MOS transistors, both at a simplified level, appropriate for hand calculations, and at a more complex level, suitable for computer simulation. In the last part of the chapter, we shall study noise performance and, finally, we shall discuss layout techniques suitable for analogue applications.
The subjects considered in this chapter form the basis for the following chapters; knowledge of these concepts is a key element in any successful analogue design.

1.1 ELECTRICAL CONDUCTION IN SOLIDS

Modern electronics is based on electrical conduction in solids. Electrical conductivity allows us to classify solids into three categories: conductors, semiconductors and insulators. All three classes of materials are used in microelectronics. However, the key component is the semiconductor. The use of metals and insulators to design integrated circuits does not require specific knowledge. Yet, semiconductor devices does require a good grounding in solid-state physics. A first fundamental notion is the band diagram. The band diagram defines the relationship between the energy of electrons and the momentum $p=mv$). For electrons in the free space the E *versus* p relationship is quite simple; it is given by

$$E = \frac{p^2}{2m} \tag{1.1}$$

derived directly from the basic laws of physics. For electrons in solids, however, the E versus k relationship is much more complex. This difficulty arises because of the interaction of electrons with the atoms of the crystalline reticule. Fig. 1.1 shows a typical E versus k diagram (for simplicity's sake we shall consider k as being unidimensional, but it actually should be represented by a vector). The diagram is drawn together with the parabola of a free electron and reveals that only given energy intervals (allowed bands) have a corresponding wave propagation vector. We can see that for large energies, the allowed bands become broad and the forbidden regions become a bit more larger. Moreover, near band limits, the diagram approaches a parabola, as with free electrons. This feature allows us to describe the motion of electrons in a solid as if they were free particles by simply replacing the mass with another quantity called the *effective mass*, m^*. This is determined by the curvature of the E versus p relationship. We may note that near the top of the bands (where the curvature is negative) the effective mass should be negative. Such a situation is not realistic and it is overcome by transferring the negative sign of the

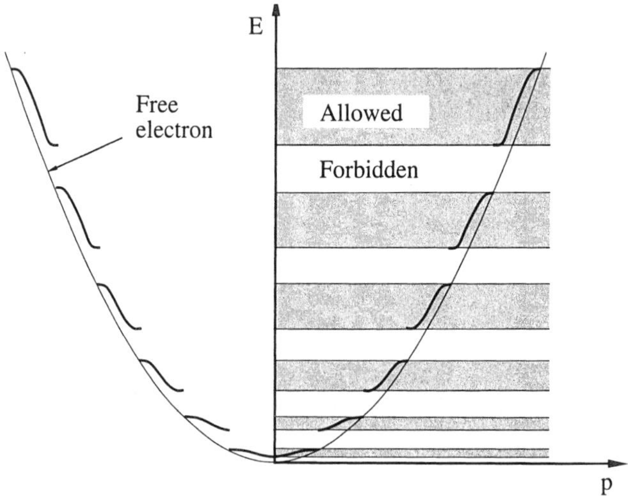

Fig. 1.1 - Simplified E versus k diagram for electrons in solids.

TABLE 1.1

Energy Gap of Solid Materials

Material	Energy gap
Metal	none
Semiconductor	0.5 - 3 eV
Insulator	> 3 eV

mass to the charge of the particle, i.e. electrons with energy near the top of the bands are represented as holes (particles with positive charge). Nevertheless, we can understand the incongruity of negative mass by thinking in terms of all the forces acting over a a particle. Often happens that if we place a solid in a liquid it seems to have a negative mass if its density is lower than the one of the liquid.

In what was derived above, we should consider that as k is a vector, so the effective mass, m^*, is a tensor. Therefore, even in approximate analysis, different directions of motion would require the use of different values of effective mass. For this reason, it turns out that the electrical properties of solids depend on their crystallographic orientation. We shall discuss the consequences of this when we look at matching in active and passive elements.

We have seen that the band diagram defines possible energy levels for electrons in a solid. Under normal conditions, they occupy all the lowest available places and fill the bands up to a given energy level. The last filled (or nearly filled) band is called the *valence band* and the first empty (or almost empty) band is the *conduction band.* We should know that an important parameter characterizing solids is the distance between these two bands called the *energy gap.* Its value allows us to distinguish between metals, semiconductors and insulators. For metals, as shown in Tab. 1.1, the energy gap does not exist, or, rather, the valence band overlaps the conduction band. For semiconductors the energy gap is in the range *0.5-3 eV.* For insulators the energy gap exceeds *3 eV.*

1.2 FERMI-DIRAC STATISTIC

At absolute zero, all electrons freeze at the lowest possible energy level. For an intrinsic semiconductor (a perfectly pure material), the bands up to the valence band are completely filled, while energy levels starting with the conduction band are completely empty. No conduction can take place under this condition: electrons in a completely filled or completely empty band are not

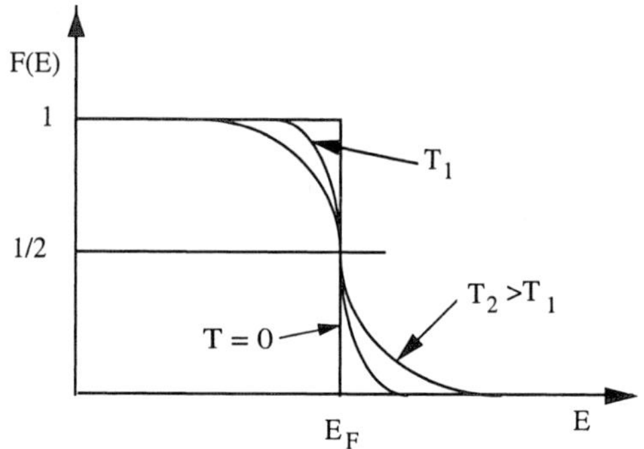

Fig. 1.2 - Fermi-Dirac distribution function.

able to move since no energy level for moving around or no electrons themselves are available. As the temperature rises, energy becomes available in the crystal. At room temperature, because of the relatively small energy gap in semiconductors, some electrons can acquire enough energy to jump from the valence to the conduction band thus causing conduction. This situation is quantitatively specified by the Fermi-Dirac statistic, which gives the probability of occupation *F(E)* of a state level with energy *E* at a given temperature *T.* We have

$$F(E) = \frac{1}{1 + e^{(E - E_F)/kT}} \tag{1.2}$$

where E_F is the Fermi energy, *k* is Boltzman's constant ($8.62 \cdot 10^{-5}$ *eV/K*) and *T* is the absolute temperature.

Fig. 1.2 plots the Fermi-Dirac distribution for three different values of temperature. At absolute zero the Fermi-Dirac distribution looks like a step function with the Fermi energy defining the threshold limit of the occupied states. As the temperature increases, the edges of the function start to round off. However, we have a significant deviation from step behaviour only within a few *kT*'s of E_F. Moreover, we observe that for temperatures above zero, the Fermi level represents that energy level whose occupation probability is *1/2*.

Electrons in the conduction band and holes in the valence band contribute to conduction. Their number can be obtained by using the Fermi-Dirac distribution, together with the state density function *Z(E)*. In a given infinitesimal energy interval *E, E+dE* the number of electrons can be expressed by $N_e(E)dE=F(E)\,Z(E)dE$. Therefore, the number of electrons in the conduction band can be calculated using the integral, extended over the conduction energy band, E_c - E_t

$$n = \int_{E_c}^{E_t} N_e(E)dE = \int_{E_c}^{\infty} N_e(E)dE = \int_{E_c}^{\infty} Z(E)\frac{dE}{e^{(E-E_F)/kT}+1} \qquad (1.3)$$

Since holes mean lack of electrons, their number at a given infinitesimal energy interval is $N_h(E)dE=[1-F(E)]\ Z(E)dE$. Therefore, the number of holes in the valence band can be calculated from the integral, extended over the valence energy interval

$$p = \int_{E_b}^{E_v} N_h(E)dE = \int_{-\infty}^{E_v} N_h(E)dE = \int_{-\infty}^{E_v} \frac{e^{(E-E_F)/kT}}{e^{(E-E_F)/kT}+1}Z(E)dE \qquad (1.4)$$

E_t and E_b are the energies at the top of the conduction band and the bottom of the valence band respectively. In the above integrals (1.3) and (1.4), we approximated the calculation to infinity (positive or negative) because the contributions well above the conduction band or much below the valence band are negligible.

Calculation of equations (1.3) and (1.4) is difficult because of the complex behaviour of the state density function in the valence and conduction bands. However, we observe that in an intrinsic material, the electrons in the conduction band can only come from the ones leaving the valence band. Therefore, we have

$$n = p = n_i \qquad (1.5)$$

where n_i is called the intrinsic concentration. It depends on a number of physical properties, the most important of which are the energy gap and temperature that produces an exponential change in the tail of the Fermi-Dirac distribution. Intrinsic concentration is expressed for silicon $(E_g=1.21\ eV)$ by the empirical relationship.

REMEMBER

Some background knowledge in solid state physics is very important for analog circuit designers. Knowing equations and parameters values help you to better interpret the results of computer simulations.

$$n_i = 3.954 \cdot 10^{16} T^{3/2} e^{-1.21q/2kT} cm^{-3} \qquad (1.6)$$

We observe that the exponential term dominates the temperature dependence. It produces, approximately, a doubling of n_i for each $10\ K$ temperature increase. At room temperature $(T=300\ K)$ n_i equals $1.42 \cdot 10^{10}\ cm^{-3}$

Example 1.1

Using equation (1.6), calculate the intrinsic concentration at 330 K and 400 K. Compare the results obtained with the approximate rule, according to which n_i doubles for each 10 K.

Solution: *At room temperature $V_T = kT/q$ holds 25.86 mV. Rising temperature at 330 K and 400 K, V_T reaches 28.45 mV and 34.48 mV respectively. Using (1.6) we have*

$$n_{i,330K} = 3.954 \cdot 10^{16} 330^{3/2} e^{-\frac{-1\ 21}{2\ 8\ 62\ 10^{-5}\ 330}} = 1.374 \cdot 10^{11} \text{cm}^{-3}$$

$$n_{i,400K} = 3.954 \cdot 10^{16} 400^{3/2} e^{-\frac{-1\ 21}{2\ 8\ 62\ 10^{-5}\ 400}} = 7.583 \cdot 10^{12} \text{cm}^{-3}$$

At 330 K the intrinsic concentration increases by a factor of 9.68 compared to the value at room temperature (a little more than the expected value 8); while at 400 K it increases by only 533 (much less than the expected figure 1024). Therefore, the given rule of doubling of n_i for each 10 K temperature increase must be kept in mind with a proper "warning".

The use of (1.5) together with (1.3) and (1.4) allows us to calculate the position of the Fermi level for an intrinsic material. Again, the result depends on the specific behaviour of *Z(E)*. However, for silicon, the semiconductor most used in microelectronics, we assume that, at a first approximation, the Fermi level is very close to the middle of the energy gap.

MEMORY HINT

The atomic density of solids is around 5 multiplied 10 exponent 22; the intrinsic concentration of electrons at 330 K (57 °C or 134 °F) is 10 exponent 11, half of the exponent of atomic density.

We have seen that the intrinsic concentration of silicon, n_i, at room temperature is extremely small, in fact, it is approximately *13* orders of magnitude smaller than its atomic density. Consequently, conductivity in intrinsic materials (silicon, for our purposes) is rather poor. Only doping can improve conductivity: it is termed *p-type* if we add atoms from the third group of the periodic system; or *n-type* when atoms from the fifth group are used. The doping atoms do not modify the crystallographic structure of silicon: they take substitutional positions (that is, the same places normally occupied by atoms of the basic material in an intrinsic crys-

TABLE 1.2

Activation energy of doping elements in silicon

III Group	Activation Energy	V Group	Activation Energy
B	0.045 eV	P	0.045 eV
Al	0.067 eV	As	0.054 eV
Ga	0.072 eV	Sb	0.039 eV
In	0.160 eV		

tal). Therefore, the band diagram changes only locally. For n-type doping, the energy level required to accommodate an extra-electron is achieved by locating an additional level in the forbidden gap. For p-type doping, the lack of an electron pushes a localised level from the valence band into the forbidden band. The distance of these localised levels from the bottom of the conduction band or from the top of the valence band is very small, in the order of a few tens of milli *eV*. This energy is termed the activation energy. Table 1.2 gives the values for the doping atoms most commonly used in silicon technology.

At zero temperature, the extra electrons contributed by the *n-type* doping freeze into the localised states. For *p-type* doping, the "pushed up states" are empty because no energy is available to ionize them. As the temperature increases, the energy available in the crystal (kT for each degree of freedom) favours the activation of electrons both inside and outside of the localised states (Fig. 1.3). Because of the very low activation energy, at room temperature ($kT=0.026\ eV$) almost all the localised levels are ionised. Moreover, since the doping concentration N_D or N_A is much higher than the intrinsic concentration n_i, we have

$$n \cong N_D \quad \textit{for n-type} \tag{1.7}$$

$$p \cong N_A \quad \textit{for p-type} \tag{1.8}$$

At thermodynamic equilibrium, electron and hole densities are linked by the mass-action law

$$n \cdot p = n_i^2 \tag{1.9}$$

Therefore, after using the equation in (1.7) or (1.8), we may conclude that only one of the two carriers dominates the conduction. This carrier is called the majority carrier, while the other type is termed the minority carrier.

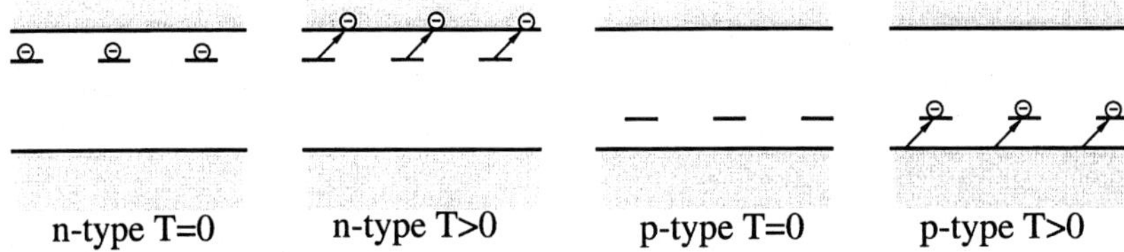

Fig. 1.3 - Electron occupation of energy levels at zero temperature and T>0

The imbalance between electrons and holes in a doped material determines an upward shift of the Fermi level for n-type materials and a downward shift for p-type materials. This shift modifies the probability of level occupancy in conduction (or in the valence band) almost exponentially. Therefore, at a first approximation, the shift of the Fermi level ($E_F=q\Phi_F$), compared to the intrinsic position, can be expressed by

$$\Phi_F = -\frac{kT}{q} ln\frac{n_i}{N_A} \quad for\ p\text{-}doping;\ [V] \tag{1.10}$$

$$\Phi_F = -\frac{kT}{q} ln\frac{N_D}{n_i} \quad for\ n\text{-}doping;\ [V] \tag{1.11}$$

> **NOTE**
>
> The position of the Fermi level inside the energy gap changes with doping. It varies by one $V_T = kT/q$ (*26 mV* at room temperature) for each variation of the doping by a factor *e* (*2.71*).
>
> The doping that brings the Fermi level into the valence or conduction band (degeneration) is more than $10^{20} cm^{-3}$. It is around *0.8%* of the atomic density!

Considering that the energy gap of silicon is *1.21 eV* and that *kT/q*, at room temperature is approximately *26 mV*, the above equation allows us to assert that a doping concentration larger than $4 \cdot 10^{20}$ cm^{-3} (that is e^{24} times the intrinsic one, n_i) will take the Fermi level close to the boundary of the forbidden gap. Such a doping condition, capable of pushing the Fermi level inside the conduction or the valence band, is usually referred to as *degeneration*. Degeneration means that the semiconductor state is no more accomplished the Fermi level being in the conduction band as it happens for metals.

Example 1.2

The doping concentrations on the two sides of an integrated p-n junction are $5 \cdot 10^{15}$ cm^{-3} *and* $8 \cdot 10^{18}$ cm^{-3} *respectively. At room temperature, calculate the Fermi level difference between the p-side and the n-side of the junction.*

Solution: *The position of the Fermi level on the p-side and on the n -side, referred to the middlegap, is calculated by (1.10) and (1.11) respectively*

$$\Delta\Phi_{F,p} = -8.62 \cdot 10^{-5} \cdot 300 \cdot \ln\frac{1.42 \cdot 10^{10}}{5 \cdot 10^{15}} = 0.330\text{V}$$

$$\Delta\Phi_{F,n} = -8.62 \cdot 10^{-5} \cdot 300 \cdot \ln\frac{8 \cdot 10^{18}}{1.42 \cdot 10^{10}} = -0.521\text{V}$$

subtracting the two terms we obtain

$$\Delta\Phi_F = 0.330 + 0.521 = 0.851\text{V}$$

the achieved value is typical for p-n junctions used in CMOS technology

The concepts presented in this section and in the previous paragraph help us to better understand the electrical operation of semiconductor devices; in particular, they allow us to estimate their key property: *conductivity.* This is expressed by the well known equation

$$\sigma = \sigma_n + \sigma_p = q(n\mu_n + p\mu_p) \tag{1.12}$$

where μ_n and μ_p represent the mobility of electrons and the mobility of holes, respectively.

Conductivity depends on the concentration of electrons and holes, that for doped materials are related to their doping concentration, and their respective mobility. The latter parameter depends in turn on the reticular structure of the semiconductor which determines a larger or smaller mean free path. However, mobility itself also depends on temperature and doping concentration. Doping atoms, in fact, change the cross section, so that, by increasing density, they reduce mobility. Therefore, the doping level influences conductivity with a significant extent. Shortly, we shall observe the dependance of electron and hole mobility on doping.

1.3 PROPERTIES OF MATERIALS

The key material used in microelectronics technology is silicon, which is associated with other elements to produce solid state devices or, more generally, integrated circuits. Important compounds used in microelectronics are silicon dioxide, polysilicon and silicon nitride. Packaging and power dissipation issues require several other organic and inorganic components to be used. In this section we shall consider the physical and technological properties of only those compounds which are specifically used in integrated circuit technology. For the other elements, the reader should refer to more specific literature.

1.3.1 Silicon

The substrate on which an integrated circuit is built and the basic material for microelectronics is monocrystalline silicon. Only special applications involve use of a substrate other than silicon (sapphire, for example). In this case a thin layer of silicon is grown on the top of the substrate using the epitaxy technique. If the substrate is a monocrystal with a reticular constant equal to (or very close to) that of silicon we achieve a monocrystal and we use it as a bulk material. Some technologies also use an epitaxial layer of silicon grown on the top of silicon substrate.The doping of the substrate material and that of the epitaxial layer are done differently to optimise performance (speed, latch-up, etc.).

The monocrystalline silicon that we use, whose physical properties are summarised in Table 1.3, is in the form of a thin ($300\text{-}1000\mu m$) slice or *wafer*. The wafers are made by slicing a single crystal ingot whose diameter ranges from *4* to *12* inches. This silicon is extremely pure (*electronic grade*). It contains less than one part per billion of impurities. Its doping (deliberately introduced) depends on the technology to be adopted. For a CMOS p-well, silicon is lightly n-doped, for CMOS n-well or twin wells it is slightly p-doped.

The slices are oriented in the crystallographic orientation defined as <*100*> and <*111*> by the Miller indexes. One or two *flat* grounds identify the different type of doping and crystallographic orientation. A wafer with crystallographic orientation <*111*> which is also p-doped, has only one flat, while <*111*> n-doped, <*100*> p-doped, and <*100*> n-doped wafers have a secondary flat oriented to the first at *45°*, *90°*, or *180°*, respectively.

The mobility value depends on whether conduction takes place on the surface (or rather, on the oxide-semiconductor interface) or in the bulk. A large carrier scattering is produced at the crystal's surface by localised trapping lev-

TABLE 1.3

Physical properties of silicon

Property	Value	Dimension
Atomic density	$5 \cdot 10^{22}$	atoms/cm^3
Density	2.33	g/cm^3
Atomic weight	28.1	g/mole
Reticular constant	0.543	nm
Thermal Conductivity	1.41	Ω/cm °C
Intrinsic resistivity (@ 300 °K)	$2.5 \cdot 10^5$	Ω–cm
Relative dielectric constant, ε_r	11.9	-
Absolute dielectric constant, ε_0	$8.858 \cdot 10^{-14}$	F/cm

els. It brings surface mobility below bulk mobility by a factor ranging from two to three. Carrier mobility also depends on temperature and doping concentration. Fig. 1.4 shows the behaviour of electron and hole mobility on silicon as a function of doping. We observe a decrease in about one order of magnitude when the doping goes from a light (*10^{14}* cm^{-3}) to a strong concentration (*10^{19}* cm^{-3}). This also determines a greater decrease in the resistivity, as shown in Fig. 1.5. We observe that a typical value of resistivity, for medium doped sili-

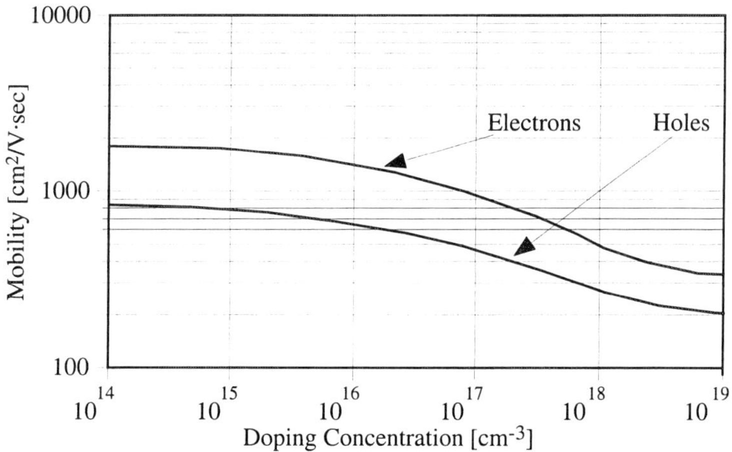

Fig. 1.4 - Mobility of electrons and holes in bulk silicon (T=300°K)

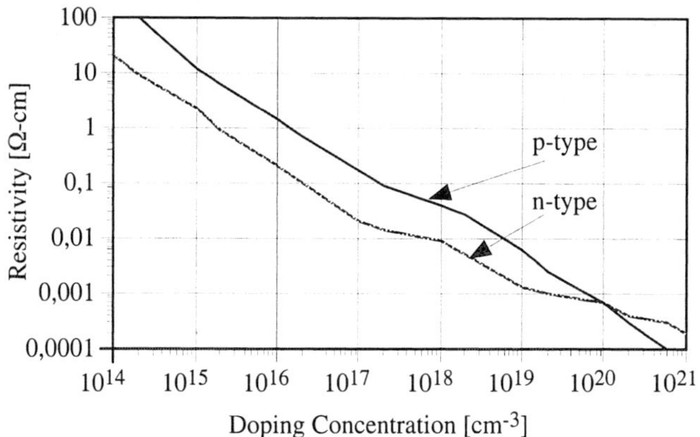

Fig. 1.5 - Resistivity of bulk silicon versus doping concentration.

con, is *1 Ω-cm* while for heavily doped silicon it can be as low as *1 mΩ-cm*.

Up until now we have discussed the properties of homogeneous silicon. However, an integrated circuit never uses homogeneous materials, allowing different structures to be used which ensure a proper change of doping in given regions. Different doping is achieved by selective masking associated with ion implantation and thermal diffusion. These two steps modify the doping into thin layers (from a fraction to a few microns) close to the surface. The doping profile looks like a Gaussian or an error function. This type of doping can enhance or reverse the nature of the substrate doping. In the latter case, the added dopants exceed the doping of the substrate and make it complementary to the substrate. This occurs up to a given limit, where the added dopants equal the substrate concentration. Beyond this, the material is doped as much as the substrate, thus making a *p-n* junction. This specific point is called the metallurgical junction of the *p-n* structure.

Doped layers are used to obtain active devices, namely, source and drain of MOS transistors, and in making resistors as well. For the source and the drain of transistors, the lowest possible specific resistance is desired (to speed up performance). In contrast, if we want to create a resistor, we have to exploit the resistivity features of diffused layers. When using a diffused layer to make

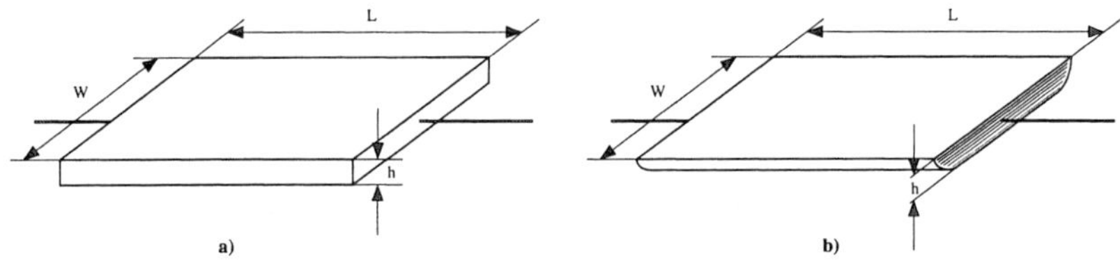

Fig. 1.6 - Diffused resistor: a) uniform resistivity b) graded resistivity.

a resistor, the designer must estimate the resistance value achieved. At a first approximation, we can schematize a diffused resistor with a parallelepiped having uniform resistivity and contacts at two opposite sides (Fig.1.6-a) Its resistance can be expressed by

$$R = \frac{\rho L}{A} = \frac{\rho L}{hW} = R_{\square}\frac{L}{W} \tag{1.13}$$

The quantity $R_{\square}$ is called the specific (or sheet) resistance of the material. It is determined by the resistivity and the thickness of the layer. In all technologies the process steps are carefully optimised. Many trade off determine the resistivity and thickness of all the resistive layers. Therefore, the circuit designer cannot normally modify these parameters but must only define width and length. Going back to (1.13), we note that $R_{\square}$ is the resistance of a parallelepiped of material with a squared top ($W/L = 1$); for this reason, it is named *resistance per square* and is measured in $\Omega/\square$.

In diffused layers, the resistivity is not constant but increases with depth and tends towards infinity at the metallurgical junction. Therefore, for a more realistic calculation, we have to describe the resistor as the parallel connection of infinitesimal elements (having, for simplicity, the same width). The total resistance is achieved by

$$G = \frac{1}{R} = \int_0^h dG = \int_0^h \sigma(z)\frac{W}{L}dz = \frac{1}{R_{\square}}\frac{W}{L} \tag{1.14}$$

Even in this case, therefore, it is possible to define the specific resistance of the layer, $R_{\square}$. For typical CMOS technology, Fig. 1.7 plots the sheet resistance of a layer with two different thicknesses ($1\ \mu m$ and $5\ \mu m$) as a function of the

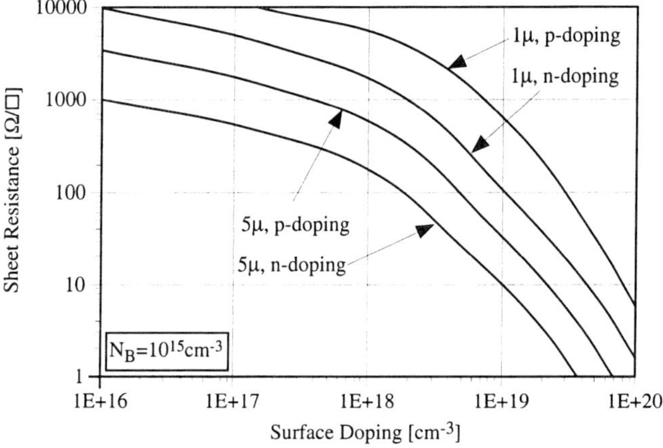

Fig. 1.7 - Sheet resistance of silicon diffused layers

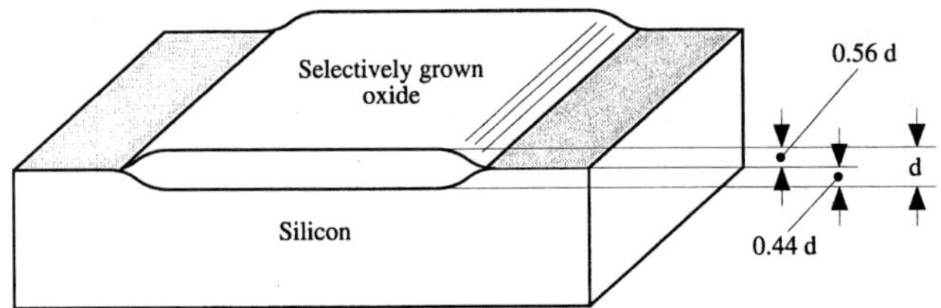

Fig. 1.8 - Selective growth of silicon dioxide

surface doping. The substrate doping is $10^{15}\ cm^{-3}$. Moreover, we assume that starting from a given surface doping, the doping in the depth has a Gaussian profile. Because surface doping ranges from 10^{17} to 10^{20} in microelectronics technologies, typical sheet resistance extends from a few $\Omega/\square$ to several $k\Omega/\square$.

The typical sheet resistance of an *N-well* diffusion is 1-2 $k\Omega/\square$. The sheet resistance of an *N-diff* or a *P-diff* is between $40\ \Omega/\square$ and $80\ \Omega/\square$. These numbers are large compared with those of conductors. Nevertheless, it is worth remembering that even the metal lines used in semiconductor technology have a finite specific resistance. It ranges between *75* and $200\ m\Omega/\square$.

1.3.2 Silicon dioxide

Silicon dioxide is an excellent insulator. Microelectronics technology uses it widely either to make active devices or to ensure insulation. Silicon dioxide is grown by thermal oxidation of silicon or by chemical vapour deposition (*CVD*). Thermal growth is achieved in dry or wet conditions (i.e., in the presence of oxygen or water) at a temperature that ranges from *800°C* to *1100°C*. The oxidation reactions are.

$$Si + O_2 \rightarrow SiO_2 \tag{1.15}$$

$$Si + 2H_2O \rightarrow SiO_2 + 2H_2 \tag{1.16}$$

The growth of silicon dioxide involves the following silicon consumption: if *d* is the thickness of the oxide grown, *0.44 d* is the thickness of the silicon consumed. Therefore, in the case of selective oxidation, the surface of the defined pattern is not flat, as shown in Fig. 1.8. The step formed by the transition from silicon to silicon dioxide can become a serious problem in small line-width technologies. Here, reliability problems can arise because layers that grow on non-flat surfaces can be affected by micro-cracks when their thickness is less than the surface curvature. Because of this problem, modern

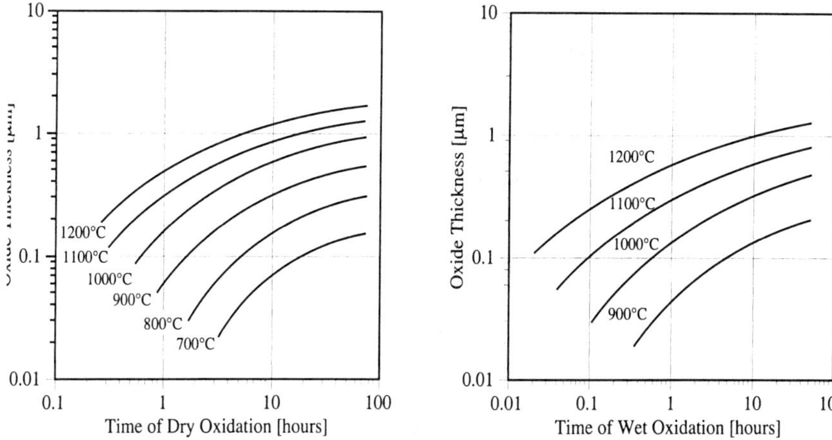

Fig. 1.9 - Oxide thickness as a function of the reaction time in dry (a) and wet (steam) ambient (b) for several oxidation temperatures

technologies widely use planarization techniques.

The growth speed of thermal oxide depends on the temperature of the reactor; for wet oxidation it is about one order of magnitude greater than for dry oxidation. Fig. 1.9 shows the thickness of the growth layer as a function of the oxidation time for different temperatures. The growth speed is not constant, but decreases as oxide thickness increases. This is due to an increasing electrochemical barrier against oxygen that is created by the already grown insulator.

It is also possible to obtain silicon dioxide by putting either silicon or oxygen on top of the growth surface. This process, called chemical vapour deposition (*CVD*), is achieved by pyrolitic decomposition of silane in the presence of oxygen

$$SiH_4 + 2O_2 \rightarrow SiO_2 + 2H_2O \tag{1.17}$$

The temperature ranges from *300°C* to *500°C*. The reaction (1.17) can take place at low pressures (*LP-CVD*) or at atmospheric pressure (*AP-CVD*). The growth speed of *CVD* oxide is higher than that of thermal oxide by about one order of magnitude.

The major advantage of grown *CVD* is that the reaction takes place at a relatively low temperature. This is very important because high temperatures must be avoided after aluminium deposition, otherwise they would lead to eutectic silicon-aluminium that causes permanent degradation of the silicon junctions. In addition, low temperatures avoid further doping diffusions thus preserving the doping profiles.

CVD oxide is widely employed for surface protection. This kind of use not only requires a mechanical defence but a chemical barrier as well. The major reliability problem derives from a deterioration of the aluminium used in pads and interconnections because of electromigration produced by the combined action of current and aggressive ions. Suitable electrochemical barriers can keep away the undesired ions (like the sodium present in moisture). For this reason, the oxide is often doped with boron.

HOW THICK IS THE OXIDE?

The thickness of silicon dioxide used depends on the specific fabrication step.

Keep note of the following typical figures

- For the gate of transistors: 5 - 15nm.
- For poly-metal insulation: 500 nm.
- For field protection: 600 nm

The specific capacitance of a 8 nm thick oxide (ε_r=3.6) is 4 fF/μm^2

Table 1.4 lists the most important physical properties of silicon dioxide. The value of the dielectric constant varies in a relatively large range, depending on the fabrication conditions. An important figure to note is dielectric strength; it spans from *2* to *8 MV* per centimetre, with the lower value pertaining to polysilicon oxide and the higher to monocrystalline silicon oxide. With such dielectric strength it is possible to support from 6 *V* to 24 *V* in a *30 nm* layer (a typical gate oxide thickness used in modern technology).

TABLE 1.4

Physical properties of silicon dioxide

Property	Value	Dimension
Density	2.22	g/cm^3
Dielectric strength	2-8 10^6	V/cm
Resistivity (@ 300 °K)	10^{15}- 10^{17}	Ω cm
Relative dielectric constant	3.4 - 4.2	

1.3.3 Polysilicon

Modern technologies use polysilicon to create MOS transistor gates, to make capacitor plates and for short interconnections. Polysilicon is grown on top of silicon dioxide by pyrolitic decomposition of silane (SiH_4) at about *600°C*. Since the substrate is amorphous, a monocrystalline structure is not achieved. We have instead a conglomerate of monocrystal grains with a size in the range of $0.1\text{-}1\ \mu m$. The thickness of polysilicon layers usually ranges from *200 nm* to *600 nm* with a typical standard deviation of *2%*.

> **NUMBERS TO REMEMBER**
>
> Specific resistance of polysilicon layers:
>
> - high resistive poly: 0.5-2 kΩ/□.
> - low resistive poly: 20-40 Ω/□.
> - Ti silicate 1-5 Ω/□

The mobility of carriers in polysilicon is very poor ($30\text{-}40\ cm^2/V\text{-}sec$). This degradation is due to the grain borders, where the crystallographic structure changes sharply. These borders create energy barriers for electrons and holes which makes crossing them difficult. An acceptable conductivity can be achieved by heavily doping the polysilicon ($10^{20}\text{-}10^{21}\ atoms/cm^3$). A fraction of the dopants atoms clusters close to the grain borders and reduces the localised energy barriers. As a result, the sheet resistance of polysilicon layers becomes around *20-40 Ω/□*. For a number of applications, this value is not small enough: high resistance produces noisy operations and, more notably, high resistive gates limit transistor speed. In demanding situations, the sheet resistance of polysilicon is improved by using sandwiched layers. Typically, a layer of refractory metal silicate (WSi_2, $MoSi_2$, $TiSi_2$) covers a thin film of polysilicon (*200 nm*). This sandwich has a typical sheet resistance of *1 - 5 Ω/□*.

1.3.4 Silicon Nitride

Silicon nitride is another material employed in silicon technology. Its major use is to protect surfaces. In addition, since it is not etched by etchants acting with silicon dioxide, it is also used in intermediate technological steps. Silicon nitride is grown by decomposing dichlorosilane and ammonia at *700-800°C*. The basic reactions are

$$3SiH_4 + 4NH_3 \rightarrow Si_3N_4 + 12H_2 \tag{1.18}$$

$$3SiH_2Cl_2 + 4NH_3 \rightarrow Si_3N_4 + 6HCl + 6H_2 \tag{1.19}$$

Its speed of growth is excellent, ranging from *10* to *20 nm* per minute. Its insulating properties are also good (resistivity is $10^{14}\text{-}10^{16}\ \Omega/cm$) with the relative dielectric constant ranging from *4* to *9*. This figure is better than for silicon dioxide. However, silicon nitride is rarely used to make capacitors because it is normally grown with a rather high thickness. As its dielectric strength is similar to silicon dioxide (*5-10 MV/cm*), even relatively small thicknesses (fractions of a micron) are capable of sustaining the voltages used in modern microelectronics.

1.4 CMOS TECHNOLOGY

MOS technology integrates both n-channel and p-channel transistors on the same chip. If the substrate of the circuit is *p*-doped, the *n*-channel transistors sit directly on the substrate, whereas the *p*-channel devices need a *well.* This is a diffused layer with complementary doping compared to the substrate. The well is also called *tube*, while the technology is termed *n*-well technology. For a *n*-type substrate the arrangement is complementary: the *p*-channel transistors are made in the substrate and the *n*-channel transistors sit inside the *p-well.*Modern technologies use twin-wells to make the two types of transistors inside wells regardless of substrate doping (normally *p*-type). This approach allows us to optimise electrical behaviour at the expense of additional fabrication steps.

Fig. 1.10 shows the cross section of a typical CMOS technology. The substrate is *p*-type (often the substrate consists of a *p*-epi grown on a p^+ *Si* substrate). The *p*-well and the *n*-well are achieved with complementary mask using a phosphorous and boron implant respectively followed by a drive-in diffusion. The active area defines the region where transistors are located. On the top of the active area we have a thin oxide (for *0.5 μm* technology it is *9-12 nm* for a shorter channel-length the oxide is thinner; for example with a *0.35 μm* technology it is *5-7 nm*). Around the active area there are structures made by highly doped region they achieve the so-called channel stop to avoid lateral current leakage. Over the thin oxide there is the silicon gate that also separates the source and drain. Because of the lateral diffusions the source and the drain extend under the silicon gate leading to an overlap between the source (and the drain) and the silicon gate. In addition, the effective channel length is shorter than the designed length by a given extent.

Fig. 1.10 also shows a poly-poly capacitor. It is achieved on top of the thick

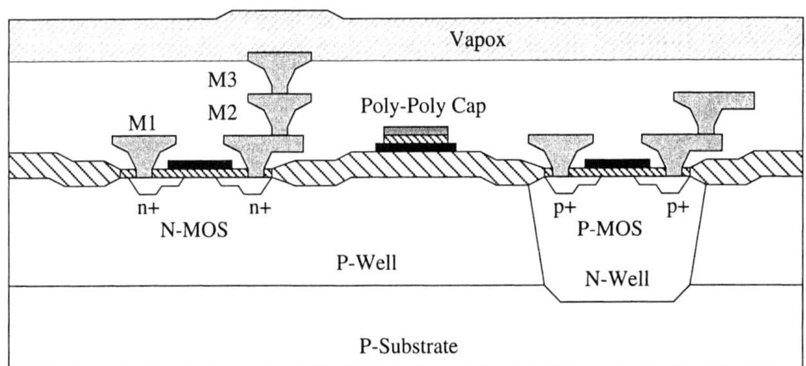

Fig. 1.10 - Cross section of a CMOS p-well circuit containing an n-channel and a p-channel transistor

oxide using poly 1 and poly 2. The thin oxide between the two poly layers ensures a good specific capacitance. Of course the poly plates are connected to the circuit using the poly itself or, better, with metal lines. The figure refers to a triple-metal technology (advanced technologies often offer a number of metal layers, some of which are achieved using copper). The cross section shows a stack of the three metals interconnected by via.

On top of the structure we have a *vapox* layer (oxide achieved with chemical vapour deposition). It protects the circuit from chemical aggression and is not present only where the pad for pin connection is located.

The above described technology refers to only one possible process. We have alternative solutions for each step of the fabrication process. Combining them leads to solutions that optimize one or the other design target. For analog applications a good output resistance of transistors in saturation and the availability of well matched, linear and reliable capacitors and resistors is very important. For mixed analog-digital solutions, instruments providing proper defence against digital noise are very essential.

1.5 MOS THRESHOLD VOLTAGE

Conduction between source and drain in MOS transistors takes place under the control of the source-to-gate voltage. The current flow does not begin sharply, but it is conventionally assumed that conduction occurs if the gate voltage exceeds a given value called the threshold voltage, V_{Th}. The analytical expression of this important parameter comes from studying the *MOS* structure or, rather, a structure where the *M*, metal, is replaced by a heavily doped polysilicon (although metal-gate technologies have been almost completely abandoned and are now used only in some high-voltage processes, metal gate solutions are again being considered for decananometer technologies). The first step in the threshold study is to consider the band diagram. It is shown in Fig. 1.11 (n-channel transistor and, therefore, p-doped substrate) for two different biasing of the gate. The wave propagation vector is not relevant to the present study, therefore the diagrams in Fig. 1.11 (as well as the one in the following Fig. 1.12) only show the Fermi level and the two limits of the forbidden gap across the three regions of the structure.

The polysilicon of the gate is strongly n-type doped, enough to degenerate the material; its Fermi level E_F is very close to the conduction band (in some cases inside it). The semiconductor is p-type and, far from the surface, the Fermi level is close to the valence band. The distance from the middle of the energy gap is Φ_{FS} (subscript *S* means silicon). Moreover the energy gap, E_g, in silicon is *1.21 eV* while in SiO_2 it is around *8 eV*.

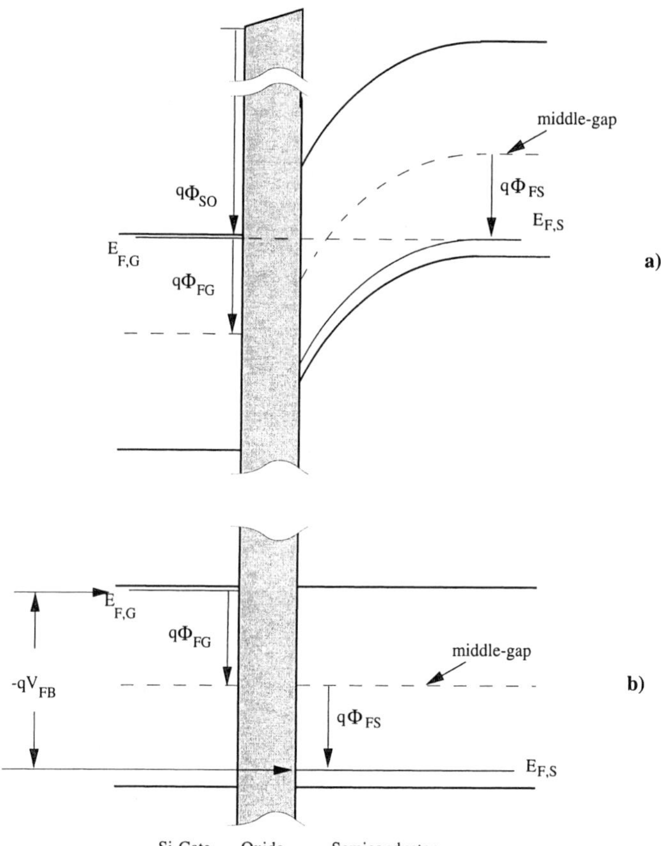

Fig. 1.11 - Band diagram of an MOS with zero biasing and for flat band conditions

Zero voltage applied to the structure gives rise to the band diagram shown in Fig. 1.11 a): the Fermi levels on the gate side and the silicon side have the same value. Compensation between the different Fermi levels on the two sides of the structure (contact potential) is achieved by band bending in the silicon and biasing the oxide. The band bending deriving from surface depletion also provides the charge required for oxide biasing. The equal and opposite charge on the gate side does not produce any significant bending because of the silicon gate strong doping.

Any difference between the Fermi levels on the two sides of the structure reflects an applied voltage. It determines larger or smaller depletion (or accumulation) on the silicon surface with a corresponding biasing of the oxide. Fig. 1.11 b) shows the special case of zero voltage across the oxide. In this condition the bands are flat and the associated applied voltage is termed *flat band voltage*, V_{FB}

$$V_{FB} = \Phi_{FG} + \Phi_{FS} \cong \frac{E_g}{2q} + \Phi_{FS} \quad (1.20)$$

In real devices, during the fabrication of an MOS structure, a certain amount of charge remains trapped at the oxide-semiconductor interface. This charge, Q_{SS}, depends on fabrication conditions and crystallographic orientation. Typical values are

$$Q_{ss} \cong 2 \cdot 10^{-8} coul/cm^2 for\langle 111\rangle \quad (1.21)$$

$$Q_{ss} \cong 4 \cdot 10^{-9} coul/cm^2 for\langle 100\rangle \quad (1.22)$$

This interface charge calls an equal and opposite charge to the gate side, thus biasing the oxide. To achieve the flat band condition an additional voltage must be applied across the oxide. After accounting for the Q_{SS} effect, the flat band voltage becomes

$$V_{FB, real} = \frac{E_g}{2q} + \Phi_{FS} - \frac{Q_{ss}}{C_{ox}} \quad (1.23)$$

TAKE NOTE

Charges trapped at the semiconductor-oxide interface lead to localised energy level inside the energy gap. Those levels can favour the scattering of electrons from the valence to the conduction band.

Localised energy levels may increase the low frequency noise.

Fig. 1.12 shows the band diagram for positive voltage applied to the gate in a very specific case. The band bending is such that the Fermi level at the oxide-semiconductor interface is taken to the upper side of the energy gap exactly symmetrically opposite the bulk position. This condition is called limit of the strong inversion: the surface of the semiconductor becomes equivalent to an n-type doped material whose carrier concentration is roughly equal to the bulk doping. Therefore, any region with n-type doping placed at the side of the structure will be able to freely exchange electrons and create a conductive channel. The voltage required to obtain this condition is termed the threshold voltage, V_{Th}. (It is worthwhile observing that the value of the threshold voltage in Fig. 1.12 is quite low, in this specific case being equivalent to a small fraction of the energy gap. This is the so-called *native threshold,* i.e. the value achieved before any step is taken to adjust the threshold to a more convenient level)

To calculate the analytical expression of V_{Th} we start from the flat band con-

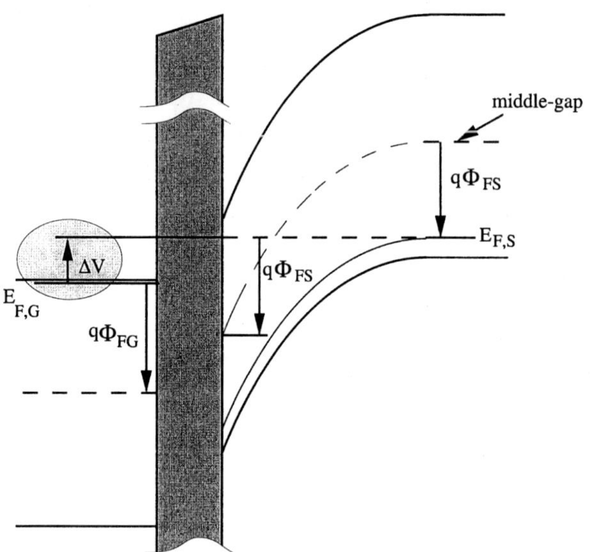

Fig. 1.12 - Band diagram of an MOS structure for flat band conditions

ditions. From the corresponding biasing, we move to strong inversion by inducing a surface depletion capable of shifting the Fermi level by $2q\Phi_{F,S}$. The extent of the depletion is

$$x_d = \sqrt{\frac{2\varepsilon}{qN_A}|2\Phi_{FS}|} \tag{1.24}$$

The charge associated with the depletion biases the oxide, hence requiring a voltage given by

$$V_{ox} = \frac{1}{C_{ox}} qN_A x_d \tag{1.25}$$

After taking into account all the above mentioned contributions and using (1.23), (1.24), and (1.25), the threshold voltage becomes

$$V_{Th,0} = V_{FB,real} - 2\phi_{FS} + V_{ox} \tag{1.26}$$

$$V_{Th,0} = \frac{E_g}{2q} - \Phi_{F,S} - \frac{Q_{SS}}{C_{ox}} + \frac{\sqrt{2q\varepsilon N_A}}{C_{ox}} \sqrt{|2\Phi_{F,S}|} \tag{1.27}$$

Subscript *0* indicates that the substrate is not biased compared to the source.

Example 1.3

Calculate the threshold voltage for a MOS structure whose substrate doping is $N_A = 4\ 10^{16}\ cm^{-3}$, and where the concentration of trapped charge is $Q_{SS}=2\cdot 10^{-8}\ cm^2$, the oxide thickness is 15 nm and $\varepsilon_{r,ox}= 3.6$. Compare the contributions of the various terms achieved.

Solution:

Using equation (1.10) the Fermi level shift for doping $N_A= 10^{16}\ cm^{-3}$ is $\Phi_{F,S}= 0.336V$.

The oxide capacitance per unit area is calculated by

$$C_{ox} = \frac{\varepsilon_0\varepsilon_{r,ox}}{t_{ox}} = \frac{8.858\cdot 10^{-14}\cdot 3.6}{15\cdot 10^{-7}} = 0.212\mu F/cm^2$$

Note that the above equation t_{ox} is expressed in cm. The threshold voltage becomes

$$V_{Th,0} = V_{Th,nat} = \frac{E_g}{2q} - \Phi_{F,S} - \frac{Q_{SS}}{C_{ox}} + \frac{\sqrt{2q\varepsilon_0\varepsilon_r N_A}}{C_{ox}}\sqrt{|2\Phi_{F,S}|}$$

$$= \frac{1.21}{2} - 0.336 - \frac{2\cdot 10^{-8}}{0.212\cdot 10^{-6}} +$$

$$+ \frac{\sqrt{2\cdot 1.6\cdot 10^{-19} 8.858\cdot 10^{-14}\cdot 11.9\cdot 10^{16}}}{0.212\cdot 10^{-6}}\sqrt{2\cdot 0.336} =$$

$$= 0.605 - 0.336 - 0.095 + 0.274\cdot\sqrt{0.672} =$$

$$= 0.605 - 0.336 - 0.095 + 0.225 = 0.399V$$

The second term of the expression corresponds to the Fermi shift. We can observe that its value for the given doping is a fraction of half of the energy gap; it is weakly affected by N_A because of the logarithmic behaviour in (1.10). By looking at the other contributions we can observe that the effect of the trapped charges is (in this example) quite limit. By contrast, the voltage across the oxide is significant: it results from the coefficient 0.274 multiplied by the square root of $2\Phi_{FS}$.

Before discussing the effect of a given substrate bias on the threshold voltage it is worth observing that the threshold voltage achieved may not be suitable for circuit design. It may be too low (below *0.2 - 0.5V*), producing rather large leakage currents in the off-state, or, by contrast, it may end up being too high (above *1.5 V*) becoming a problem for the bias voltages used today. Fortunately, thresholds can be corrected with an additional implant in the channel region before polysilicon gate deposition. The implanted ions, Q_{imp}, have an effect equivalent to the charge at the oxide-semiconductor interface, Q_{SS}. Therefore, with this additional implant the expression of the threshold voltage becomes

$$V_{Th,0} = \frac{E_g}{2q} - \Phi_{F,S} - \frac{(Q_{SS} + Q_{imp})}{C_{ox}} + \frac{\sqrt{2q\varepsilon N_A}}{C_{ox}}\sqrt{|2\Phi_{F,S}|} \tag{1.28}$$

Example 1.4

Calculate the implant dose needed to increase the threshold voltage calculated in Example 1.3 up to 0.85 V

Solution: *By comparing (1.27) and (1.28) we see that the only difference comes from the term* Q_{imp}/C_{ox}*. Therefore,*

$$V_{Th} - V_{Th,nat} = 0.85 - 0.19 = -\frac{Q_{imp}}{C_{ox}}$$

$$Q_{imp} = -0.66 \cdot 0.212 \cdot 10^{-6} = -1.4 \cdot 10^{-7} cm^{-2}$$

which corresponds to a light implant of donors.

Let us consider now the case of reverse biasing the source bulk junction, $V_{SB}<0$. In the light of the previous discussion, we can see that further strain to equalize the Fermi level between source and channel is necessary, thus requiring higher band bending. Therefore, surface depletion must increase to

$$x_d' = \sqrt{\frac{2\varepsilon}{qN_A}|V_{SB} - 2\Phi_{F,S}|} \tag{1.29}$$

Which, in turn, leads to a larger biasing of the oxide. The threshold voltage then becomes

$$V_{Th} = \frac{E_g}{2q} - \Phi_{F,S} - \frac{(Q_{SS} + Q_{imp})}{C_{ox}} + \gamma\sqrt{|V_{SB} - 2\Phi_{F,S}|} \tag{1.30}$$

where we have defined a new parameter γ called *body effect coefficient.* It is expressed by

$$\gamma = \frac{\sqrt{2q\varepsilon N_A}}{C_{ox}} \tag{1.31}$$

Using equation (1.25) the threshold voltage can be rewritten as

$$V_{Th} = V_{Th,0} + \gamma\{\sqrt{|V_{SB} - 2\Phi_{F,S}|} - \sqrt{|2\Phi_{F,S}|}\} \tag{1.32}$$

This equation better shows the dependence of the threshold voltage on the substrate biasing by means of the body effect coefficient γ. Since well doping can be higher than the doping of bulk material (to ensure the necessary doping compensation), the threshold of the transistors placed in the well may be more sensitive to substrate biasing than transistors made directly in the substrate. Moreover, the expression of γ shows that the body effect diminishes as technology moves towards thinner oxides.

Example 1.5

Calculate the body effect coefficients of an n-channel and a p-channel transistor integrated with n-well technology. The process features are:
$N_{A,Sub} = 2\ 10^{16}\ cm^{-3}, N_{D,Well} = 4\ 10^{16}\ cm^{-3}, C_{ox} = 1.3\ fF/\mu^2, \Phi_{F,n} = 0.45\ V, \Phi_{F,p} = 0.36\ V.$
Calculate the p-type and n-type threshold variation for a substrate biasing $|V_{SB}| = 2V$.

Solution: *The expression of the body effect coefficient is given by (1.31). Remembering that* $\varepsilon = \varepsilon_0\varepsilon_r = 11.9 \cdot 8.858 \cdot 10^{-14} = 1.05 \cdot 10^{-12}\ F/cm$
and that $1.3\ fF/\mu^2 = 1.3\ 10^{-7}\ F/cm^2$ *we obtain*

$$\gamma_n = \frac{\sqrt{2q\varepsilon N_{A,Sub}}}{C_{ox}} = \frac{\sqrt{2 \cdot 1.6 \cdot 10^{-19} 1.05 \cdot 10^{-12} 2 \cdot 10^{16}}}{1.3 \cdot 10^{-7}} = 0.63 \cdot V^{1/2}$$

Since the doping concentration of the well is twice that of the substrate, we have

$$\gamma_p = \frac{\sqrt{2q\varepsilon N_{D,Well}}}{C_{ox}} = 0.892 \cdot V^{1/2}$$

with the above results, using (1.32) and $V_{SB} = 2\ V$ *we obtain:* $\Delta V_{Th,n} = 373\ mV$ *and* $\Delta V_{Th,p} = 563\ mV$. *These figures are signifi-*

cant when compared with the value of the threshold voltage. For present technologies, this is the case being the threshold normally less than 1V.

1.6 I-V CHARACTERISTICS

We study the I-V characteristics of an MOS transistor by distinguishing between three different regions of operation: the *weak inversion*, whose gate voltage is lower than the threshold ($V_{GS} = V_{Th}$), the *saturation*, whose drain current is almost totally controlled by the gate voltage, and the *linear* (or triode) region, with parabolic-like *I-V* characteristics. The transition points between the triode and saturation regions are determined by the condition $V_{DS} = V_{GS} - V_{Th}$.

The use of three different approximations to study the I-V characteristics is quite a rough approximation. The transition between adjoining regions may be discontinuous. This situation must be avoided in computer programs because of possible unrealistic results and convergency problems.

The typical I-V characteristics of an MOS transistor are shown n Fig. 1.13. The linear region and the saturation regions of operation are separated by a dividing line. The weak inversion region is very close to the *0* current axis. The equations that we are going to obtain shortly are simple and are useful only for hand calculations made by the designer to get a feeling for circuit operation. These equations describe the electrical behaviour in each region with a reasonable smooth transition between adjacent regions.

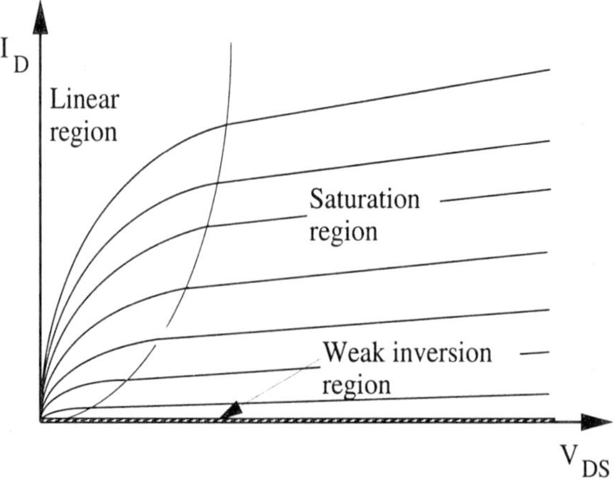

Fig. 1.13 - I-V static characteristics of an MOS transistor.

1.6.1 Weak Inversion Region

Fig. 1.14 shows the band diagram in the transversal cross section of a transistor. Both V_{GS} and V_{DS} are set to zero. Moreover, the source-substrate and drain-substrate junctions are reversely biased (this better ensures that source and drain are isolated from the substrate) and the V_{SB} voltage is increasing the energy barrier at the two sides of the channel. Looking at Fig. 1.14 we can recognise two back to back p-n diodes in the source-channel-drain structure. Any voltage applied between source and drain will fall almost completely across the reversely biased diode, and the transistor current will be given by its reverse saturation current. As is known, this current depends exponentially on the height of the energy barrier.

As the gate-source voltage increases, a fraction of it, say $((n-1)/n)$, drops across the gate oxide while the part, $1/n$, diminishes the barrier.The equivalent increase in the reverse saturation current results in

$$I_S = I_{D0} e^{qV_G/nkT} e^{-qV_B/nkT} \tag{1.33}$$

When accounting for the slight dependence on the V_{DS} voltage, which acts as the reverse biasing voltage, the drain current I_D is evaluated as

$$I_D = I_{D0} e^{qV_G/nkT} e^{-qV_B/nkT} [1 - e^{-qV_{DS}/kT}] \tag{1.34}$$

This result is quite important: it shows that in weak inversion the *I-V MOS* characteristic changes exponentially with variations in the control voltage, V_G, similarly to a bipolar transistor. The only difference is given by the factor n. For a typical technology the value of n is between *1.5* and *3*.

REMARK

A MOS transistor in the weak inversion region (sub-threshold) behaves like a bipolar transistor. However voltage **continues** to be the electrical quantity that controls the device

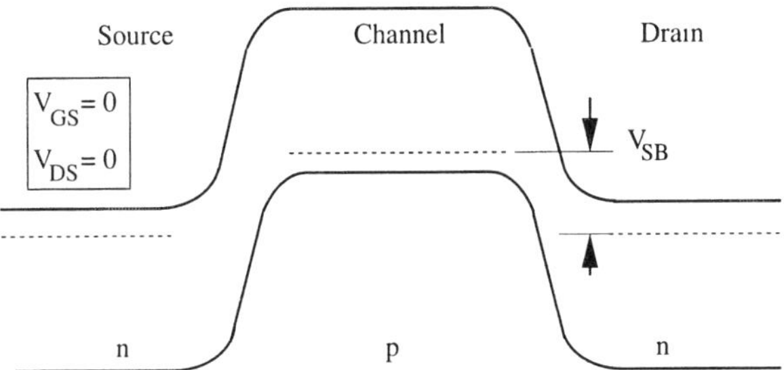

Fig. 1.14 - Band diagram of the longitudinal cross section of an MOS transistor

1.6.2 Linear (or Triode) Region

Any V_{GS} which is larger than the threshold voltage leads the oxide-silicon interface to strong inversion. We have seen that, conventionally, strong inversion begins when V_{GS} equals V_{Th}. When a larger voltage is applied, the exceeding part produces an accumulation of mobile charges at the oxide-semiconductor interface, thus forming the so-called *inversion layer*. For a generic position x, its charge per unity area, Q_{inv}, is given by (Fig. 1.15)

$$Q_{inv}(x) = C_{Ox}\{V_{GS} - V(x) - V_{Th}(x)\} \tag{1.35}$$

which depends on the drop voltage, $V(x)$, along the channel and on the resulting threshold voltage variation.

We can now calculate the resistance across an infinitesimal element, dx, of the channel using the equation

$$dR = \frac{dx}{\sigma A} = \frac{dx}{Q_{inv}(x)\mu W} \tag{1.36}$$

where μ is surface carrier mobility and W is the effective width of the transistor. The drop voltage across this element results as

$$dV = I_D dR = \frac{I_D dx}{C_{ox}[V_{GS} - V_{Th} - V(x)]\mu W} \tag{1.37}$$

We have to remember that the threshold voltage changes along the channel because of the body effect, according to

$$V_{Th} = V_{Th,0} + \gamma\{\sqrt{|V_{SB} - 2\Phi_F - V(x)|} - \sqrt{|2\Phi_F|}\} \tag{1.38}$$

Therefore, using equation (1.36) in equation (1.37) and integrating along the channel, we get

$$\begin{aligned} I_D = \mu C_{ox}\frac{W}{L}\Big\{&(V_{GS} - V_{Th,0} - \gamma\sqrt{|2\phi_F|})V_{DS} - \frac{1}{2}V_{DS}^2 + \\ &+ \frac{2}{3}[|V_{SB} - 2\Phi_F|^{3/2} - |V_{DB} - 2\Phi_F|^{3/2}]\Big\} \end{aligned} \tag{1.39}$$

The above equation is difficult to use for hand calculations. It can be simpli-

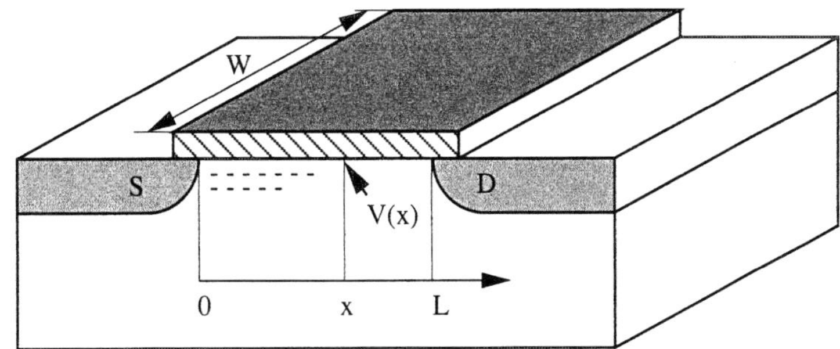

Fig.

fied by neglecting the effect of (1.38) to account for the variation of the threshold voltage along the channel. In this case we get

$$I_D = \mu C_{ox}\left(\frac{W}{L}\right)\left[(V_{GS} - V_{Th})V_{DS} - \frac{1}{2}V^2_{DS}\right] \tag{1.40}$$

We observe that both equations (1.39) and (1.40) contain the term $\mu C_{ox}(W/L)$; therefore, the drain current is larger with electrons acting as mobile carriers (their mobility is larger than that of holes) and for thinner gate oxides. In addition, the drain current is proportional to the aspect ratio of the gate, (W/L). Moreover, note that (1.40) represents a parabola in the I_D-V_{DS} plane whose maximum is achieved in

REMEMBER

In the triode region the current is linearly proportional to the voltage exceeding the threshold level (overdrive voltage).

$$V_{GS} - V_{Th} = V_{DS} \tag{1.41}$$

We observe that when the above condition is used in equation (1.35) the charge in the inverted layer at the drain end ($x=L$; $V(L)=V_{DS}$) goes to zero. Since the inversion layer cannot be reversed, any V_{DS} larger the one given by equation (1.41) will lead to unrealistic situations. Condition (1.41) hence defines the applicability limit of the above formulation and establishes the limits of the triode region.

It should be noted that for a very small V_{DS} (V_{GS}-V_{Th} >>V_{DS}) the *I-V* curve approximates a straight line whose slope is proportional to the gate-to-source voltage, V_{GS}. This feature is often exploited to achieve a voltage controlled resistor.

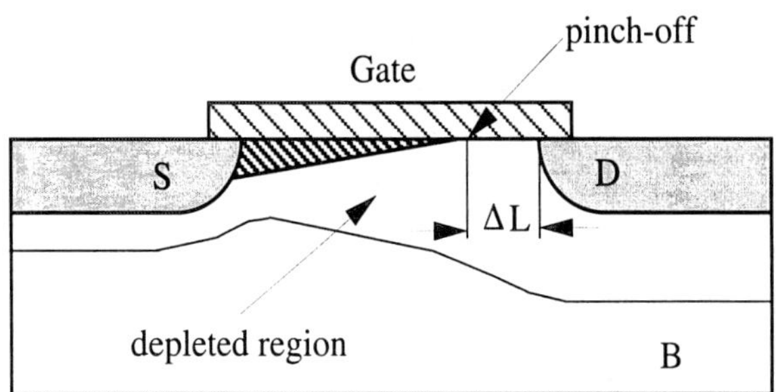

Fig. 1.16 - Cross section of an MOS transistor in the saturation region

1.6.3 Saturation Region

The transistor leaves the linear region and enters the saturation region when the voltage at the gate is larger than the threshold and when V_{DS} exceeds the so-called saturation voltage, V_{sat}. It is defined by means of equation (1.41) as

$$V_{sat} = V_{GS} - V_{Th} \tag{1.42}$$

In saturation, the inversion layer vanishes inside the channel and a portion, ΔL, of the channel itself becomes completely depleted (Fig. 1.16). Therefore, the saturated transistor becomes the connection series between a transistor biased at the limit of the saturation and a piece of depleted material, ΔL. The voltage across the depleted part is

$$\Delta V = V_{DS} - V_{sat} \tag{1.43}$$

The width ΔL can be approximated by

$$\Delta L = \sqrt{\frac{2\varepsilon}{qN_A}(V_{DS} - V_{sat})} \tag{1.44}$$

We can estimate the current flowing in the transistor by the suitable use of equations (1.40) and (1.41)

$$I_D = \frac{1}{2}\mu C_{ox}\frac{W}{L-\Delta L}(V_{GS} - V_{Th})^2 = \left(\frac{1}{2}\mu C_{ox}\frac{W}{L}V_{DS}^{\ 2}\right)\frac{L}{L-\Delta L} \tag{1.45}$$

The result is the product of two terms: the drain current of a transistor with length L biased at the limit of saturation and a correcting factor $L/(L\text{-}\Delta L)$ accounting for the channel length reduction. Using equation (1.44) we have

$$\frac{L}{L-\Delta L}=\frac{1}{1-\sqrt{\dfrac{2\varepsilon}{qN_AL^2}(V_{DS}-V_{sat})}}\approx$$
$$1+\sqrt{\frac{\varepsilon}{qN_AL^2}}\sqrt{(V_{DS}-V_{sat})}\cong 1+\lambda V_{DS} \tag{1.46}$$

This defines a new parameter, λ, called the channel modulation parameter. The above result is greatly approximated: we have replaced the square root of $(V_{DS}\text{-}V_{sat})$ with V_{DS}. Moreover, the square root of the volt "missing" in the dimensions has been included in parameter λ. Thus, as we see from (1.46), the channel length modulation causes a linear increase in the drain current which is proportional to V_{DS}. A suitable (but rough) empirical expression for λ, also supported by (1.46), is the following

$$\lambda\cong\frac{10^7}{\sqrt{N_A}\cdot L} \tag{1.47}$$

where the length, L, is measured in microns and, N_A, the doping concentration, in cm^{-3}. (Of course, for complementary transistors, N_A should be replaced by N_D).

Combining equation (1.46) with equation (1.45) leads to the drain current in the saturation region

$$I_D=\frac{1}{2}\mu C_{ox}\left(\frac{W}{L}\right)(V_{GS}-V_{Th})^2[1+\lambda V_{DS}] \tag{1.48}$$

Observe that the I-V characteristics expressed by (1.40) shows a zero slope in the I_D-V_{DS} plane at the boundary between the triode and the saturation region. By contrast the slope calculated with (1.48) is λ. This discontinuity in the I-V derivative is an example of possible problems that occurs when using simplifying physical approximations.

KEEP IN MIND!

The transistors' saturation current increases as the square of the overdrive voltage.

Non-linear responses lead to harmonic distortion. Analog circuits often require a distortion-free, linear response.

The product μC_{ox} is often represented by the symbol k_n (or k_p) that is called the *process transconductance parameter.* Its value depends on the technology and the type of channel carriers. For given modern CMOS technologies we can use the following figures $t_{ox} = 12\ nm$, $\mu_n = 460\ cm^2/Vsec$ and $\mu_p = 148\ cm^2/Vsec$. Therefore, for n-channel and p-channel transistors, respectively, this specific technology leads to

$$k_n = \mu_n C_{ox} = 120 \mu A/V^2$$
$$k_p = \mu_p C_{ox} = 39 \mu A/V^2 \quad (1.49)$$

Note that if the oxide thickness is halved (as it happens when the transistor line-width is reduced by a factor two or so) the process transconductance parameter doubles.

1.7 EQUIVALENT CIRCUITS

In the previous section we derived the voltage-current relationship for an MOS transistor. The equations obtained in the previous section do not completely describe the operation of a transistor. Parasitic effects and additional parameters representing the dynamic behaviour should also be included. These effects can be described by equations. However, since circuit designers prefer to study circuits using schematics, parasitic and dynamic effects are studied by considering equivalent circuits. We shall distinguish between the large signal equivalent circuit, suitable for describing device nonlinearity, and the small signal equivalent circuit obtained from a linearization of the large signal equivalent circuit.

1.7.1 Large Signal Equivalent Circuit

Fig. 1.17 shows the large signal equivalent circuit of an MOS transistor. A cross-section of a typical physical structure, placed under the electrical network, helps us to understand the component's role. Some of the elements are non linear; others, at a first approximation, can be assumed to be linear.

Resistors R_s and R_d describe the resistive paths from the source and drain contacts to the endings of the channel. Their value depends on the specific resistance of the diffused layers and on the aspect ratio of the contact regions. Typical values of drain and source sheet resistances are around *50-200 Ω/□*. Aspect ratios range from a fraction to one square; therefore, R_s and R_d are typically *10-50 Ω*.

Two diodes, D_{BS} and D_{BD}, and their associated parasitic capacitances describe the reversely biased *p-n* junctions insulating the source and the drain

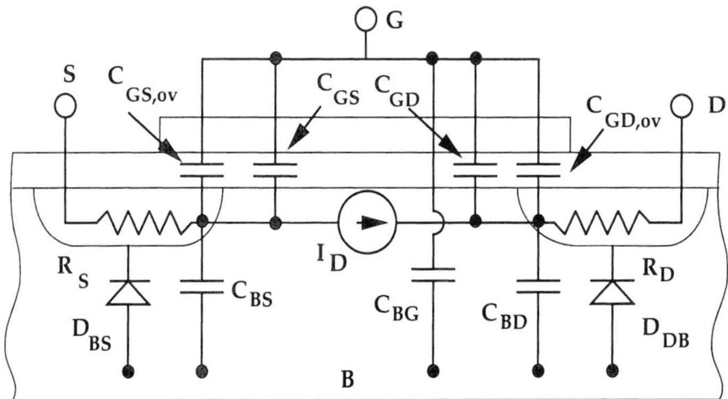

Fig. 1.17 - Large signal equivalent circuit of an MOS transistor

from the substrate. The major issue in circuit behaviour for these is the leakage current. In reversely biased conditions, this is dominated by the generation-recombination effect, I_{GR}, that is expressed by

$$I_{GR} = A\frac{qn_ix_j}{2\tau_0} \tag{1.50}$$

where A is the area of the junction, x_j is the width of the depletion region, n_i is the intrinsic carrier concentration and τ_o is the mean lifetime for minority carriers. Note that the generation-recombination current is proportional to the depletion width. Therefore, the current increases as the reverse biasing increases. Moreover, I_{GR} is proportional to n_i, making it exponentially dependent on temperature. A typical value of I_{GR}/A at room temperature ranges from 10^{-17} to $10^{-16} A/\mu m^2$.

Example 1.6

Calculate the leakage current in a p-channel drain diffusion. Use the following figures: W=100 μm, drain extent L_D=2μm, p^+-n leakage current: I_{LA}=0.023 fA/μ^2, p^+-n sidewall leakage current: I_{LS}=0.4 fA/μ.

Solution: *We note that the above figures provide the leakage parameters per unity area and per unity length in the sidewalls. Therefore, the leakage in the sidewall should correspond to an equivalent depth of 17.4 μm. This amount is much higher than any reasonable figure. We thus evince that leakage is much higher in the bent sidewalls of the diffusion than in a flat region.*
The leakage current is calculated by

$$I_{leak} = I_{LA}WL_D + I_{LA} \cdot 2(W + L_D) = 86.2fA$$

It is worthwhile observing that in 1 second for example, this current, discharges a capacitor of 1 pF by 86 mV.

Capacitances C_{BS} and C_{BD} represent the junction capacitance of the two diodes D_{BS} and D_{BD}. For step junctions, as is well known, they are inversely proportional to the square root of the reverse biasing. Moreover, they are proportional to the area of the junction. We shall see that suitable folding of the transistor layout allows us to reduce the drain and the source contact area while ensuring an acceptable value of R_s and R_d.

The most important element of the equivalent circuit is the voltage controlled current source I_D. Its value depends on the voltages applied to gate, source, drain and bulk, according to the expressions derived in the previous section for the three regions of operation: sub-threshold, saturation and linear.

Five capacitances account for the capacitive behaviour of the gate. Two of these, $C_{GS,ov}$ and $C_{GD,ov}$, are due to the overlap, x_{ov}, of the gate with source and drain caused by lateral diffusion (*0.6-0.8* times the junction depth). They are linear and can be calculated, using C_{ox} the gate oxide capacitance per unit area, by

$$C_{ov} = Wx_{ov}C_{ox} \tag{1.51}$$

The other three capacitances, C_{GS}, C_{GD} and C_{GB}, account for the non linear coupling between the gate and the other three terminals. There are several approaches to describe the behaviour of the three capacitances with regard to the gate and drain voltages. The Meyer model empirically splits the gate capacitance, $C_{ox}WL$, into varying amounts between the gate and the other terminals. Fig. 1.18 shows the dependence of the three capacitors on the V_{GS}

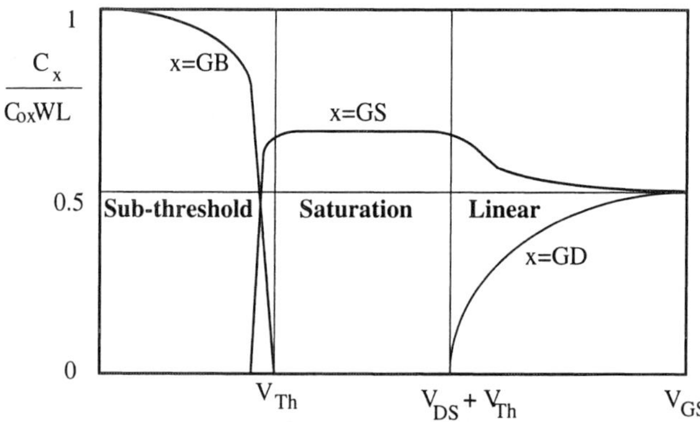

Fig. 1.18 - Gate capacitances as a function of the V_{GS} voltage

voltage.

In the sub-threshold region, the conductive channel has not yet been created. Therefore, the coupling with source and drain is very weak: C_{GB} is the only significant capacitance. By contrast, in the saturation and linear region, the conductive channel produces an equivalent shielding for the substrate; thus, C_{GB} becomes negligible. In the linear region, the channel is equivalent to a resistance beneath the gate. This is described by a distributed RC network connecting the source to the drain. However, for simplicity's sake, only two concentrated capacitors, C_{GS} and C_{GD}, are used. Their value is almost equal for large values of V_{GS}. As V_{GS} decreases and reaches the saturation limit, the drain is no longer connected to the channel, so that C_{GD} goes to zero. In saturation, the only relevant capacitance is C_{GS}, whose value is assumed as being approximately equal to $2/3\ C_{ox} WL$.

The Meyer model, discussed above, is not precise enough for accurate circuit simulations. Nevertheless, it is adequate enough for performance estimations when selecting a circuit architecture. Of course, more accurate models exist: the Ward-Dutton model calculates the device distribution of charge and introduces a non-reciprocal capacitor model. A similar approach, ensuring charge-conservation, is used in the BSIM model. Complex representations are essential for computer simulations while a simple description is sufficient for first order estimation and hand calculations.

1.7.2 Small Signal Equivalent Circuit

The linearization of the elements of the large signal model leads to the small signal equivalent circuit. This is shown in Fig. 1.19. The voltage controlled current source generates three terms proportional to the small signal voltages v_{gs},

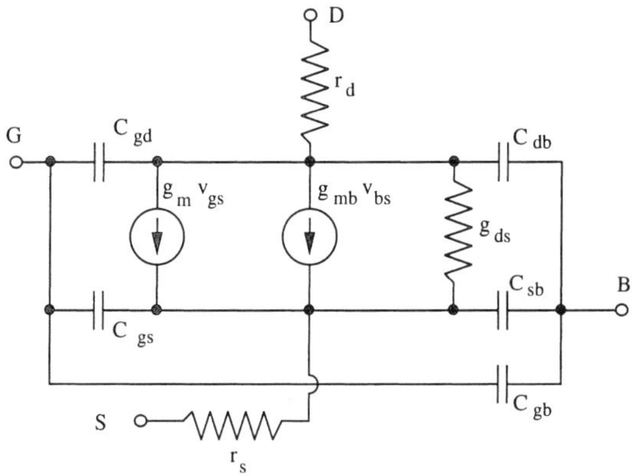

Fig. 1.19 - Small signal equivalent circuit of the MOS transistor

v_{ds} and v_{bs}. We thus have

$$i_d(v_{gs}, v_{ds}, v_{bs}) = \frac{\partial I_D}{\partial V_{GS}} v_{gs} + \frac{\partial I_D}{\partial V_{DS}} v_{ds} + \frac{\partial I_D}{\partial V_{BS}} v_{bs} \tag{1.52}$$

$$i_d = g_m v_{gs} + g_{ds} v_{ds} + g_{mb} v_{bs} \tag{1.53}$$

The two equations above define the transconductance g_m, the output conductance g_{ds}, and the substrate transconductance g_{mb}. These parameters are calculated from the *I-V* characteristics in the three regions of operation; in the weak inversion region it results that

$$g_m = -g_{mb} = \frac{I_D}{n\frac{kT}{q}} \tag{1.54}$$

$$g_{ds} = \lambda_{wi} I_D \,; \quad \lambda_{wi} \text{ being } \lambda \text{ in weak inversion} \tag{1.55}$$

In the linear region, if we use (1.40) we get

$$g_m = \mu C_{ox}\left(\frac{W}{L}\right) V_{DS} \tag{1.56}$$

$$g_{ds} = \mu C_{ox}\left(\frac{W}{L}\right)[V_{GS} - V_{Th} - V_{DS}] \tag{1.57}$$

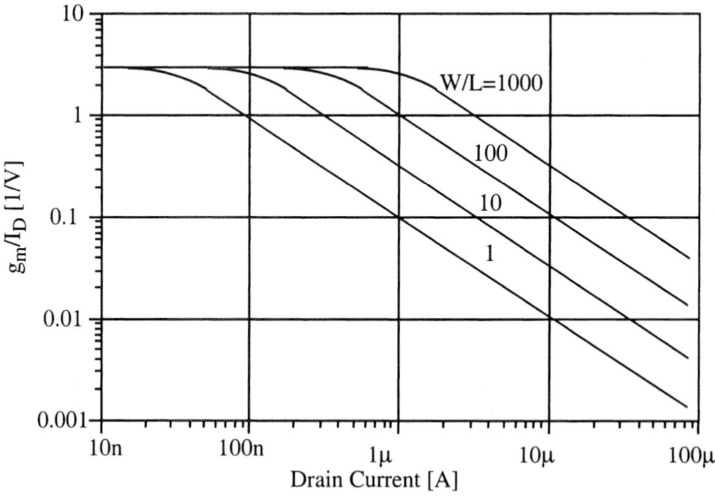

Fig. 1.20 - Transconductance versus the drain current in subthreshold and saturation

and in the saturation region, using (1.48), we have

$$g_m = \mu C_{ox}\left(\frac{W}{L}\right)(V_{GS} - V_{Th}) = \frac{2I_D}{(V_{GS} - V_{Th})} = \sqrt{2\mu C_{ox}\left(\frac{W}{L}\right)I_D} \tag{1.58}$$

$$g_{ds} = \lambda I_D \tag{1.59}$$

where the λ factor conforms the definition given in (1.46).Its dimension is the inverse of Voltage, V^{-1},

EQUATIONS TO KEEP IN MIND!

The transconductance of a MOS in the *saturation region* can be calculated using:

$$\mu C_{ox}\left(\frac{W}{L}\right)(V_{GS} - V_{Th})$$

$$\frac{2I_D}{(V_{GS} - V_{Th})}$$

$$\sqrt{2\mu C_{ox}\left(\frac{W}{L}\right)I_D}$$

The equations that we used to achieve the transconductance parameters do not depend on V_{BS} for the linear and saturation regions. Therefore, we do not obtain any expression for g_{mb}. Since the above described results help when making hand calculations and since accounting for the dependence on substrate voltage is a complex task, we usually take the effects of g_{mb} into consideration only at a higher level of accuracy that requires the use of a circuit computer simulator (like Spice). For analog applications we often use transistors operating in the saturation region. Hence it is useful to consider in some details the equations derived for such regions of operation.

From equation (1.58) we note that transconductance is proportional to the current and inversely proportional to the overdrive voltage $V_{GS} - V_{Th}$. This result, compared with the transconductance expression of a bipolar transistor ($g_{m,BJT} = I/(V_T)$), reveals that this parameter is much lower than its *BJT* counterpart, since the overdrive voltages commonly used in MOS circuits are always higher than $V_T = kT/q$.

Moreover, the same equation (1.58) shows that the transconductance is proportional to the square root of the drain current and the square root of the transistor's aspect ratio (*W/L*). Consequently, transconductance increases proportionally to the first power of the drain current in weak inversion and only proportionally to the square root when in saturation. This feature can be seen from the plot of g_m/I_D versus I_D (Fig. 1.20).

Example 1.7

We want to achieve a transconductance equal to 1 mA/V in a p-channel and n-channel transistor respectively (in saturation). The drain current is a fixed design parameter and holds 200 μA. Calculate the required aspect ratios. Use the tranconductance parameters given in (1.49).

Solution: *From the last relationship given in (1.58) we get*

$$\frac{W}{L} = \frac{g_m^{\ 2}}{2\mu C_{ox} I_D}$$

from which we obtain

$$\left(\frac{W}{L}\right)_n = 20.8; \left(\frac{W}{L}\right)_p = 64.1$$

We can observe that the same transconductance is achieved by a bipolar transistor whose collector current is only 26 μA.

When the drain current becomes too small, the transistor enters the weak inversion region and the transconductance, as stated by (1.54), becomes proportional to the drain current.The transition between weak inversion and saturation is smooth. This is physically reasonable but, if we use the simplified equations describing the *I-V* relationship in weak inversion and in saturation, it is difficult to find the transition point. As a first attempt, we can assume that the transconductance g_m does not show a discontinuity, hence, the transition between weak inversion and saturation should occur at a current $\overline{I_D}$ for which

$$\frac{\overline{I_D}}{n\frac{kT}{q}} = \sqrt{2\mu C_{ox}\left(\frac{W}{L}\right)\overline{I_D}} \tag{1.60}$$

solved in the form

$$\overline{I_D} = \left(n\frac{kT}{q}\right)^2 2\mu C_{ox}\left(\frac{W}{L}\right) \tag{1.61}$$

In the CMOS technology used for the figures in (1.49) and (1.50), at room temperature, *n=1.5* and for a minimum area transistor (*W/L=1*), we obtain the transition point at around *328 nA* and *115 nA* for an n-channel and a p-channel transistor respectively.

1.8 MORE SOPHISTICATED MODELS

The previous section dealt with a simplified model of the MOS transistor. As already mentioned, the results obtained are useful in hand calculations or, in acquiring the necessary feeling for circuit behaviour. However, designing integrated circuits requires very precise predictions of electrical performance. Consequently, we have to employ much more sophisticated models that can be handled only with computer simulation programs. The equations that these models use are pretty complex (much more than the ones derived here), allowing accurate physical phenomenons to be represented and mathematical fitting of high order effects to be described.

The most popular circuit simulator is SPICE. It was developed two decades ago at the University of California, Berkeley, and, for a number of years, it has been the basic tool for integrated circuit designers. Nowadays other commercial products are on the market. Most of them, however, have the same structure and adopt a similar philosophy. They normally include new features, such as pre- and post-processing facilities, the possibility of performance optimization, and optional tools for implementing new and customized device models using standard or proprietary (behavioural) languages.

In this section we shall not enter into the details of circuit simulator use nor the syntax that they employ to specify the circuits or the options. All this information can be easily obtained from user manuals. However, since SPICE or its commercial versions will be extensively employed in the course of this book, it will be necessary to achieve a clear idea of the relationship between the simple equations studied in this chapter and the results achieved with circuit simulators.

We know that what we derived in the previous sections corresponds to an approximate description of reality. What we achieve from a circuit simulator is still an approximation of reality, albeit a more precise one. Results should therefore be validated by the circuit designer and, when necessary, verified by experimental measurements. Therefore, what this book seeks to achieve is not to derive quantitative equations, but instead to confer the design experience necessary, on one hand, to properly assess results obtained by the circuit simulators and, on the other, to motivate correct design decisions regarding circuit definition and optimization.

Circuit simulators describe passive and active devices with a set of equations that represent the equivalent circuit. The models available may have different degrees of complexity depending on the accuracy desired. The original Berkeley version of Spice used three levels of complexity to model MOS transistors; the first, Level 1, corresponds almost exactly to the set of equations derived in the previous sections (the Schichman-Hodges model). The other two involve more sophisticated models: Level 2 uses a geometry-based approach and calculates all effects from detailed device physics (analytic model); Level 3 is more qualitative and uses observed operation to define its equations (empirical model). The last version of Spice (3) contains a fourth MOS model, called BSIM, including features for sub-micron transistor description.

All the models included in SPICE, as well as any others contained in commercial circuit simulators, follow one of the two following main philosophies: the physically-oriented description or the fitting-oriented method. In the former case, as in Level 1, the equations derive from a physical description of phenomena. In the latter, the equations are meant to achieve the best fitting of results with experimental measurements. Here, complex models use many parameters, most of which have little or no physical impact. However, they allow more precise results to be obtained and facilitate the extraction of the parameters from experimental measurements.

Just to get an idea, let us consider the parameters that must be specified in the model card (or in the transistor card) for the simplest model used by Spice: the Level 1. We have:

VTO	threshold voltage with zero bulk biasing
KP	process transconductance parameter ($\mu\, C_{OX}$)
PHI	surface potential ($2\Phi_F$)
GAMMA	body effect coefficient
LAMBDA	channel modulation parameter
IS	saturation current of the source and drain junctions
UO	carrier mobility
TOX	oxide thickness
RD	parasitic drain resistance (*in the transistor card*)
RS	parasitic source resistance (*in the transistor card*)
AD	drain area (*in the transistor card*)
AS	source area (*in the transistor card*)

When these parameters are not specified the simulator uses default values or it calculates them through other furnished parameters.

We shall not go into detail with equations because more and more silicon foundries are using proprietary (and diverse) models. What is important is assessing results and estimating their level of plausibility: before starting a design with any new technology, one should perform some simulations on simple test circuits and critically analyse the results. This will give the designer a feel for the model's accuracy and the reliability of its parameters. Regarding this, we should remember that problems often arise when transistors operate in the regions between weak and strong inversion or when the body is not connected to the source.

An important aspect in analog design is small signal analysis. The computer carries out a linearization of the large signal equations and should give accurate results to have an analog meaning. This means that we also need high precision in the derivative of the equations used. Often, models are optimised for digital use with the transient response and the large signal behaviour being carefully fitted. In contrast, the derivatives of voltages and currents have not been suitably adapted. In such situations the analog designer must encourage people in charge of design kits to properly optimised model parameters for analog needs. The transition regions between different areas of operation is particularly critical. Typically, we have trouble with moderate inversion (between strong and weak inversion): output conductance and transconductance are not well described, and with some models we also have discontinuity.

> **DESIGN HINT**
>
> Before using a new technology become familiar with it by performing some simple but essential simulations.
>
> Decide by yourself the benchmarks that you feel most appropriate and significant.

This book will adopt the SPICE simulator for designing analog circuits extensively; the basic principles that we have discussed in this section will drive Spice use for training purposes. The models employed refer to a 0.8 μm CMOS process, described by a Level 2 model and two, 0.35 μm and 0.25 μm, CMOS technology characterized by a BSIM3 model. The parameters for these two technologies are given in Appendix A, Appendix B and Appendix C respectively. We have three sets of model parameters for the p-channel and the n-channel transistor respectively: the first set concerns the so-called typical case, that is, the one achieved with a nominal process. The other two describe the fast and slow cases. They refer to a combination of process parameters that, all together, determine a higher or smaller speed of operation. Correspondingly, the fast process leads to higher currents and higher power consumption while, whereas the slow process leads to lower currents and lower power consumption.

Example 1.8

Compare the results obtained with the simple equations derived in the present Chapter and the ones achieved with SPICE. Use a 0.8 μ CMOS twin-well technology whose Spice models (level 2) are given in Appendix A (typical case). Assume that the transistors are in saturation and that their overdrive is around a few hundred millivolts. In addition, discuss the W and L effects on drain current, transconductance and output resistance.

Solution: *We can sketch many test circuits to acquire a feeling for simulator results. A typical way to place a transistor in saturation is to use the so-called diode configuration: that is, with gate and drain connected. In this way we ensure saturation: the drain voltage is larger than the overdrive by the threshold voltage.*
The circuit netlist given below includes six diode connected transistors, of which three are n-type and three p-type; all of them have V_{GS} and V_{DS} equal to 1.2V. The .OP card foresees the calculation of voltages and currents in the operating point.

```
M1 1 1 0 0 MODN W=4U   L=2U
M2 2 2 0 0 MODN W=16U  L=2U
M3 3 3 0 0 MODN W=8U   L=1U
M4 4 4 8 8 MODP W=4U   L=2U
M5 5 5 8 8 MODP W=16U  L=2U
M6 6 6 8 8 MODP W=8U   L=1U
VB 8 0 1.2
.OP
....
```

The output file provides a great amount of information that should always be considered with great attention. A first table is given with the MOS parameters: it contains figures that almost exactly reproduce the ones on the model card. Some of these come from calculations: these are, amongst others, kp *(process transconductance),* gamma, *and* phi *(=$2\Phi_F$). Other relevant figures that we acquire are the parasitic capacitances, the sheet resistance of the source and drain diffusion (*rsh*). For the given process we have 28 $\Omega/\square$ and 32 $\Omega/\square$ for n-type and p-type diffusion respectively.*

```
MOSFET Model Parameters              Temperature =    27.000 Deg C
Type            nmos       pmos
level          2.000      2.000
vto            0.840     -0.690
kp          9.93D-05   3.91D-05
gamma          0.654      0.338
phi            0.788      0.720
pb             0.870      0.870
cgso        3.70D-10   3.70D-10
cgdo        3.70D-10   3.70D-10
cgbo        1.55D-10   1.10D-10
rsh           28.000     32.000
cj          2.94D-04   3.40D-04
mj             0.470      0.500
cjsw        2.94D-10   3.08D-10
mjsw           0.340      0.230
js          1.10D-05   3.10D-04
tox         1.60D-08   1.60D-08
nsub        6.00D+16   1.60D+16
nfs         8.50D+11   5.50D+11
tpg            1.000      1.000
xj          8.20D-08   6.80D-08
ld          0.00D+00   2.40D-08
uo           460.000    181.000
```

```
ucrit      3.60D+05  2.25D+05
uexp          0.300     0.260
utra          0.000     0.000
vmax       6.40D+04  4.60D+04
neff         11.000     2.300
delta         0.220     0.820
```

From the expressions of the process transconductance parameter, eq. (1.49), of the body effect coefficient, eq. (1.51), and the Fermi level shift, eq. (1.10) or (1.11), we can calculate the corresponding approximate values. The results achieved, together with the ones given above, lead to

Parameter	Calculated	Spice
KP (n-channel)	***91.4 μA/V²***	***99.3 μA/V²***
KP (p-channel)	***35.9 μA/V²***	***39.1 μA/V²***
GAMMA (n-channel)	***0.567 V^1/2***	***0.654 V^1/2***
GAMMA (p-channel)	***0.292 V^1/2***	***0.338 V^1/2***
$2\Phi_F$ *(n-channel)*	*0.788 V*	*0.788 V*
$2\Phi_F$ *(p-channel)*	*0.720 V*	*0.720 V*

We observe that more complex models used in Spice lead to different numbers, but the spread is less than 15-20%.
Another important piece of information provided by the output file is a table with the small and large electrical signal features of the transistors used. For the present circuit, the table below is given.

```
**** MOSFETS Model Parameters

            m1         m2         m3         m4         m5         m6
model      modn       modn       modn       modp       modp       modp
id        1.14E-05   4.57E-05   5.41E-05  -1.00E-05  -4.08E-05  -4.90E-
05
vgs          1.200      1.200      1.200     -1.200     -1.200     -1.200
vds          1.200      1.200      1.200     -1.200     -1.200     -1.200
vbs          0.000      0.000      0.000      0.000      0.000      0.000
vth          0.863      0.862      0.832     -0.715     -0.711     -0.689
vdsat        0.277      0.278      0.283     -0.429     -0.434     -0.431
gm        5.89E-05   2.36E-04   2.57E-04   3.79E-05   1.53E-04   1.73E-04
gds       3.21E-07   1.29E-06   2.87E-06   6.56E-07   2.67E-06   6.94E-06
gmb       1.86E-05   7.39E-05   7.34E-05   6.15E-06   2.39E-05   2.36E-05
cbd       0.00E+00   0.00E+00   0.00E+00   0.00E+00   0.00E+00   0.00E+00
cbs       0.00E+00   0.00E+00   0.00E+00   0.00E+00   0.00E+00   0.00E+00
cgsovl 1.48E-15   5.92E-15   2.96E-15   1.48E-15   5.92E-15   2.96E-15
cgdovl 1.48E-15   5.92E-15   2.96E-15   1.48E-15   5.92E-15   2.96E-15
cgbovl 3.10E-16   3.10E-16   1.55E-16   2.15E-16   2.15E-16   1.05E-16
cgs       1.15E-14   4.60E-14   1.15E-14   1.12E-14   4.49E-14   1.10E-14
cgd       0.00E+00   0.00E+00   0.00E+00   0.00E+00   0.00E+00   0.00E+00
cgb       0.00E+00   0.00E+00   0.00E+00   0.00E+00   0.00E+00   0.00E+00
```

From the above-given numbers, we notice a slight change in the threshold voltages; these are likely to be due to the specific sizing of the transistors. We can therefore argue that SPICE includes some dependance on width and length in the threshold calculation.
Using (1.48), the threshold voltages and the estimated process transconductance parameters, we estimate:

$$I_{D1} = \frac{1}{2} \cdot k_P \cdot \left(\frac{W_1}{L_1}\right) \cdot (V_{GS1} - V_{th1})^2 =$$

$$= 0.5 \cdot 91.4 \cdot 10^{-6} \cdot 2 \cdot (1.2 - 0.863)^2 = 10.38\mu A$$

$$I_{D4} = 0.5 \cdot 35.9 \cdot 10^{-6} \cdot 2 \cdot (1.2 - 0.715)^2 = 8.44\mu A$$

which are quite close to what SPICE calculates. By looking at the transistor sizing we would expect a current in M_2 and M_5 which is four times the values in M_1 and M_4 respectively. Actually, this is what Spice nearly achieves. What we notice is that the current in M_3 and M_6: despite having the same designed aspect ratio as M_2 and M_5, has drain currents which are more than 20% larger. This significant difference should come from the different shortening of channel length with respect to the designed figure. Assuming that the lateral diffusion of source and drain is 0.8 xj, *the two actual lengths of the n-channel and the p-channel devices are diminished by* 0.13 μm *and* 0.11 μm *respectively; the modified aspect ratios justify approximately half of the above mentioned current differences.*
Now we calculate the transconductances: from (1.58) we achieve

$$g_{m1} = \frac{2I_{D1}}{V_{GS1} - V_{th1}} = 61.6\mu A/V \qquad g_{m4} = 34.8\mu A/V$$

These figures are within approximately 10% of the ones provided by Spice.
Finally, let us consider output conductance. Using the relationship (1.47) for transistors with length L=2μ, we have: λ_n=0.0204 V^{-1}, *and* λ_p=0.0395 V^{-1}. *Moreover, from the Mosfet Model parameter table we have:* $N_A=1.6 \cdot 10^{16}$ *and* $N_D=6 \cdot 10^{16}$. *Remembering that* $g_{ds}=\lambda I_D$ *we obtain*

$$g_{ds,1} = 2.12 \cdot 10^{-7}\Omega^{-1} \qquad g_{ds,4} = 3.33 \cdot 10^{-7}\Omega^{-1}$$

These results are quite different from the ones achieved with Spice, but the reader should not be surprised: output conductance is one of the most critical parameters. Spice calculates it with the help of a fitting parameter, NEFF, which accounts for second-order effects like short channel and carrier velocity saturation.

1.9 NOISE

Noise is very important for analog circuits. It accounts for unwanted and unpredictable voltage or current fluctuations in a network. Noise is due to fundamental physical principles, but also to undesired coupling between parts of the same circuit. It may also result from interaction between the circuit and the surrounding environment. Noise is represented by current or voltage sources whose amplitude varies randomly in time. Noise is described as a stochastic phenomenon so that the mathematical methods developed to handle such processes are the ones normally used. In particular, we employ the distribution function, $F\{x \leq \xi\}$, that denotes the probability that the noise x will be smaller than ξ. We also use its derivative, the density function, $f(x)$, that allows us to calculate the mean, or the expected value, $\eta = E\{x\}$, and the variance, σ^2, of a random variable x

$$\eta(x) = \int_{-\infty}^{\infty} x dF = \int_{-\infty}^{\infty} x f(x) dx \tag{1.62}$$

$$\sigma^2(x) = E\{(x-\eta)^2\} = \int_{-\infty}^{\infty} (x-\eta)^2 f(x) dx \tag{1.63}$$

Another important mathematical instrument is the autocorrelation function; it is the expected value of the quantity

$$R_x(\tau) = E\{x(t)x(t+\tau)\} \tag{1.64}$$

Under the conditions of stationarity and ergodicity, the autocorrelation is calculated by

$$R_x(\tau) = \lim_{T \to \infty} \frac{1}{2T} \int_{-T}^{T} x(t)x(t+\tau) dt \tag{1.65}$$

It is also possible to define the correlation between two random variables (x,y) by the equation

$$R_{x,y}(\tau) = \lim_{T\to\infty}\frac{1}{2T}\int_{-T}^{T} y(t)x(t+\tau)dt \tag{1.66}$$

Again under the conditions of stationarity and ergodicity, it is possible to define the power spectrum of a random variable, $S(\omega)$, by using its autocorrelation function

$$S(\omega) = \int_{-\infty}^{\infty} R(\tau)e^{-j\omega\tau}d\tau \tag{1.67}$$

From (1.67) it follows that if the autocorrelation is a delta function, the spectrum is frequency independent (or white); otherwise the spectrum is frequency dependent (or coloured).

Circuit noise is described by random voltage or current sources applied to given points of the network. Each noise source is characterised by a given amplitude and a corresponding power spectrum that can be white or coloured. A frequent problem in circuit analysis is the calculation of the global effect of several noise generators. We can distinguish between two cases: correlated and uncorrelated noise generators. Often the sources are uncorrelated, in that they are describing mechanisms causing noise in unrelated parts of a circuit or coming from unrelated physical effects. In the case of uncorrelated sources, we use the superposition principle that tells us to combine all the effects quadratically.

Each single noise generator must be applied to the circuit and the resulting effect estimated at the output. Assuming that a given network includes either voltage and current noise sources, the noise spectrum of the noise voltage at the output results from the following relation

$$v_{out}^2(f) = \sum_i v_{in,i}^2(f)|H_i(f)|^2 + \sum_j i_{in,j}^2(f)|Z_j(f)|^2 \tag{1.68}$$

where $V_{in,i}(f)$ and $I_{in,j}(f)$ are the noise generators while $H_i(f)$ and $Z_j(f)$ are the transfer function and transfer impedance between the noise source inputs and the output. From equation (1.68) we observe that the circuit modifies the noise power spectrum through the voltage and the current noise transfer functions, $H_i(f)$ and $Z_j(f)$. Therefore, an input white noise can become coloured because of the filtering action of the circuit.

Example 1.9

Consider the circuit in Fig. 1.21. It contains an RC network with the white noise voltage generator associated to the resistor; its spectrum is white and holds $v_n^2 = 4kTR$. *Calculate the noise spectrum at the output noise and its total power*

Solution:

The spectrum of the output voltage is calculated by (1.68)

$$v_{n,out}^2(\omega) = v_{n,in}^2 \left| \frac{1}{1 + j\omega RC} \right|^2$$

As expected, the spectrum of the output noise is coloured. Its integral over the entire frequency axis gives

$$V_{n,out}^2 = v_{n,in}^2 \int_{-\infty}^{\infty} \frac{1}{1 + \omega^2 R^2 C^2} \frac{d\omega}{2\pi} = v_{n,in}^2 \frac{1}{2RC} = \frac{2kT}{C}$$

We observe that the result achieved does not depend on the value of the resistance: if the resistance increases, the spectrum of its white noise augments as well but at the same time, the low pass filtering of the RC network becomes more efficient, thus making total power independent of the resistor value. We have to observe that the dimensions of the spectrum $v^2_{n,in}$ *are* V^2sec; *hence, the dimensions of the output power are* V^2.

The previous part of this section outlined the basic elements of noise analysis and calculation. Now we shall discuss noise effects in MOS transistors, which are:

- thermal noise
- flicker noise
- avalanche noise

The first of these comes from carrier fluctuations in the channel because of a

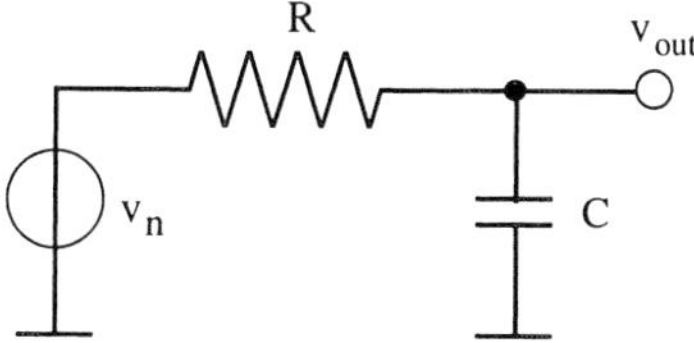

Fig. 1.21 - The simple RC network used in Example 1.9

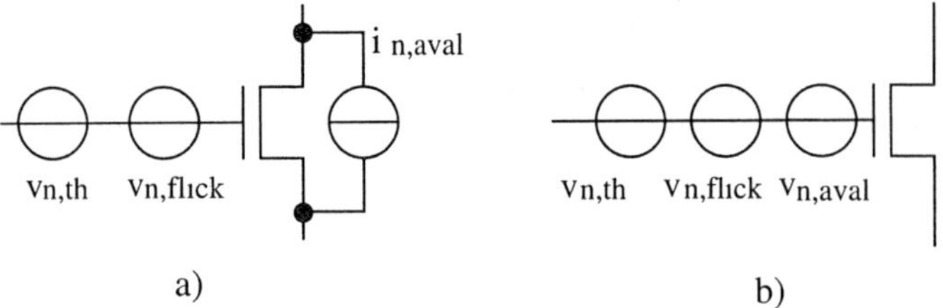

Fig. 1.22 - Equivalent circuit of noise components. a) with avalanche noise at the output, b) with all the noise sources referred to the input

non-zero operation temperature. A voltage generator in series with the gate represents this term (Fig. 1.22). The spectrum of the voltage noise generator referred to the input is given by

$$S_{v^2_{n,th}} = \frac{8KT}{3g_m}\Delta f = \frac{8KT}{3}\frac{q}{q}\frac{q}{g_m}\Delta f \tag{1.69}$$

that shows that the spectrum is white and is inversely proportional to the transconductance of the transistor. Therefore, to reduce such an important noise source, tranconductance should be maximized.

> **LEARN OFF BY HEART**
>
> The thermal noise of an MOS transistor at room temperature can be calculated by
>
> $$S_{v^2_{n,th}} = \frac{11 \cdot 10^{-16}}{g_m}\Delta f$$
>
> where g_m is expressed in *mA/V*.
>
> If g_m= *1 mA/V*, $v_{n,th}$ = 33 nV/√Hz

The second noise contribution comes from the mechanism of generation and recombination of carriers activated by localized energy levels in the forbidden gap. These energy levels either represent localized impurities or they reflect the discontinuity of the band diagram at the MOS surface. As they are spread over a wide energy interval, for this reason, the associated carrier's mean life time is widely distributed. In turn, the noise generator has a coloured power spectrum with shaping very close to *1/f*. It is represented by a voltage generator in series with the gate whose spectrum is

$$S_{V^2_{n,1/f}} = \frac{K_f}{\mu C_{ox} WL}\frac{\Delta f}{f^{\alpha}} \tag{1.70}$$

where K_f is the flicker noise coefficient and exponent α is very close to *1*. The equation shows that flicker noise is inversely proportional to oxide capacitance and gate area. Therefore, good performance can be achieved with thin oxide technologies and a large gate area.

The last contribution to noise comes from the avalanche component of the drain current. Its effect becomes significant only when the drain to source voltage is very high and determines a large electrical field near the drain. This term can be represented by a noise current generator in parallel with the output terminals (source and drain); its spectrum is white

$$S_{I^2aval} = 2qI_{aval}\Delta f \tag{1.71}$$

where q is the charge of the electron and I_{aval} is the avalanche current. We can, of course, refer the contribution to the input terminal by dividing by the square of the transconductance gain

$$\frac{v_{n,aval}^2}{\Delta f} = \frac{2qI_{aval}}{g_m^2} = \frac{qI_{aval}}{\mu C_{ox}(W/L)I_D} \tag{1.72}$$

if we want to estimate the impact of avalanche noise, we can compare it to the thermal noise contribution

$$\frac{v_{n,aval}^2}{v_{n,th}^2} = \frac{qI_{aval}}{2KTg_m} = \frac{V_{GS}-V_{Th}}{4(KT/q)}\frac{I_{aval}}{I_D} \tag{1.73}$$

It turns out that the avalanche noise becomes important when the avalanche current becomes a non-negligible fraction of the drain current. This condition occurs at a limit of operation for transistors rarely used in analog circuits.

1.10 LAYOUT OF TRANSISTORS

The last step in integrated circuit design is physical description. This consists of defining the masks to be used for processing. An MOS transistor is

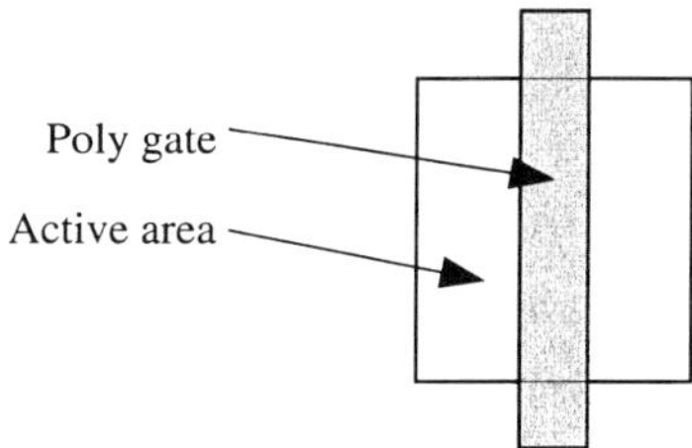

Fig. 1.23 - The two layers that achieve a transistor

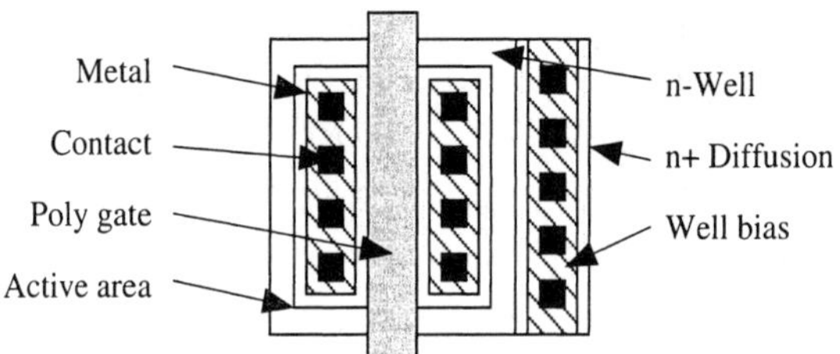

Fig. 1.24 - Layout of a p-channel transistor (inside an n-well)

achieved by the simple overlap of two rectangles: one defining the active area and the other defining the polysilicon gate (Fig. 1.23). The parts of the active area that are not protected by the gate originate the source and the drain, while the part protected by the gate forms the transistor channel. To ensure that the source and the drain are separated, even in presence of fabrication inaccuracies, the gate overlaps the active area to a given extent, its value being defined by the design rules of the technology used.

Of course, the physical description of a transistor is not limited to the masks of the active area and polysilicon. When the transistor must be realized in the well, a suitable pattern must be defined. Moreover, it is necessary to arrange the connections of source, drain and gate together with the substrate and well biasing. A typical layout of a MOS transistor (sitting in the well) is shown in Fig. 1.24. It represents a pattern typical of analog circuits: the aspect ratio (W/L) of the transistor is not at a minimum, as is usually the case for analog designs.

The key points to consider when we draw transistor layouts are the following:

- parasitic resistances at source and drain must be kept as low as possible
- parasitic capacitances should be minimized
- matching between paired elements is very important

Concerning the first condition, we should remember that drain and source diffusions have a given sheet resistance. With only a few squares, we can achieve hundreds of ohms of resistance: even with a current as low as a few tens of μA we can have drop voltages of millivolts. Therefore, as shown in Fig. 1.24, we must use multiple contacts on top of the source and drain regions to avoid parasitic transversal drop voltages. Designers prefer multiple contacts placed at a minimum distance instead of using a single large contact. Many contacts placed close to each another make the surface of metal connections smoother than when using only one contact; this prevents microcracks in the metal that can be a source of failure.

Parasitic capacitances derive from the reverse source-substrate or drain

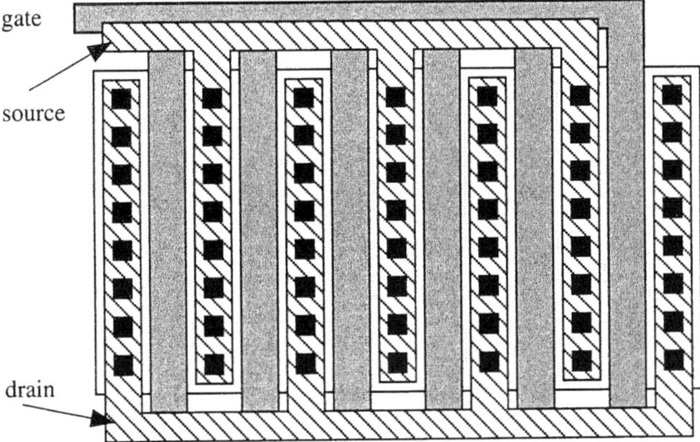

Fig. 1.25 - Interdigitized transistor.

substrate diodes. We have just seen that this is useful in establishing good contacts. Hence, the source and drain area must be large enough to accommodate contacts and to fulfil the design rules. However, it is possible to reduce the source and drain area and, consequently, reduce the parasitic capacitances. This is achieved using the layout shown in (Fig. 1.25).

> **DESIGN HINT**
>
> Reduce to a minimum the area of the diode source-substrate (or drain-substrate). This will procure minimum parasitic coupling and minimum leakage current.

The transistor is split into a given number of parts that are connected in parallel. We can see that most of the source and drain area is used doubly allowing the left and right parts of the transistors to be connected. It follows that the parasitic capacitance can be reduced up to a factor 2. We can obtain a further reduction with the waffle solution shown in Fig. 1.26. Here we see that a source or drain contact serves four pieces of gate. The parasitic capacitance is now reduced by a factor of four. However, since we have an unclear situation at the corners, the aspect ratio (W/L) together with width and length are not well-controlled. The waffle layout is rarely used; the compactness of the given solution is exploited only when the accuracy of the aspect ration is not an important issue.

Matching is very important when we have to design current mirrors and differential pairs. In general, bad matching produces high offset. Therefore, we have to use layouts that optimize matching. This is achieved by providing the best symmetrical conditions. Transistors with different orientation (Fig. 1.27 a) match badly. Moreover, we can suffer mismatch if the current in transistors is flowing in opposite directions (Fig. 1.27 b). In addition, we can effect a change

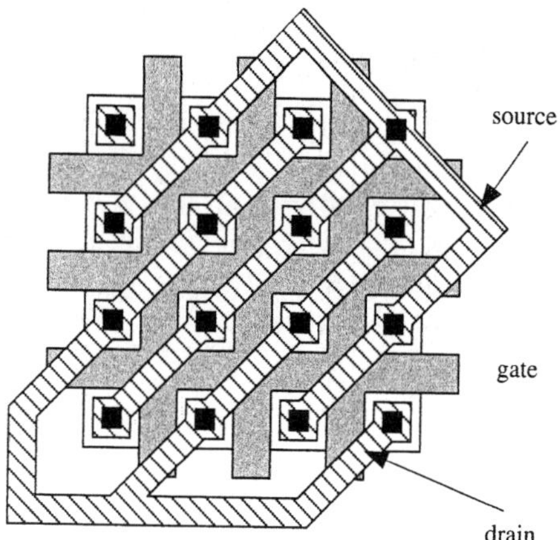

Fig. 1.26 - Layout of a waffle transistor (to use only when the accuracy of the aspect ratio is not a relevant issue)

in physical and technology parameters in points of the chip that are relatively far away.

> ***LAYOUT TIPS***
>
> - In the analog layout of the entire chip use transistors with the same orientation.
> - Split transistors into parallel elements and secure layout flexibility
> - Solidly and repeatedly bias substrate and well.

The best methods of achieving good matching are shown in Fig. 1.28. We assume that the two transistors that should match have one of the terminals (source or drain) in common so that we can use an interdigitated arrangement. Each transistor is split into four equal parts; they are interleaved in two by two's so that for one pair of pieces of the same transistor we have currents flowing in opposite directions. The layout has a slight dissymmetry; the centroid of transistor M_1

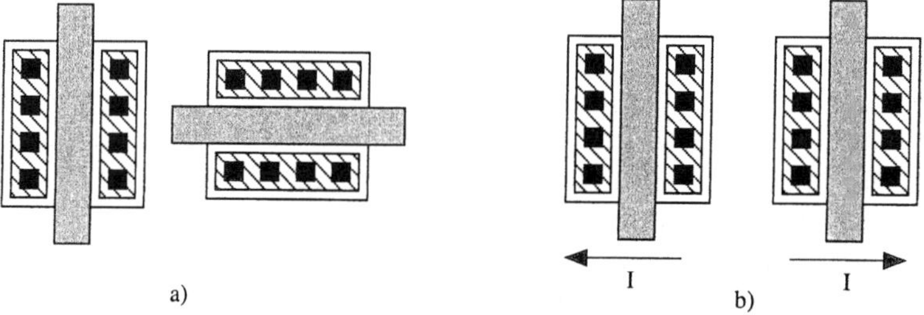

Fig. 1.27 - Badly matching transistor: a) bad orientation b) with opposite current flow

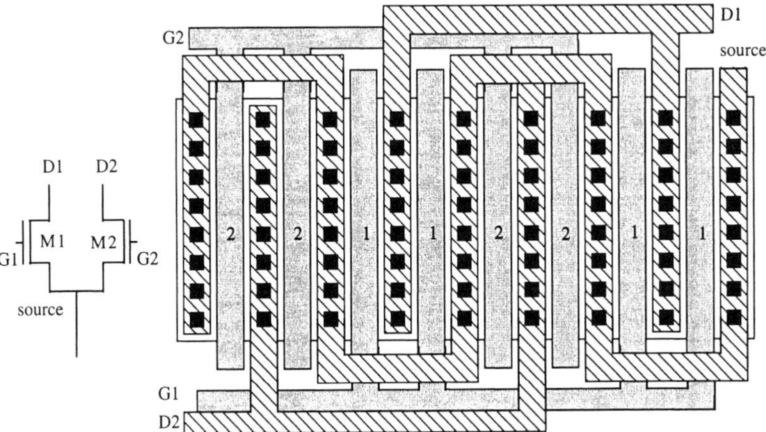

Fig. 1.28 - Layout of a matched transistor pair

is shifted to the right compared to the centroid of M_2. This dissymmetry can be corrected using two structures like the one shown inwith one on top with two pieces of M_2 on the left and the other on the bottom and with two pieces on M_2 at the right. The two structures are connected in parallel, using a cross connection. In addition to this point, it is worth anticipating a potential problem that will be discussed in detail in the next chapter: asymmetries associated with the undercut effect in the presence of different boundary conditions. This can result in the small differences between the width of boundary transistors in the stack shown in Fig. 1.28, for example.We shall that this problem can be in general alleviated using dummy elements.

A final point worth discussing concerns the biasing of substrate or of the well. This is a very important issue: we have to ensure that the biasing is as close as possible to the active devices. Any noisy signal affecting the substrate or the well should be sunk by the biasing and should not affect the circuit itself. For this reason, any possible silicon space should be used for biasing purposes.

1.11 DESIGN RULES

We have already mentioned the need to design the layout according to given rules. They are a set of geometrical recommendations (actually, they are restrictions) that guide the correct realization of the layout. The design rules take into account the limited accuracy of the technology. Among them, the limits deriving from mask alignment, mask non-linearities, lateral diffusion, etching undercut, optical resolution and diffraction, and others. The design rules pursue an optimum trade off between performance and yield. If for example, a given tech-

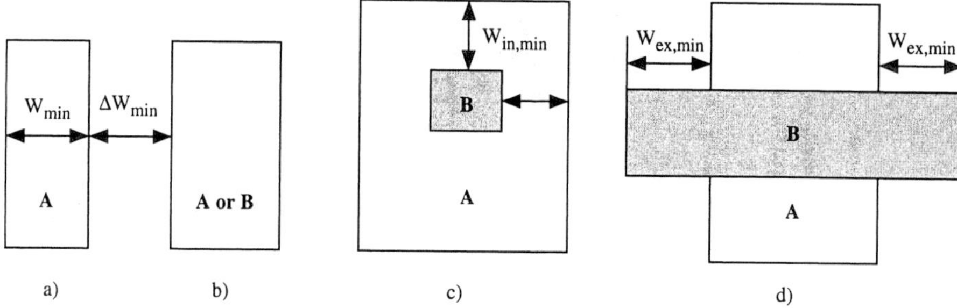

Fig. 1.29 - The four categories of design rules: a) Element width, b) Elements spacing, c) Inner overlap, d) External extension.

nology allows the contact and the metal masks to be aligned within a given accuracy, the width of the metal line placed over the contact must be suitably wide to ensure that the contact is overlapped even in the worst case of misalignment: just one missing contact in the circuit can be responsible for a catastrophic failure. Similarly, for metal etching to be effective we have to remove a suitable amount of material to prevent short or resistive parasitic connections between two unrelated metals.

The design rules define geometrical relations referring to the following four possibilities (see Fig. 1.29):

- Element width, W_{min}: it is the minimum (or the maximum) width allowed for a given element. It avoids possibly opening or vanishing of the element. For example we have a rule defining the minimum width of the poly gate and a rule defining the minimum size of poly-metal contacts.
- Elements spacing, ΔW_{min}: this is the minimum distance between two elements. This rule avoids shortening. The elements can be of the same kind (for example, metal-metal) or of a different kind. For example, we have a rule defining the minimum distance between two metal lines or the minimum distance between a poly line and an unrelated diffusion.
- Inner overlap $W_{in,min}$: is the minimum separation between two elements that we design one inside the other. This rule avoids the two elements detaching. For example, we have a rule defining the inner overlap of the contact over or below a metal.
- External extension, $W_{ex,min}$: is the minimum extension of an element overlapping another element. This kind of rule ensures that the two elements are fully overlapped. For example, we use a design rule to ensure that the poly gate always crosses the active area.

The above four categories of design rules are specified for all the possible layers used by the technology. The above description of possible rules refer to the minimum spacing. However, some rules require the assigned figure to be "exact". Therefore, excepting the latter case, the designer can exceed the minimum spacing by an extent considered appropriate.

The design rules are normally specified in an ad-hoc document: the so-called "Design rules Manual". The DR Manual is normally provided by the semiconductor house together with other informations (like the Spice models, technology features, etc.). All these documents constitute a package called "Design kit".

It is very important to check that the design rules are not violated. This operation is accomplished with suitable computer tools: the Design Rules Checker, (DRC). They are normally included in the same computer package for the physical design and the data base layout generation.

1.12 REFERENCES

- J. P. McKelvey, *Solid State and Semiconductor Physics*, Harper & Row, New York, N.Y., 1966
- A. J. Dekker, *Solid State Physics*, Prentice-Hall, Englewood Cliff, New Jersey, 1964
- J. E. Franca, Y. Tsividis, *Design of Analog-Digital VLSI Circuits for Telecommunications and Signal Processing*, Prentice-Hall, Englewood Cliff, New Jersey, 1994
- R. L. Geiger, P. E. Allen, N. R. Strader, *VLSI, Design Techniques for Analog and Digital Circuits*, McGraw-Hill Publishing Company, New York, N.Y., 1990
- Sze, *VLSI Technology*, McGraw-Hill Publishing Company, New York, N.Y., 1988
- R. Muller, T. Kamins, *Device Electronics for Integrated Circuits*, Wiley, New York, N.Y., 1972
- G. W. Roberts, A. S. Sedra, *Spice*, Oxford University Press, New York, N.Y., 1997

1.13 PROBLEMS

1.1 Calculate the intrinsic concentration at 100 K, 200 K and 300 K. Calculate the number of times the concentration doubles in the two intervals.

1.2 Calculate the concentration of minority carriers at room temperature in a material which p-type doping is $5 \cdot 10^{16}$ cm^{-3}.

1.3A wafer of silicon is placed for *1* hour in a dry oxidation fumace at *1000°C*. Afterwards, the temperature is cooled down to *800°C* for 10 hours. Estimate the thickness of the oxide grown.

1.4 Calculate the capacitance of a *50 μm* x *5 μm* MOS structure. The oxide is a *12 nm* silicon dioxide layer. Use *3.4* as dielectric constant of silicon. Estimate the maximum voltage that we can safely use.

1.5 A thin layer of n-doped silicon ($h = 1\ \mu m$, $N_D = 10^{17}\ cm^{-3}$) is used to realize a resistor. Find the sheet resistance with the help of Fig. 1.4. Calculate the sheet resistance for an equal structure achieved using p-type material.

1.6 Repeat the previous problem for a *p*-doped layer with $h = 0.5\mu m$ and $N_A = 10^{16}\ cm^{-3}$.

1.7 Given an MOS transistor with substrate doping $N_A = 10^{17}\ cm^{-3}$, interface charges $Q_{SS} = -10^{-9}\ cm^2$, oxide thickness *8 nm* and $\varepsilon_{r,ox} = 4$; determine the threshold voltage. Calculate the variation of threshold caused by an additional interface charge of $+\ 5\ 10^{-8}\ cm^{-2}$. Find the body effect coefficient.

1.8 Determine the current in an n-channel and a p-channel transistor (use the transconductance parameters given in (1.49)) with $(W/L) = 10$, $n = 1.5$, $V_{th} = 0.8V$, $V_{GS} = 1.3\ V$ and $V_{GS} = 0.75V$. Calculate the small signal parameters g_m, g_{ds}. Assume $N_A = N_D = 10^{15}\ cm^{-3}$, $L = 0.5\ \mu m$.

1.9 Repeat the previous problem with $V_{GS} = 0.95V$, $N_A = 10^{16}\ cm^{-3}$, $N_D = 10^{17}\ cm^{-3}$. What happens when V_{DS} rises to $3V$.

1.10 The source contact of a transistor with width *100 μm* is *1.5 μm* large to accommodate a *0.5 μm* contact and allow a *0.5 μm* spacing between contact and gate (or diffusion). The leakage currents of the source substrate diode are the ones in Example 1.6. Calculate the leakage current for a linear layout and the reduction that we have when the transistor is split in two parts with the source in the middle.

1.11 Determine the transconductance of an MOS transistor with $K_n = 112\ \mu A/V^2$, $(V_{GS} - V_{th}) = 0.5V$, $I_D = 400\mu A$, $(W/L) = 16$. Compare the result with the equivalent transconductance of a bipolar transistor. Assuming that the transconductance increases as the square root of the aspect ration *(W/L)*, calculate the aspect ratio that we have to use to achieve with the MOS the same transconductance of the corresponding BJT.

1.12 A p-channel MOS ($K_p = 40\mu\ A/V^2$) has an aspect ratio $W/L = 20$ and a drain current $I_D = 1\mu\ A$. Find the overdrive voltage, $(V_{GS} - V_{th})$. Determine the current at which the transistor enters the sub-threshold region.

1.13 A p-channel MOS ($K_p = 40\mu\ A/V^2$) has a drain current $I_D = 4\ \mu\ A$. Find the aspect ratio at which the transistor is at the sub-threshold region limit.

1.14 Using the two set of SPICE models in Appendix A, determine the λ factor for an n-channel and a p-channel transistor. Compare the result with the approximate equation (1.47). Use $L = 5\mu;\ 3\mu;\ 1\mu;\ 0.5\mu$.

1.15 Determine, using SPICE, the I-V characteristics of a "diode connected" transistor $V_{GS} = V_{DS,}$ $(W/L = 10)$. Find the current for $V_{GS} - V_{th} = 0.3V$.

1.16 Repeat Example 1.8 using the BSIM models given in Appendix A. Use lengths ranging from 0.4μ to 1μ.

1.17 Simulate the input/output characteristics (I_D versus V_{DS} with V_{GS} as a parameter) of an MOS transistor $W/L = 1;\ 10;\ 50$). Determine, using (1.61) the boundary between the weak inversion and the saturation region. Estimate, with computer simulation, the variation of transconductance around given point.

1.18 Using SPICE, determine the capacitance $C_{GS,}$ C_{GD} and the overlap terms in triode and in saturation conditions. Compare the achieved results with the gate capacitance $C_{ox}\ WL$. The oxide thickness is provided by the SPICE parameter tox.

1.19 Determine the output power noise considered in Example 1.9 in the case of noise generated by the thermal effect of the resistance itself. What is the value of $V^2_{n,out}$ with $C = 100\ pF;\ 10\ pF;\ 1\ pF;\ 0.1\ pF$?

1.20 Sketch the layout of four transistors having the source in common (connected to ground and the following dimensions: $W_1 = 40\ \mu m$, $W_2 = 80\ \mu m$, $W_3 = 120\ \mu m$, $W_4 = 160\ \mu m$, $L_{1,2,3,4} = 0.35\ \mu m$. The transistor can be split in a given number of parts and they must occupy the minimum area. The drain connections must be on the top of the layout.

Useful Constants

Parameter	Value	Unit
Boltzman constant	$8.62 \cdot 10^{-5}$	*eV/K*
Electron charge	$1.6 \cdot 10^{-19}$	*Coul*
Carrier intrinsic concentration @room temperature	$1.42 \cdot 10^{10}$	cm^{-3}
Si atomic density	$5 \cdot 10^{22}$	$atoms/cm^3$
Si reticular constant	*0.543*	*nm*
Si thermal conductivity	*1.41*	*Ω/cm °C*
Si intrinsic resistivity (@ 300 °K)	$2.5 \cdot 10^5$	*Ω–cm*
Si relative dielectric constant, ε_r	11.9	-
Absolute dielectric constant, ε_0	$8.858 \cdot 10^{-14}$	*F/cm*
SiO_2 dielectric strength	$2\text{-}8\ 10^6$	*V/cm*
SiO_2 resistivity (@ 300 °K)	$10^{15}\text{-}\ 10^{17}$	*Ω cm*
SiO_2 relative dielectric constant	*3.4 - 4.2*	-
Specific capacitance of a 16 nm thick oxide (ε_r=3.6)	*2*	fF/mm^2
Typical transconductance parameter t_{ox} = 12 nm, μ_n = 460 cm^2/Vsec	*120*	$\mu A/V^2$
Typical transconductance parameter t_{ox} = 12 nm, μ_p = 148 cm^2/Vsec	*39*	$\mu A/V^2$

Chapter 2

RESISTORS, CAPACITORS, SWITCHES

An analog integrated circuit is made of active and passive components. Active elements (n-channel or p-channel transistors) have already been described in the previous chapter. This chapter examines the basic properties of integrated passive components, which are: resistors, capacitors and analog switches. Although, as we will see, analog switches are made of one or a couple of MOS transistors (which are active elements), they are here classified in the passive component category because in reality they do not implement any active function. In fact, a switch, is merely a time-variant resistance. The performance of passive elements in monolithic form is quite different from that of the corresponding ideal or discrete versions. Therefore, before starting to consider the design techniques of integrated circuits, we should know their major characteristics and limits.

2.1 INTEGRATED RESISTORS

A resistor in integrated technology is made of a thin strip of resistive layer. Fig. 2.1 shows its typical structure: a long sheet of resistive material is con-

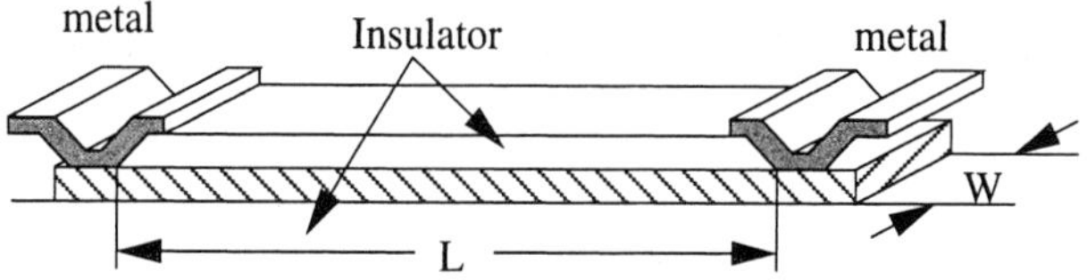

Fig. 2.1 - Typical integrated resistor

nected to metal terminals by two ohmic contacts. The body of the resistance is electrically insulated from its surroundings by an oxide layer or by a reversely biased junction. If we assume that the sheet resistance is $R_{\square}$, the total resistance, R, is given by

$$R = 2R_{cont} + \frac{L}{W}R_{\square} \tag{2.1}$$

where R_{cont} is a localized resistance describing the endings and metal connection contacts. The contact resistance value depends on the specific case; it can range from 10Ω to 50Ω. The L/W ratio expresses the aspect ratio; sometimes it is referred to as the number of squares since it is actually the number of squares that we see when looking down at the top of the structure.

We learnt in the previous chapter that resistive elements are mainly achieved by diffused layers (n^+, p^+, and well), or by polysilicon (first poly or, when available, second poly). Moreover, we know that specific resistances range from a few tens of $\Omega/\square$ to a few $k\Omega/\square$. Therefore, it is not difficult to achieve resistors up to a few $k\Omega$, while it becomes a serious problem reaching hundreds of $k\Omega$ or $M\Omega$. Fig. 2.2 shows cross-sections of various CMOS integrated resistor structures. Fig. 2.2 a) and b) use highly-doped diffusions. Both refer to *n-well* technology; the first utilizes a n^+ diffusion achieved directly on the substrate, the other has a p^+ diffusion sitting inside the well. For both structures, insulation is ensured by reversely biased junctions. We can observe that the n^+ and p^+ layers contact the metal lines directly. In contrast, the

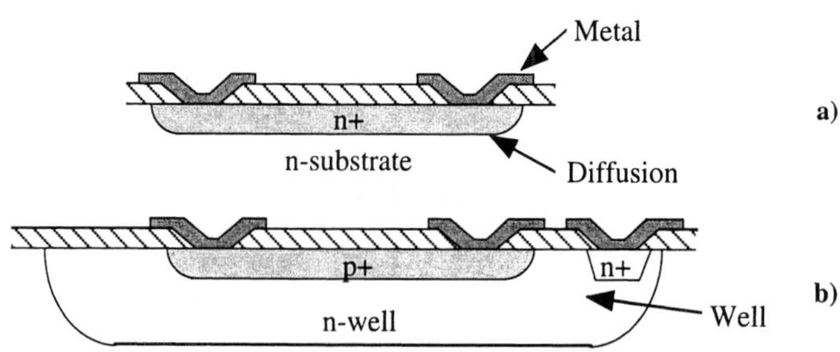

Fig. 2.2 - Cross section of integrated resistances: a) diffusion b) diffusion into well

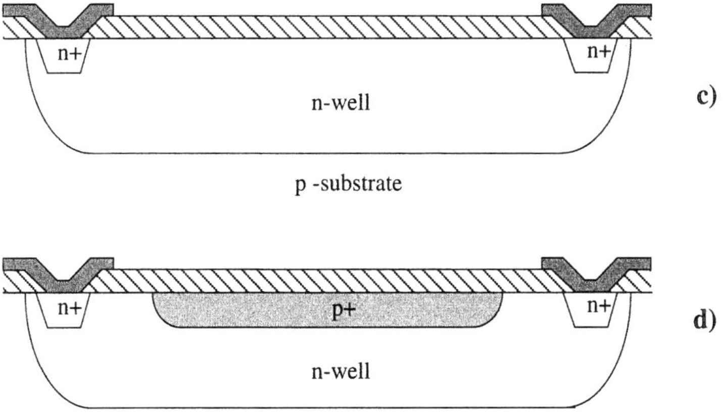

2.2 (c and d) - Cross section of integrated resistances: c) well d) pinched well

lightly doped diffusion of the well has interposed a n^+ diffusion. This happens because a lightly doped silicon to metal junction realizes a rectifying structure. We achieve an ohmic contact only with heavily doped silicon-to- metal junctions.

The reversely biased diode used to ensure the insulation of the resistance generates a capacitive coupling between the body of the resistor itself and the substrate (or the well). Such a parasitic capacitance, in normal cases, does not produce a significant frequency limitation. However, it is necessary to deal with it because of the noise that could affect the resistor. We have seen that spur signals often affect the substrate: through the parasitic coupling, the noise reaches the resistance and worsens circuit performance. In this respect, the resistor in Fig. 2.2 b) that we place in the well is better protected since the well itself provides local shielding.

Well layers offer a specific resistance that is larger than p^+ or n^+ diffusion, therefore, they are preferred when designing large resistors. Fig. 2.2 c) shows the basic structure of a well resistor. In Fig. 2.2 c) and Fig. 2.2 d) we have a p-type substrate, unlike what was used in Fig. 2.2 a) and Fig. 2.2 b). This is just to remind the reader that in integrated technologies, we can find either *n-type* (with *p-well*) and *p-type* (with *n-well*) substrate. Going back to the well resistors, again we note that we have two highly doped diffusions that ensure ohmic

KEEP IN MIND!

The substrate noise is one of the most important limit to resolution in mixed signal integrated circuits. Integrated resistors are very sensitive elements. We have to carefully protect them against substrate noise coupling.

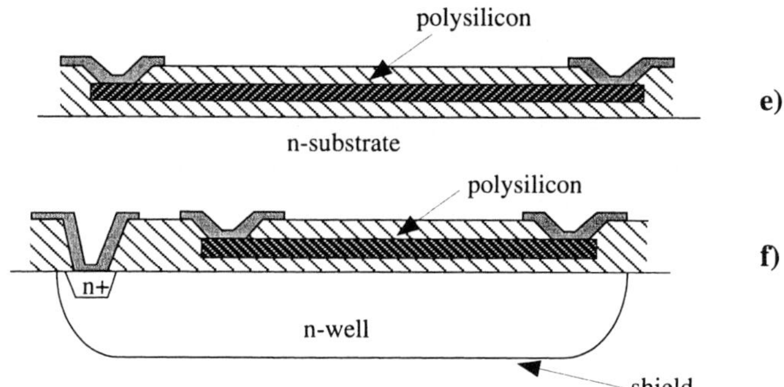

2.2 (e and f) - Cross section of integrated resistances: e) polysilicon f) with a well shielding

contacts at the terminals. Moreover, the insulation of resistors from their surroundings is achieved by the reversely biased diode: the well-to-substrate. For this kind of resistor, we have the same problems of coupling with substrate as discussed above. However, any shielding to limit noise injection is, of course, impossible. The only possibility that we have is to create a tight substrate biasing placed around the well that is likely to drain spur signals before they reach the well.

The specific resistance of wells is quite high. Nevertheless, for getting further highly resistive layers, we can put a complementary diffusion on the top of the well itself to reverse the doping at the surface. The resistance is achieved with a buried layer, which means the doping concentration is much higher than at the surface. The resulting structure in Fig. 2.2 d), called *pinched well*, shows a significant increase in specific resistance (from a few to several $k\Omega/\Box$). Moreover, the conduction takes place in the bulk of the material. We therefore have a reduction of the low frequency noise ($1/f$) as well, mainly produced by surface defects.

We can also design resistors with polysilicon. Fig. 2.2 e), Fig. 2.2 f), Fig. 2.2 g) and Fig. 2.2 h) show their typical structures. We can use polysilicon 1 and, when available, polysilicon 2. The use of polysilicon allows us to better confront the problem of shielding. Since polysilicon is not in the direct vicinity of the substrate, we have a weaker coupling compared with diffusions. Moreover, we can employ single or double shielding. The former is achieved by using the well (Fig. 2.2 f) or by making the resistor with poly 2. The latter can be carried out by using two levels of shielding with poly 1 and well (Fig. 2.2 g). Of course, special care should be taken to ensure the quietness of the voltage at which the shields are connected.

In some situations (for example, when the circuit uses plastic packages), coupling with a noisy environment can also be derived from the top of the structure.The plastic used for the package is slightly conductive (especially at

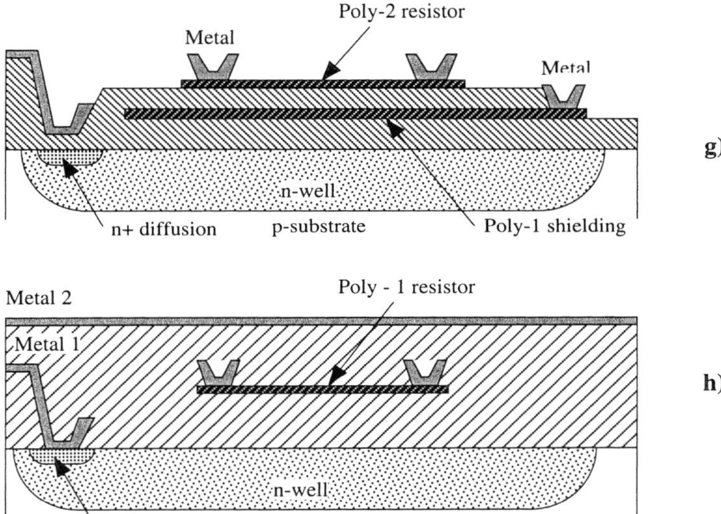

2.2 (g and h)- Cross section of integrated resistances: g) polysilicon h) with well shielding

high frequencies) and since it is in close proximity to silicon, it can lead to spur signals. In this case we should ensure shielding from the top. This can be achieved using, for example, metal 2 on the top of the entire structure (Fig. 2.2 h). Another case for which we have to shield the resistor on the top is when noisy lines over-cross the resistors. For this, we need two levels of conductors: one for shielding (metal 1) and one for the over-crossing (metal 2).

2.1.1 Accuracy of Integrated Resistors

In the previous section we discussed how to make resistors in integrated technology. In this section we consider both the achievable absolute accuracy and the achievable relative accuracy (matching accuracy). For this purpose, we make the assumption that the resistance of the strip dominates the resistance of the entire structure: contacts should be designed to have negligible resistance. We have

$$R = \frac{L}{W} R_{\square} = \frac{L}{W} \cdot \frac{\bar{\rho}}{x_j} \tag{2.2}$$

where $\bar{\rho}$ is the mean resistivity, L and W define the number of squares and x_j is the average thickness of the strip.

We observe that the value of the resistor depends on four parameters that are achieved with independent technological steps. Therefore, to estimate their accuracy we assume them to be statistically independent. The standard devia-

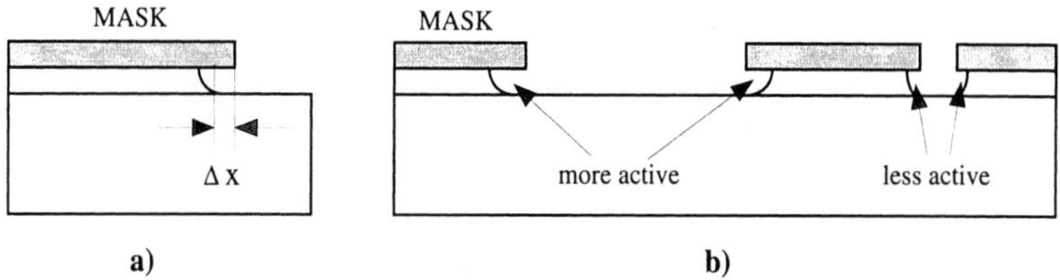

Fig. 2.3 - Undercut effect and its boundary dependence

tion of the resistor is thus related to the standard deviation of parameters by

$$\left(\frac{\Delta R}{R}\right)^2 = \left(\frac{\Delta L}{L}\right)^2 + \left(\frac{\Delta W}{W}\right)^2 + \left(\frac{\Delta \bar{\rho}}{\bar{\rho}}\right)^2 + \left(\frac{\Delta x_j}{x_j}\right)^2 \tag{2.3}$$

Often, the structures used in practice have a much greater length than width. This is because we normally need values higher than the specific resistance. It follows that the contribution of the length is negligible compared to the contribution of the width. Therefore, when we design resistors, we should keep the width under control much more than the length. To this end, we should remember that the accuracy of geometrical parameters mainly depends on the lithographic steps used to define patterns and on the successive technological processing handling. The major resulting errors are:

- undercut effect
- boundary mismatch
- side diffusion

A pattern defined by etching does not correspond to the designed pattern. We have errors in the fabrication of masks and errors in the optical apparatus; these limits are not listed above because the most important limitation in etched strips comes from the undercut effect. This is due to the anisotropic action of the etch: to be sure that patterns are well-defined, the etching action is continued beyond what is required and erosion under the protection (undercut) results (Fig. 2.3 a)). The outcoming pattern is smaller than the designed pattern by an amount, Δx, which depends on the etched material and on the etching process. For wet etching of polysilicon, Δx ranges from *0.2* μm to *0.4* μm with standard deviation in the order of *0.04 - 0.08* μm. Undercut is smaller for reactive ion etching, where Δx is around *0.05* μm with standard deviation *0.01* μm.

A further source of inaccuracy is the boundary dependence of undercut etching. As shown in Fig. 2.3 b), etching is more active when the region to be removed is wide. By contrast, it is less effective if the strip to be removed is narrow. For this reason, the undercut on the strip in the middle in Fig. 2.3 b) results differently on the two sides because the boundaries are different. Thus, when a number of parallel closed strips must be designed, as occurs in serpentine resistances, the two terminal elements undergo larger etching on the external side. We thus have a smaller width inducing a larger resistor value in the terminal elements and, in turn, an error in the resistor value. We can solve the problem by the use of dummy strips placed around the structure. Dummy elements do not have any electrical function. They are used exclusively to define symmetrical geometrical boundary conditions, thus giving the same undercut effect on the two side of the etched strip. We will see the layout of this kind of structure shortly.

Example 2.1

An integrated resistance is made like the serpentine structure shown in Fig. 2.6 a). The specific resistance of the material used is 30 Ω/□; the width of the strips is 3 μm; each element is 42 μm long and the distance between the centers of the two elements is 6 μm. Find the nominal resistance, the actual resistance and the inaccuracy coming from the geometrical contribution. The undercut is 0.2 μm with standard deviation 0.04 μm. The inaccuracy at the corners is 30% of the nominal contribution (1/2 square).

Solution:
Analyzing the figure, we see that we have 6 vertical parts of 12 squares, five horizontal parts of 1 square and 10 corners. Since the corners contribute to half of a square, the nominal resistance is

$$R_{nom} = R_{\square}N_{squares} = 30(12 \cdot 6 + 5 + 10/2) = 2460\Omega$$

In reality, because of the undercut effect, the effective width of the strip is 2.6 μm. In contrast, the length of the various parts increases: the vertical elements (excluding the corners) become 36.4 μm and the horizontal ones 3.4 μm. The squares at the corner becomes smaller (2.6 × 2.6 μm) but they are still squares. Therefore, accounting for these effects we have

$$R_{actual} = R_{\square}N'_{squares} = 30(14 \cdot 6 + 5 \cdot 1.3 + 10/2) = 2865\Omega$$

which is 16% larger than the nominal figure.
From the equations shown above, we observe that the squares achieved account for 1715 Ω while the corners account for 150 Ω. Because of the standard deviation of the undercut effect, the width of the strip can vary from 2.56 to 2.64 μm and the resulting error of the resistance is ±1.5% (±25.8 Ω). Since the inaccuracy deriving from the corners is 30% of their resistance, we have a possible error of ±45 Ω. Assuming that the two contributions are fully corre-

lated, we have a total error of ±70.8 Ω corresponding to 2.6% of the actual resistance. By contrast, in case of fully uncorrelated contributions we have an error of ±51.9 Ω.

Another element contributing to geometrical inaccuracy which at the same time causes difficulties in calculating the resistance, is lateral diffusion under the protecting mask. This effect occurs when we make any diffusion. We depose dopants through the protection mask and we activate their diffusion into the material with a thermal process. The diffusion proceeds in the three dimensions and we change the doping even under the protection to a given extent. If the depth of the metallurgical junction is x_j, the side diffusion results about $0.6\text{-}0.8\ x_j$ (Fig. 2.4) with a standard deviation in the order of $0.05\ x_j$. This effect corresponds to an equivalent enlargement in the width of the strips. However, the increase achieved is not equal to the side diffusion because its resistivity is higher than in the body of the strip. Typically, we have to account for only 30-*50%* of the side diffusion.

We see now how the accuracy of a resistor depends on the accuracy of its mean resistivity, $\bar{\rho}$. It can be affected by the following sources of error:

- polysilicon grain size
- crystal defects
- doping dose
- stress
- temperature

Polysilicon is made of a conglomerate of monocrystalline grains. Its resistivity depends on a complex mechanism of conduction which is strongly affected by discontinuity at the grain borders. Since the number of these discontinuities is related to the poorly-controlled dimension of the monocrystalline grains, polysilicon resistivity has a limited accuracy. We achieve better controlled resistivities with diffused resistors where the conduction takes place in monocrystalline material and does not suffer from discontinuity. However, this is not wholly true: even over much smaller areas, the bulk material is never perfect. During crystal growth, some defects are introduced into

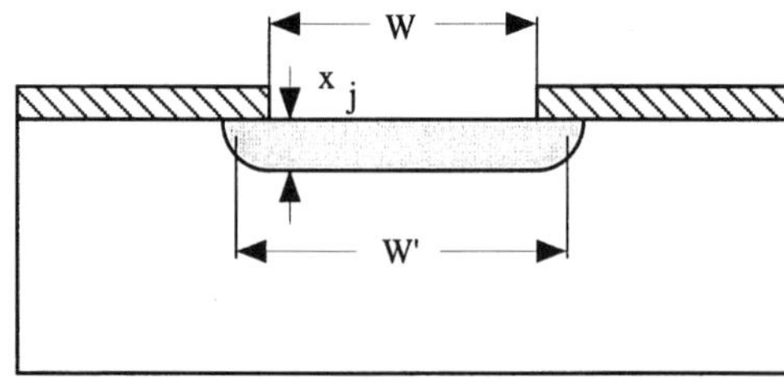

Fig. 2.4 - The side diffusion enlarges the width

the crystallographic structure. From the electrical point of view, theireffect is to perturb conduction. They introduce localized potential barriers to the carriers whose mobility is reduced locally, and, as a global result, the resistivity is not accurately defined.

In addition, inaccuracy in resistivity is determined by the limited control of the doping dose. Depending on the process, doping is determined by the flow of a given gas, the concentration of dopant on it, the temperature in the furnace and many other parameters. All of these are affected by uncontrolled fluctuations that globally determine an inaccuracy in the doping level.

> **REMARK**
>
> The absolute and matching accuracy of diffused layers are superior to those of the polysilicon layers. However, it is easier to shield polysilicon layers since they are "distant" from the substrate.

An important factor contributing to error in resistivity is stress. Plastic packages are made by injecting fluid plastic into a mould. When it becomes solid, its volume diminishes and produces a stress on the chip surface. The pressure achieved is very high (many hundreds of atmospheres), close to the fracture limit of silicon. Under such high pressure, a change of the material's surface structure results and, in turn, we have a modification in resistivity. With silicon slices in the <100> direction, we have an anisotropic response: resistivity in one crystallographic direction changes differently than in the orthogonal direction. Since the minimum occurs at *45°*, this orientation is sometimes used in the layout to minimize stress contribution to inaccuracy (for <111> materials we isotropic behaviour).

We have seen in Chapter 1 that temperature strongly affects the resistivity of

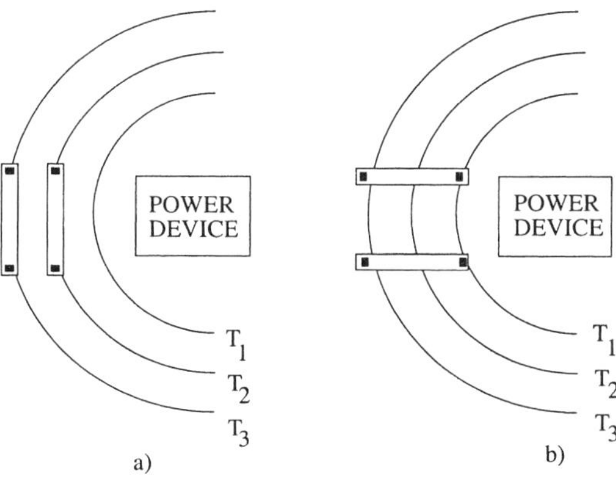

Fig. 2.5 - The temperature in the resistor's centroid is different in case a) and equal in case b)

TABLE 2.1:

Resistor features

Type of layer	Sheet Resistance $\Omega/\Box$	Accuracy %	Temperature Coefficient ppm/°C	Voltage Coefficient ppm/V
n + diff	30-50	20-40	200-1K	50-300
p + diff	50-150	20-40	200-1K	50-300
n-well	2K-4K	15-30	5K	10K
p-well	3K-6K	15-30	5K	10K
pinched n-well	6K-10K	25-40	10K	20K
pinched p-well	9K-13K	25-40	10K	20K
first poly	20-40	25-40	500-1500	20-200
second poly	15-40	25-40	500-1500	20-200

diffused layers and, in turn, that of integrated resistors, particularly for lightly-doped materials. The temperature coefficient can be as large as *0.5-1%/°C*, corresponding to a significant increase of the resistor's value for modest increases of temperature. Such large temperature coefficients can produce resistor mismatch even inside the chip. When the circuit contains a power device, its dissipation determines a temperature gradient. Therefore, the mean temperature of resistors can be different, as shown in Fig. 2.5 a). Moreover, temperature changes at different power dissipations inside the chip. In this case, a layout symmetrical to the power devices is recommended, as shown in Fig. 2.5 b).

Finally, we have to consider the last parameter that affects the accuracy of integrated resistors: the thickness of the resistive layer. For this element, we account for two sources of errors: the implant dose and the deposition rate.

Concerning the first, we must note that the implant dose does not only affect resistivity, but also the depth of the metallurgical junction, whose position determines the equivalent thickness of the diffused layer. The second source of error derives from anomalies in the deposition rate, which depends, in turn, on temperature fluctuations in the reactor and on alterations in the incoming gas concentrations.

Table 2.1 summarises the most important properties of resistors. It is worthwhile noting that the worst performance is exhibited by resistors with very high sheet resistance. Therefore, accurate elements with large resisitve value are not feasible in integrated technology. In addition, we see that the temperature coefficients and voltage coefficients are rather high; much higher than in normal discrete components. We must remember these features when designing integrated systems and, accordingly, we should not rely on resistors when we have to design very precise circuits. Nevertheless, the matching that we can obtain between nominally equal resistors or resistors whose ratio is a relatively small number, can be quite good. Since resistors of the same kind are achieved with the same technological steps, mismatch derives only from the variation of process parameters along the chip. We have good control of the process for short distances and we can obtain matching accuracies better than *0.1 - 0.2%* if we are very careful in the layout. Some hints for an appropriate layout of integrated resistors are provided in the next section.

2.1.2 Layout of Integrated Resistors

The layout of an integrated resistor contributes to its important performance parameters (absolute value, matching accuracy and stability). From Table 2.1 we see that the specific resistance of good quality layers (namely, with low temperature and voltage coefficients) is limited to a few tens of $\Omega/\square$. So, high value resistors with good quality must be designed with a very large L/W aspect ratio. An efficient use of the chip area is obtained by arranging very long strips into a serpentine fashion, as shown in Fig. 2.6. The resistance of the squares at the corners is different from that in the rectilinear stretch. This happens because in proximity to the bend, the current density becomes non-uniform. The value

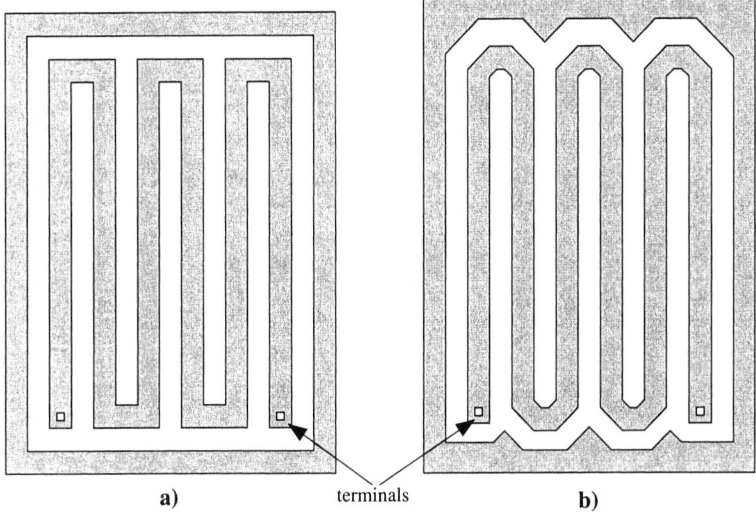

Fig. 2.6 - Typical resistor layout with dummy strips placed around it

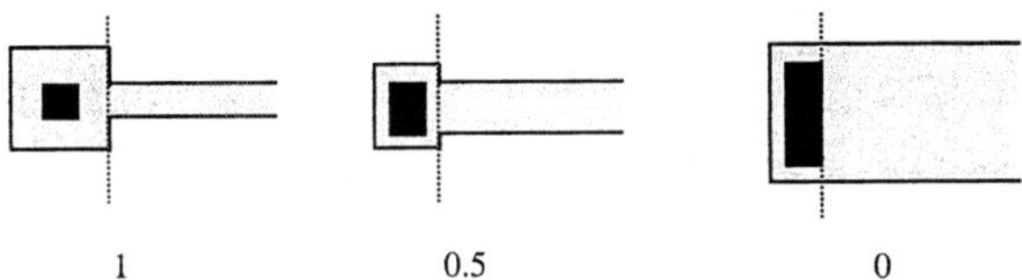

Fig. 2.7 - Possible types of resistor bending and their estimated effect (in resistor □)

of the resistance at the corner is, in first approximation, accounted for as one half of a square. This can be accepted for limited accuracies; for precise applications it is better to use rounded or 45° bending. This solution is shown in Fig. 2.6 b). Both layouts in Fig. 2.6, in addition to the resistive pattern, use dummy strips around the resistance. As already mentioned, the dummy strips define a uniform boundary around the body of the resistor. Therefore, with a constant undercut effect we achieve optimum control of width reduction.

> ***TAKE HEED!***
>
> The layout of integrated resistors significantly affects the performance of the circuit that uses them. The analog designer should consider layout design as important as circuit design (or more).

Endings contribute to the resistance. Their effect involves the contact resistance between the metal and the resistive layer. In addition, if they are not correctly designed, a localized resistance will result due to a disturbance to the current laminar flow. Fig. 2.7 shows some examples of good and bad endings. In case a) the narrowing of the current passage establishes a localized resistance roughly estimated as equivalent to one square; In case b) we are in the middle, in case c) the additional contribution is negligible. The numbers provided are not exact values, they are used merely to give an idea of the possible effect in the case of bad endings.

Matching between resistors is always required to obtain precise results when their ratio is important, such as in voltage references, A/D and D/A converters, amplifiers, etc. For a precise absolute accuracy with this kind of applications, in addition to the points recommended, it is necessary to control (or to compensate) the gradients of the physical parameters, which are the major factors responsible for resistor mismatch.

Good practice in compensating gradients at the first order can be gained from using interdigitized or common centroid structures. An example of interdigitized matched resistors is shown in Fig. 2.8. Two nominally equal resistors, R_1 and R_2, are split into five equal parts connected together by metal connections. At the two sides of the structure, two dummy strips make the boundary of the terminal elements uniform. Any change of resistivity along

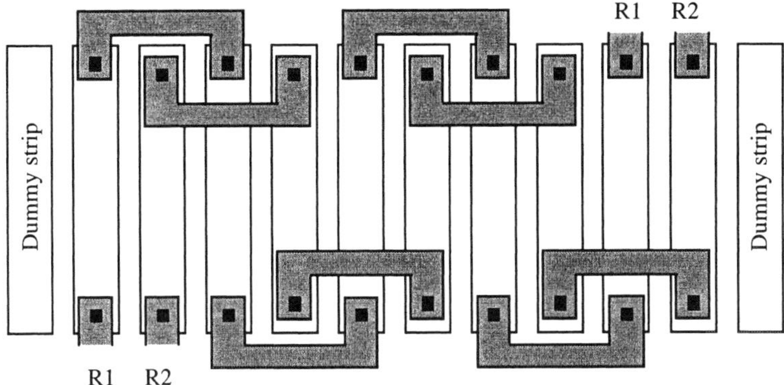

Fig. 2.8 - Interdigitized resistors with dummy elements

the structure produces small variations in closed elements. We therefore have a global mismatch limited to the local inaccuracy between two adjacent strips. Common centroid structures are a two-dimensional extension of the interdigitized approach.

Example 2.2

The two interdigitized resistors shown in Fig. 2.8 are made by 25 equivalent squares of a 35 Ω/□ polysilicon. The resistance of the metal-poly contact is 5 Ω with 2% standard deviation. The specific resistance gradient ∇_x in the horizontal direction is 200 ppm/μm; the width of strips is 6 μm and they are 3 μm apart. Calculate the matching accuracy between the two resistors.

Solution:

The horizontal distance, D_x, between the two centroids is 9 μm; therefore, the resistance mismatch contributed by the poly strips is

$$\Delta R_{poly} = n_{sq} R_{\square} \nabla_x D_x = 25 \cdot 35 \cdot 2 \cdot 10^{-4} \cdot 9 = 1.57\Omega$$

We have 10 poly-metal contacts; their total resistance is 50 Ω and their contribution to error is $\Delta R_{cont} = 5 \cdot 0.02 \cdot 10 = 1\ \Omega$. Since the two sources of error are statistically independent, we combine them quadratically.

$$\Delta R_{tot} = \sqrt{\Delta R_{poly}^2 + \Delta R_{cont}^2} = \sqrt{1.57^2 + 1} = 1.87\Omega$$

Each resistance is 925 Ω; therefore, their matching accuracy is about 0.2%.

A final important point on the layout of resistors concerns minimizing the noise injected from the substrate. For this, it is useful to put substrate biasing completely around the resistor: it defines a sink for spur signals. Moreover, for well resistors it is suggested that substrate bias should also be placed between the single branches of resistors. This better guarantees transversal insulation. Indeed, the lateral diffusion of the well, in worst case conditions, may produce a slight overlap of diffusion tails, resulting in transversal shunting resistances. The substrate diffusion employed compensates the doping between adjacent strips and firmly defines the p-n junctions. The substrate diffusion must always be frequently contacted and connected to a metal line. This ensures an equipotential voltage of the substrate which establishes a guard ring around the resistor.

For poly or highly doped diffusion, we achieve protection using wells (Fig. 2.9). The biasing of the well goes all the way around the structure to catch spur signals as they cross the first ring of substrate biasing. With the solution in Fig. 2.9, we achieve a double ring of rejection of noise injected from the substrate.

2.2 INTEGRATED CAPACITORS

A capacitor, in CMOS technology, is fabricated like a parallel plate structure. The electrodes are achieved using conductive layers made available by the technology (metal, polysilicon, diffused layers). Insulation is established by silicon dioxide, polysilicon dioxide or, rarely, *CVD* oxide. In the structure

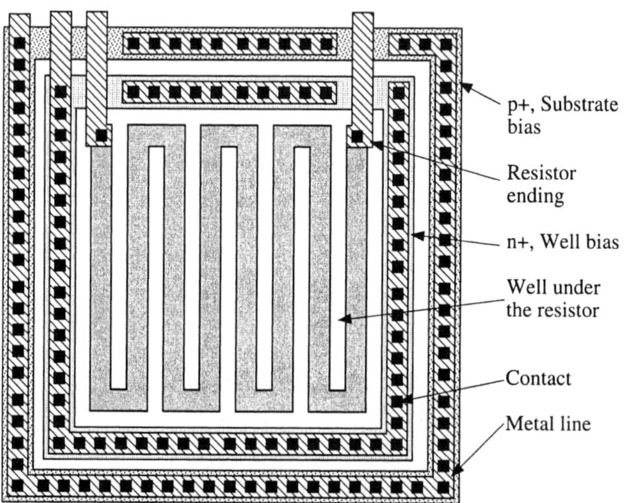

Fig. 2.9 - Integrated resistor with double protection against spur signals

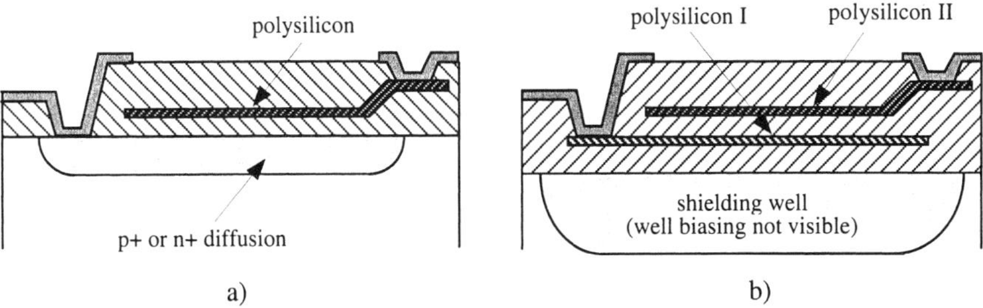

Fig. 2.10 - Two possible cross sections of integrated capacitor

of Fig. 2.10 a) the bottom plate is made by a p^+ or n^+ diffusion while the top plate is built up by polysilicon. The contribution to capacitance in the thick oxide area is usually neglected. The area of the capacitor is determined by the overlapping of the two plates in the thin oxide region. Alternatively, the bottom plate of the capacitor can be made by any p^+ (or n^+) diffusion inside the well, which, biased with a quiet voltage, operates as a shield against the substrate. For both the above-mentioned possibilities we have to use a special process: n^+ or p^+ layers under polysilicon are not achieved with conventional CMOS processes. With normal technologies, in fact, the polysilicon gate protects the material underneath from implantation. This is, incidentally, the technique used to separate the source and the drain of transistors. Therefore, in order to achieve diffused plates it is necessary to use additional technological steps for an n^+ or a p^+ implant before depositing polysilicon.

The structure of Fig. 2.10 b) uses two polysilicon layers to achieve both the bottom and the top plate of the capacitor. Again, an optional well under the structure allows us to shield the capacitor from the substrate. Even for this kind of structure, it is necessary to use additional technological steps for the deposition of the second level of polysilicon. Since cheap, digital processes do not foresee extra implants or two poly layers, the technologies that admit effective capacitors (like the structures in Fig. 2.10) are normally accredited as being "analog".

The capacitance, neglecting fringing and parasitic effects, is given by

$$C = \frac{\varepsilon_0 \varepsilon_r}{t_{ox}} WL \tag{2.4}$$

where ε_r is the relative dielectric constant and t_{ox} is the oxide thickness, W and L are the geometrical dimensions of the plates, assumed as being rectangular.

The oxide thickness of typical technology today is around *10 nm*. Assuming the relative dielectric constant ε_r as equal to *3.8* (the nominal value for SiO_2), the capacitance of the thin oxide per unit area results $3.36\ fF/\mu m^2$. This means

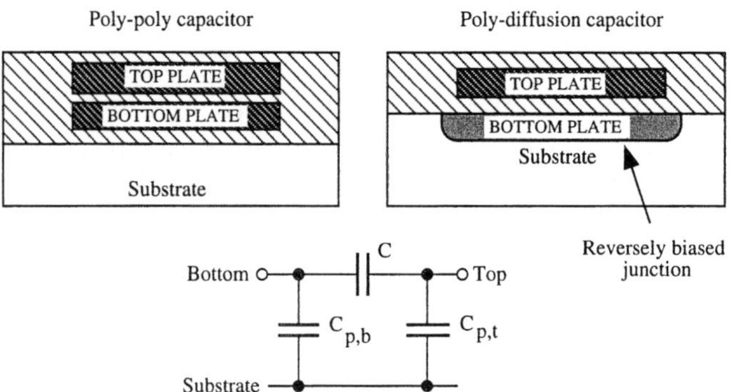

Fig. 2.11 - Equivalent circuit of an integrated capacitance

that we design a capacitor of *2 pF* with plates as large as $600\ \mu m^2$ (about *1* $mill^2$). Therefore, in integrated technology, *0.1 pF* is a small capacitance while *100 pF* is a large one.

The capacitors in Fig. 2.10, as well as in any other possible integrated structure (like sandwiches of metal layers), have both the top and the bottom plates in close proximity to the substrate. Therefore, a non-negligible parasitic coupling results. Because of this, the equivalent circuit of any integrated structure is not a simple capacitor but, at a first approximation, it is given by the network shown in Fig. 2.11. In addition to the designed capacitance, we have two parasitic elements describing the coupling between the bottom and the top plate with the substrate. Any resistive effect (due to the finite conductivity of the plates) or leakage contributions are ignored here.

> **REMEMBER!**
>
> An integrated capacitance is not, strictly speaking, a capacitor. Its equivalent circuit includes parasitic capacitance. Moreover, for high frequency applications, we also have to account for the distributed resistance of the plates used.

In the case that the bottom plate is made of a diffused layer, the parasitic $C_{p,b}$ comes from a reversely biased junction. It is a non-linear element with a value around a few percent of integrated capacitors. For the bottom plate made with polysilicon, the parasitic $C_{p,b}$ is normally smaller, as the bottom plate sits upon relatively thick oxide; its value holds a fraction of a percent of the integrated capacitor. For both structures, the top plate parasitic, $C_{p,t}$, is clearly smaller than the bottom plate parasitic; it is around one order of magnitude less than the value of $C_{p,b}$.

2.2.1 Accuracy of Integrated Capacitors

The accuracy of an integrated capacitor is determined by the accuracy of the parameters that we have in equation (2.4). Again, as for resistors, we observe that since they derive from different technological steps, all the parameters can be assumed as being statistically independent. Therefore, we have

$$\left(\frac{\Delta C}{C}\right)^2 = \left(\frac{\Delta \varepsilon_r}{\varepsilon_r}\right)^2 + \left(\frac{\Delta t_{ox}}{t_{ox}}\right)^2 + \left(\frac{\Delta L}{L}\right)^2 + \left(\frac{\Delta W}{W}\right)^2 \tag{2.5}$$

Of course, we have not accounted for ε_o which is an absolute physical constant.

It is now useful to consider the factors that produce inaccuracy in the four parameters involved. As far as the relative dielectric constant is concerned, the most important effects that we have to account for are:

- oxide damage
- impurities in the oxide
- stress
- bias history (mainly for CVD oxide)
- temperature
- bias conditions.

Oxide is an amorphous material, we do not have any regular structure. However, it is possible to identify damages that change the dielectric properties of the insulator locally. In addition, we have impurities, mainly deriving from the technological steps that follow oxide growth, which degrade the oxide as well. Even stress modifies the relative dielectric constant, which, as we already mentioned, can be very large for plastic packages. Concerning the type of oxide used, there are some differences between the oxide grown thermally and the CVD. For the latter, we use a relatively low temperature. It is therefore possible to have some anisotropy that results in possible dielectric polarization. This, in turn, produces a dependence of the dielectric constant on bias history. Finally, temperature and bias also modify the dielectric constant. We account for their effect in the temperature and voltage coefficients.

The second parameter in equation (2.5) is oxide thickness. Its accuracy mainly depends on the precision of the growth rate which, in turn, is related to the kind of starting material. When we use oxide polysilicon grain size affects the growth rate. With oxide from monocrystalline silicon we have a better starting material and therefore better accuracy. In any case, thermal oxide grows very slowly and we can control its thickness very accurately. Therefore, the contribution of oxide thickness to capacitance inaccuracy can be controlled well.

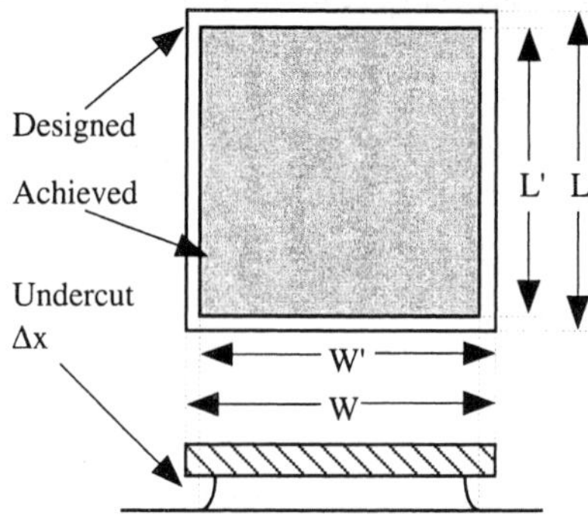

Fig. 2.12 - Reduction in capacitor area due to the undercut effect

The other two parameters that we have to account for are the width and the length of the capacitor plate. We observe that we have an intrinsic advantage compared to the resistors: the value of the capacitor is proportional to the product of the two geometrical quantities. Therefore, we can use comparable dimensions (normally equal) to optimize capacitor accuracy. As we have already discussed for integrated resistors, the accuracy of geometrical dimensions depends on:

- etching definition
- boundary mismatch
- lateral diffusion

The undercut effect should be carefully accounted for, especially when we design matched capacitors. As we can see from Fig. 2.12 the actual area A' is smaller than the designed area, A. We have

$$A' = W'L' \cong WL - 2(L + W)\Delta x = A(1 - \Delta x P/A) \tag{2.6}$$

thus, the relative reduction is proportional to the undercut extent, Δx, and the ratio between perimeter, P, and the designed area, A. Therefore, if we want to design well-matched capacitors, we should use the same perimeter-area ratio in matched elements.

Boundary mismatch, as we have already discussed for resistors, produces different undercuts. Therefore, for good accuracy, we should keep the boundary of all the sides of the capacitor plates uniform. This can be achieved with dummy elements, as we shall see in next section that looks at layout problems.

When we estimate the value of capacitance, we consider only the area with thin oxide. In fact, we should also account for the capacitance of thick oxide areas, which slightly increases the calculated value. This source of error should be carefully considered when designing very demanding circuits. In addition, we have to remember that we assume the capacitor to be a parallel plate structure. Strictly speaking, this is not true: near the periphery of the plates, the fringing effect bends the electric field, thus reducing capacitance value in a strip region whose width is about the same as the oxide thickness. This corresponds to a reduction in the total capacitor value by a factor proportional to t_{ox} *P/A*. To make this error negligible, we should use a much larger *L* and *W* than the oxide thickness, t_{ox}.

Example 2.3

We have two poly-poly capacitors whose plate area is defined by poly-2 etching patterns. The nominal sizes of the two capacitors are 10 ×10 μm and 20× 20 μm respectively. The undercut is 0.05 μm with a standard deviation 0.005 μm; the capacitance per unity area is 3.36 fF/μm². *Calculate the effective capacitance and the errors.*

Solution:
Accounting for the undercut effect, the capacitor sizes become 9.9×9.9 μm and 19.9×19.9 μm respectively, thus corresponding to 98 μm² and 396 μm². The percentage reductions of area for the two capacitors are 2% and 1%. The achieved values of capacitance are

$$C_{10\times 10} = 98 \cdot 3.36 = 329\text{fF} \quad C_{20\times 20} = 396 \times 3.36 = 1330\text{fF}$$

Note that the ratio between these two figures is 4.04 instead of the nominal ratio of 4. The absolute accuracy contributed by width and length comes from the undercut standard deviation. We have

$$\Delta C_{10\times 10}(W, L) = 2(W_{eff} + L_{eff})\Delta x_j C_{ox} =$$

$$= 39.8 \cdot 0.005 \cdot 3.36\text{fF} = 0.67\text{fF}$$

while for the bigger capacitance we have

$$\Delta C_{20\times 20}(W, L) = 79.8 \cdot 0.005 \cdot 3.36\text{fF} = 1.34\text{fF}$$

The two calculated errors are 0.2% and 0.1% of the achieved value respectively.

The performance of different types of integrated capacitors are summarised in Table 2.2. We can observe that, in general, integrated capacitor performance is better than resistor performance. This is because the relative dielectric constant is more accurate than resistivity, the thickness of the oxide is better controlled than the thickness of the resistive layers, and the width and length of the capacitors are normally designed with the same dimensions. The shadowed part of the table reports features of other capacitor structures that can be realized

TABLE 2.2:

Capacitor Features

Type	t_{ox} nm	Absolute Accuracy %	Temperature Coefficient ppm/°C	Voltage Coefficient ppm/V
poly-diff.	6-20	7-14	20-50	60-300
poly I-poly II	8-25	6-12	20-50	40-200
metal-poly	500-700	6-12	50-100	40-200
metal-diff.	1200-1400	6-12	50-100	60-300
metal I-metal II	800-1200	6-12	50-100	40-200

with CMOS processes. We see that, independently of other performance parameters, the great oxide thickness makes the metal-poly, metal-diffusion, metal I-metal II structures almost impractical for the large silicon area required.

2.2.2 Layout of Integrated Capacitors

Remember that capacitors are important elements for precise analog signal processing. They are frequently used as matched elements to define precise capacitor ratios. Moreover, often they are connected to the virtual ground of an operational amplifier. From these two characteristics we deduce the need for careful layout matching and for good shielding to limit the injection of spur signals that, through virtual grounds, are directly transferred into the circuit.

The layout of a capacitor depends on the layers used to realize the two plates and on whether or not the contact of the top plate can be designed on the thin oxide region. Fig. 2.13 shows three possible layouts of a single capacitor made with a double poly technology. In the first two cases, the area of the capacitor is determined by poly II. Poly I is wider and is contacted by a suitable metal connection. To ensure better control at the corners they have a 45° shaping in poly II. The overlap of poly I and poly II corresponds to thin oxide. In the first case, the contact of poly II on top is allowed; in the second it is forbidden. This because of reliability issues: a contact on the top of the thin oxide can be produce damages on the oxide itself in some technologies. We there-

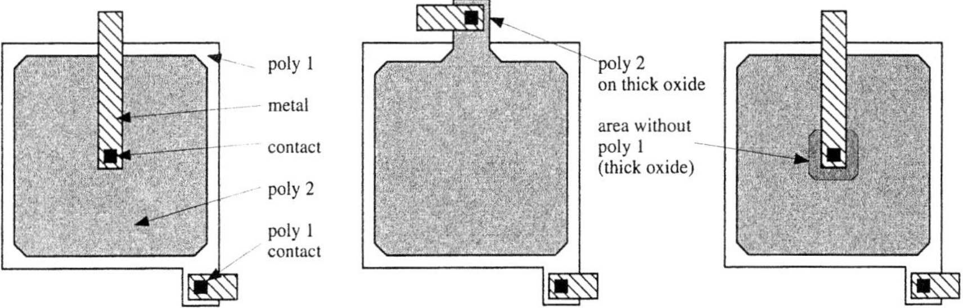

Fig. 2.13 - Three possible layouts of double poly capacitors

fore have to lead poly II out of the thin oxide area to create the contacts. This solution can be expensive in terms of silicon area. The third layout in Fig. 2.13 provides a solution to this problem: poly I has a donut-like shape: in the inside hole we have thick oxide that is suitable for the poly I contact. This solution is effective from the point of view of area consumption. However, the accuracy now depends on the pattern precision of both poly I and poly II.

> **TAKE NOTE**
>
> The most important issue for the layout of precise capacitors is the undercut effect. To compensate it remember to keep constant the area perimeter ratio!
>
> Another problem is the noise from the substrate. Remember to use a shielding well under the capacitive structure.
>
> The etching on the corner is also a problem to keep in mind. The best is to use 45° shaped corners

We have seen that the capacitor value depends on the undercut effect and, based on this, on the perimeter-to-area-ratio. Therefore, when precise matching of two or more capacitors is required, it is absolutely necessary to stick to the rule of constant perimeter-to-area-ratio. This condition is easily fulfilled when we have to design capacitors with an integer proportion. The smallest capacitor is assumed to be "unity" while the others are merely the parallel connections of a suitable number of unity elements. As an example, Fig. 2.14 shows the layout of three capacitors, C_1, $C_2=3C_1$ and $C_3=4C_1$. Since the total capacitance is $8C_1$, 8 unity elements have been arranged into a 2×4 array. We see that three elements on the top are connected in parallel and the four at the bottom are in parallel as well. It is clear that the perimeter-to-area-ratio of all the capacitors is the same. Moreover, since we have dummy elements all around the array, we create the same boundary conditions at the periphery. With this (already discussed) artifice, matching accuracy is further improved. When the ratio between matched capacitors is not an integer number it is not possible to use

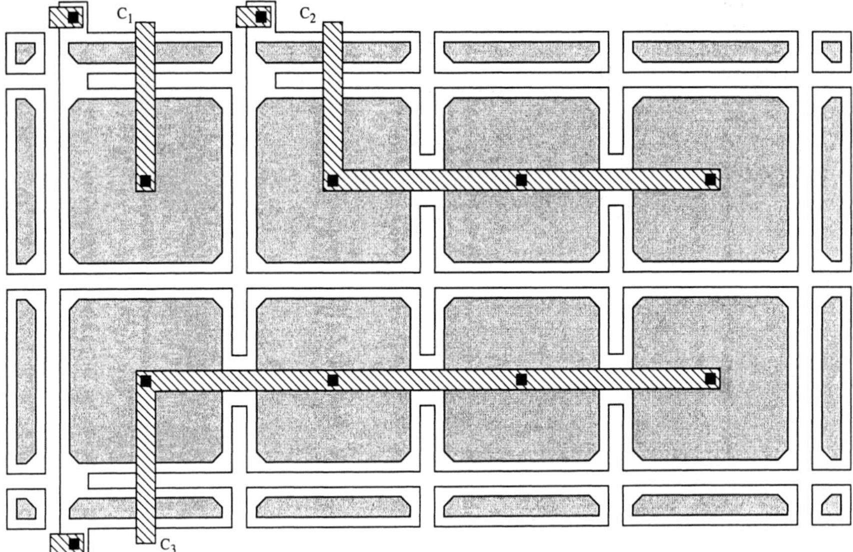

Fig. 2.14 - Three matched capacitors: C_1, C_2=3C_1, C_3=4C_1 Note that the area-perimeter ratio is kept constant. Moreover, dummy elements around the structure ensure the same undercut.

the approach discussed above. In this case we connect in parallel a number of unity elements with an additional rectangular capacitor whose value, C_k, is between *1* and *2*. A suitable choice of length and width of the rectangular element permits us to achieve the same area perimeter ratio as the unity squared capacitor. If the size of the plate of the unity capacitance, C_u, is L_u and the dimension of the rectangle making C_k are L_r and W_r, we have

$$\frac{C_k}{C_u} = \frac{L_k W_k}{L_u^2}$$
$$\frac{L_u}{4} = \frac{L_k W_k}{L_k + W_k} \tag{2.7}$$

That leads to a real solution for $C_k > C_u$.

A limitation not yet accounted for is the limit derived from any potential gradient in technological or geometrical parameters. The problem is resolved, at a first evaluation, using an interdigitized or a common centroid arrangement. Earlier, when we discussed resistor layout, we presented an interdigitized ordering. Now, in Fig. 2.15 we have a common centroid arrangement. It is a binary array of *5* capacitors (C_1, C_2= C_1, C_3= $2C_1$, C_4= $4C_1$, C_5= $8C_1$) with one terminal in common. The total number of unity elements is *16*; they

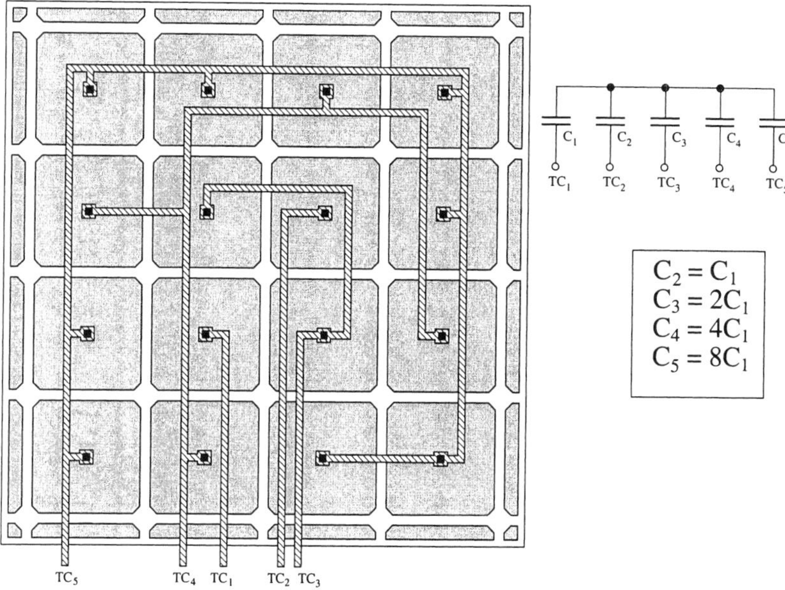

Fig. 2.15 - Common centroid layout of a capacitor array

are placed in a 4×4 array. The two unity capacitors, C_1 and C_2, are positioned in the middle of the array; Capacitor C_3 completes the inner core. The remaining capacitors are arranged in the periphery: one element of C_4 interleaves two elements of C_5. The common terminal is achieved with a large first poly plate. The parallel connection of the second poly plates is done with metal lines. Again, dummy elements ensure the same boundary conditions. One limit of the given arrangement comes from the parasitic effect of metal connections. To attenuate this effect, metal lines are designed with minimum width.

Shielding is widely used to protect against substrate noise or parasitic couplings. Normally we use a well under the capacitor (or the capacitor array) and, in addition, we establish tight substrate biasing all around the well. Both well and substrate biasing should be connected to quiet voltages. Even shielding on the top can be necessary at times. Shielding is achieved properly by using the conductive layers available on the top of the capacitor structure. The parasitic capacitance between the top plate of the capacitive structure and the shield used must be accounted for and included in the equivalent circuit (Fig. 2.11).

2.3 ANALOG SWITCHES

A switch is a short circuit in the "on" state and an open circuit in the "off"

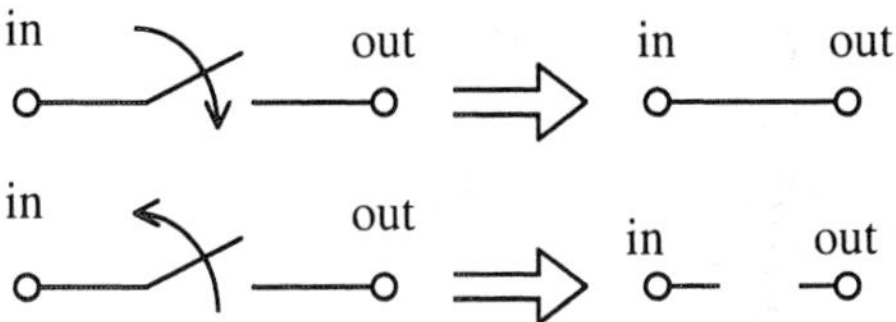

Fig. 2.16 - Equivalent circuits of an ideal switch in the "on" and the "off" state

state (Fig. 2.16). For many analog applications, a switch is used merely to transfer a charge from one node to another and to set a voltage on a high impedance point. For this kind of use, after a possible transient, the current in the switch goes to zero or is negligible. These operative conditions are easily implemented by one MOS transistor or a pair of complementary transistors. The typical circuit that uses a MOS transistor as the switch is shown in Fig. 2.17: the capacitor C_1 is charged by the input voltage through a switch. The "on" and "off" states are ensured by a proper voltage applied to the gate of M_1. Fig. 2.17 uses an *n-channel* device; thus, we have to apply a positive control to drive it in the on-state. In contrast, a voltage as negative as possible must be used to achieve the off-state. The opposite is necessary for *p-channel* devices. Normally, the gate of transistors is controlled by the highest voltage available on the chip (V_{DD}) or by the lowest one.

With the switch in Fig. 2.17 in the on state, after a transient, we have $V_{out}=V_{in}$, hence the drain to source voltage of M_1 goes to zero. The MOS enters the linear region and its on-resistance is approximated by

$$R_{on} = \frac{1}{g_{ds}} = \frac{1}{\mu C_{ox}\frac{W}{L}(V_{GS}-V_{Th})} \tag{2.8}$$

The value of the on-resistance depends on the overdrive voltage, $V_{ov}=V_{GS}-V_{Th}$ and on the aspect ratio, *W/L*, through the transconductance parameter, μC_{ox}.

The rail voltages used in modern technologies are *3.3 V* or *2.4 V*; therefore, *1 V* of overdrive (or less) is what we can typically expect. Under this condition, the on-resistance of a minimum area switch (*W/L* = *1*), using the figures given in (1.49) becomes

$$\begin{aligned} R_{on,n}(W/L = 1; V_{ov} = 1) &= 8.4k\Omega \\ R_{on,p}(W/L = 1; V_{ov} = 1) &= 25.6k\Omega \end{aligned} \tag{2.9}$$

Using a n-channel and the p-channel transistor respectively.

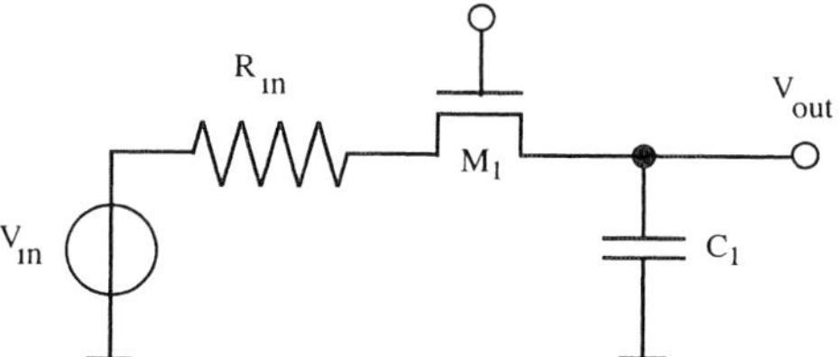

Fig. 2.17 - Typical circuit with a CMOS switch used to charge a capacitor

> **NOTE**
>
> We achieve good analog switches when we transfer voltage signals. The achievable on-resistance can be kept to a low value and off-current and leakage current are quite low.
>
> When we want to transfer a current we always have a drop voltage across the switch.

Now we have to assess whether or not the achieved resistances are suitable for typical integrated applications. For this, let us assume that capacitor C_1 is $2\ pF$ (a value which is rather large for integrated circuits). Moreover, let us assume that the switch is driven by a *2MHz* clock and that it remains in the on-state for *50%* of the clock period (*250 nsec*). The resulting RC time constants are *16.4 nsec* and *51.6 nsec* for the n-type and the p-type switch respectively. This means that we have *15.2* and *4.84* time constants available. Assuming an exponential response of the circuit (we neglect any operation in the saturation region), the output voltage reaches *0.9999997* and *0.992* of the final voltage respectively. The former result is good enough for any analog application, the latter one corresponds to an error of *0.8%*, which is normally not acceptable for precise requirements. The calculations above lead to the following rules of thumb:

- a minimum area n-channel switch is capable of driving *2 pF* up, running at a few *MHz* clock
- a minimum area p-channel switch is capable of driving *2 pF* with a clock control not exceeding 1MHz.

for higher frequencies of operation (or for larger capacitors), one must use transistors with a *W/L* aspect ratio larger than one. This, as results from (2.8), leads to a lower on-resistance while maintaining the same overdrive voltage.

The above calculations assume a proper overdrive of transistors (like *1 V*) when operating in the on-state. Such an operation, of course, depends on the value of the source voltage, which is equal to the input signal, and on the gate voltage used. When the input approaches the gate control, the overdrive diminishes and, at a given level, vanishes. At this point, the on-resistance goes to infinity and the switch no longer operates properly. This occurs for an n-chan-

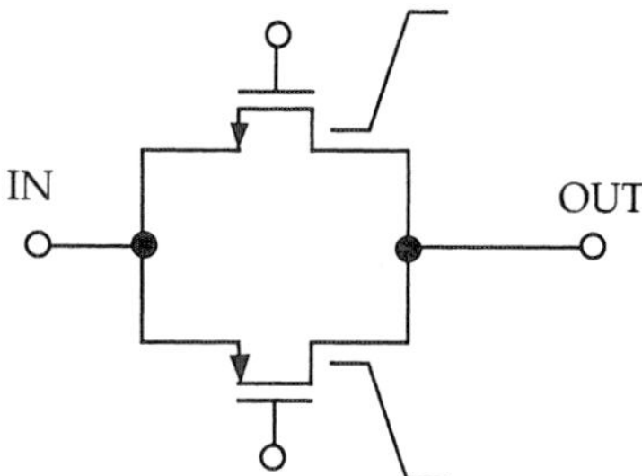

Fig. 2.18 - Switch achieved with complementary transistors.

nel transistor when the input voltage approaches by the n-channel threshold the positive rail (V_{DD}). By contrast, a p-channel transistor malfunctions when driven close to the negative rail.

> ***TAKE CARE!***
>
> In low voltage applications driving switches is a problem. We need to drive the gate of the switch with a voltage that is at least one threshold higher (for n-type, lower for p-type) than the switched signal.

A suitable solution to the problem discussed above is by the use of complementary transistors as shown in Fig. 2.18. We have an n-channel and a p-channel element whose switching from the on to off state is controlled by two complementary phases. It is obvious that when the n-channel transistor is not working properly because the input signal is too high, the p-channel enjoys a rather large overdrive and ensures a satisfactory on-resistance. A similar behaviour is displayed by the n-channel transistor when the input voltage approaches the negative rail. We can distinguish between three regions of operation: $V_{in} < V_{Th,p}$, where only the n-type transistor works, $V_{DD} - V_{in} < V_{Th,n}$, where only the p-type element functions, and the region in between where both transistors contribute with a finite on-resistance.

We can observe that the region in which both transistors operate is reduced when we move from $V_{DD} = 5\ V$ to $3.3V$ or below. The situation worsens with a lower supply and becomes below a given level. When the supply voltage goes down to $V_{Th,n} + V_{Th,p}$, the region in which both transistors operate becomes just a point ($V_{in} = V_{Th,n}$). Here the "on-resistance" goes to infinity. For supply voltages below $V_{Th,n} + V_{Th,p}$ we have a "gray" region where the switches can't open. Nevertheless, the two-transistor solution is always better than the one-transistor solution. It is definitely advisable when the input signal varies with swings that approach positive or negative rails. The price that we have to pay, apart from the additional transistor, is an additional control signal generator and, more importantly, the addition of another control line.

Example 2.4

Determine the on-conductance of a switch as a function of the input voltage. The switch is made by two complementary transistors with a minimum aspect ratio. Use the following parameters: V_{DD}=3.3V, $V_{th,n}$=0.86 V, $V_{th,p}$=0.72 V, $\mu C_{ox,n}$=100 $\mu A/V^2$, $\mu C_{ox,p}$=38 $\mu A/V^2$. Determine the change of on-resistance when the p-channel aspect ratio increases.

Solution:

The gate of the n-channel transistor is biased at 3.3 V while the gate of the p-channel transistor is set to 0V. For V_{in}<0.72 V the control of the p-channel gate does not exceed the threshold: the on-conductance comes only from the n-channel transistor

$$G_{on} = \mu_n C_{ox}\left(\frac{W}{L}\right)_n [V_{DD} - V_{in} - V_{Th,n}] \quad \text{for } V_{in} < 0.72V$$

Similarly for V_{in}>2.44 V, the overdrive of the n-channel vanishes and we only have to account for the p-channel contribution:

$$G_{on} = \mu_p C_{ox}\left(\frac{W}{L}\right)_p [V_{in} - V_{Th,p}] \quad \text{for } V_{in} > 2.44V$$

for 0.72V <V_{in}< 2.44V both transistors contribute to the conductance

$$G_{on} = \mu_n C_{ox}\left(\frac{W}{L}\right)_n (V_{DD} - V_{in} - V_{Th,n}) + \mu_p C_{ox}\left(\frac{W}{L}\right)_p (V_{in} - V_{Th,p})$$

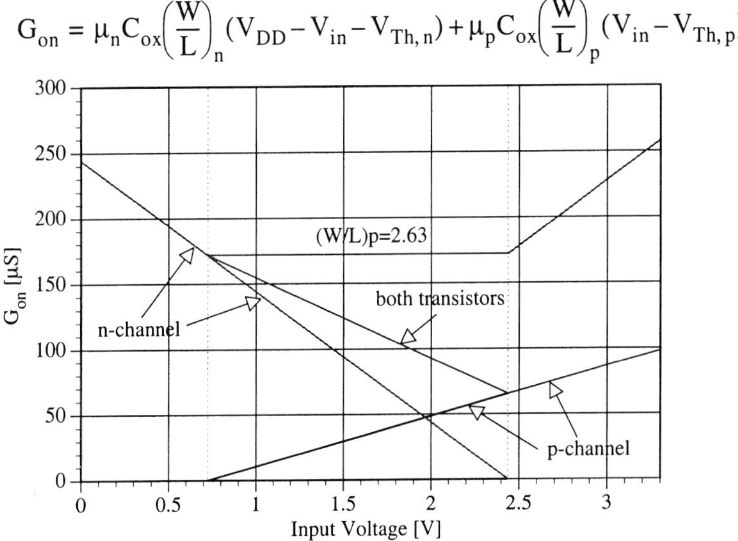

Fig. 2.19 - On-conductance of the complementary switch transistor in Example 2.4

From the equations above, it turns out that the conductance versus V_{in} (Fig. 2.19) comes from the combination of two straight lines.Their slope is $\mu_n C_{ox}(W/L)_n$ and $\mu_p C_{ox}(W/L)_p$.
With minimum aspect ratios, the lower value of conductance is achieved at V_{in} = 2.44 V (the point where the contribution of the n-channel vanishes). G_{on} is 92.78 μA/V and the on-resistance R_{on} = $1/G_{on}$ = 15,3 kΩ.
Moreover, we observe that when the p-channel aspect ratio increases the slope of the straight line, its conductance increases as well. As we can observe from the figure, if

$$\mu_n C_{ox}\left(\frac{W}{L}\right)_n = \mu_p C_{ox}\left(\frac{W}{L}\right)_p$$

the conductance becomes constant in the interval $0,72V < V_{in} < 2.44V$. Using numerical values, we have: $(W/L)_p$ = 2.63. For a larger value of $(W/L)_p$ the minimum of conductance is swapped at V_{in} = 0,72V.

2.3.1 Charge Injection

One of the major problems coming from the use of CMOS switches is the so-called *charge injection* (and *clock feedthrough*). This effect mainly results from the generation and the dissolution of the conductive channel sitting under the gate when the transistor is in the on state. In addition, we have to account for parasitic capacitive couplings. These two sources of error are such that if one of the terminals of the switch is connected to an high impedance node, any injected charge remains trapped on the capacitance associated with the high impedance node, thus changing its voltage.

We have seen in Chapter 1 that the channel of an MOS transistor working in the triode region houses the following amount of charge

$$Q_{ch} = WL_{eff}C_{ox}(V_{GS} - V_{Th}) \quad for \ V_{DS} = 0 \tag{2.10}$$

Where we use the effective length of the channel $L_{eff}=L-2x_{ov}$, with x_{ov} being the extent of source and drain overlap. In addition to the channel charge, we have to remember the charge in the overlap capacitance

$$Q_{ov} = Wx_{ov}C_{ox}V_{GS} \tag{2.11}$$

When the switch is turned off, the charge of the channel disappears and the charges in the overlap capacitances vary according to the gate voltage swing.

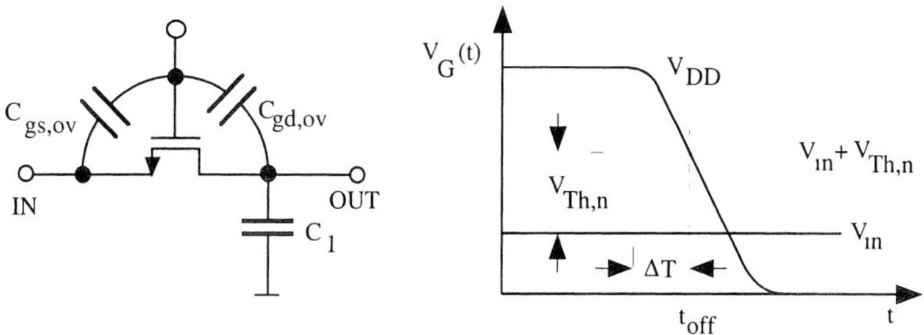

Fig. 2.20 - Charge injection in a single transistor switch.

What we want to study is where the charges go and what effect they produce. It is quite clear that the only nodes where charges can go are the two terminals of the switch. A fraction of the total charge will affect one terminal, the rest will influence the other. The splitting will depend on the electrical boundary conditions and on the speed of variation of the gate voltage. Let us go into this point now in more detail. We refer to Fig. 2.20 showing a simple "sample and hold" with a switch made by an n-channel transistor; the control voltage changes from V_{DD} to ground. The specific input level, V_{in}, allows the gate control to cross the turning- off level $(V_{in} + V_{Th,n})$ at time t_{off} with a delay ΔT after the beginning of the turning-off phase. Incidentally, we note that if V_{in} increases, both the channel charge and the delay ΔT decrease.

In the short period ΔT, the on-resistance of the transistor increases from a rather low value up to infinity: at time t_{off}, source and drain are electrically disconnected. While the on-resistance increases, part of the channel charge near the output node may flow through the channel and reach the low impedance node (V_{in}). The fractional part pulled by the input node depends on the control speed and on the equivalent resistance seen at the input node. The analysis of the transient is difficult even with circuit simulators. However, it is obvious that for very slow switching, nearly all the channel charges can reach the low impedance point. In contrast, for quick switching, the channel is closed instantaneously and an almost equal splitting between source and drain results. We can assume that a fraction α of the charge from the channel affects the output node and is integrated in the store capacitor.

> **REMEMBER**
>
> Clock-feedthrough is the result of a complex process. It leads to a voltage offset that depends on many factors: the technology, the sizing of transistor used, the input voltage, the clock phase and its timing.

Similarly, we analyse the charge injection from the overlap capacitances. The voltage swing at the gate produces an injection of charge in the input and output nodes because of the parasitic couplings $C_{gs,ov}$ and $C_{gd,ov}$. We distinguish between the injection during the ΔT period and what happens afterwards. When the channel is still existent, the low impedance node pulls part of the charge: we assume that a fraction, β, remains in the storing capacitor. After t_{off}, we have no interacting injections on the two sides.

Summarizing the points above, we can calculate the total charge that remains in the storing capacitor

$$Q_{inj} = \alpha\{WL_{eff}C_{ox}(V_{DD}-V_{in}-V_{Th})\} +$$

$$+\beta\left\{\frac{Wx_{ov}C_{ox}C_1}{Wx_{ov}C_{ox}+C_1}(V_{DD}-V_{in}-V_{Th})\right\}+ \tag{2.12}$$

$$\frac{Wx_{ov}C_{ox}C_1}{Wx_{ov}C_{ox}+C_1}(V_{in}+V_{Th})$$

This charge, divided by the value of the storing capacitance itself, gives the voltage error produced by charge injection. The second and third term are normally referred to as *clock feedthrough*. For the small capacitors used in integrated circuits, this error can become a design problem.

Example 2.5

The sample and hold in Fig. 2.20 is made by an MOS transistor with W=2μm,L=0.8μm C_{ox}=2 fF/μ^2 and overlap extent x_{ov}=0.1μm. The storing capacitance is 0.5 pF. Determine the voltage offset produced by switching-off of the transistor under the following conditions: VDD=5V, Vin+Vth=2.7V, α=0.4, β=0.8, C_1 = 2 pF.

Solution:
The gate capacitance and the overlap capacitance are respectively

$$C_{gate} = WL_{eff}C_{ox} = 2\cdot 0.6\cdot 2 = 2.4\text{fF};\ C_{ov} = Wx_{ov}C_{ox} = 0.4\text{fF}$$

Using equation (2.12) we obtain

$$Q_{inj} = 2.208 + 0.736 + 1.08 = 4.024\text{fF}$$

for the three charge injection contributions, which produces an offset voltage of 2.012 mV.

2.3.2 Charge Injection Compensation

We have seen that the injection of charge and the one called clock feedthrough can result in a considerable error. This is particularly problematic for high resolution data converters where the accuracy required is below the so-called quantization step: $V_{Ref}/2^N$, (V_{Ref} being the reference voltage used and N the number of bits). For $V_{Ref} = 1\ V$ and 12 bits of resolution, the quantization step is as low as $0.25\ mV$. It is therefore necessary to reduce the charge injection drastically. Unfortunately, it is quite difficult to fully cancel this source of error. The commonly used techniques are based on compensation strategies which are only capable of alleviating the problem. The most popular methods can be classified as:

- use of dummy switch
- use of parallel switches
- use of complementary transistors
- use of compensation networks
- use of fully differential solutions.

Use of dummy switch

A direct way of cancelling the effect of a given charge injected into a capacitor is to compensate it by injecting an equal and opposite amount of charge. We have seen that the charge injection depends on a combination of many parameters, including the transistor's parameters. Therefore, the best way to achieve compensation is to use a transistor as well. Fig 2.21 shows the circuit diagram. A transistor M_2 with source and drain shorted (the dummy switch) is connected to the storing capacitor, C_1, and is driven by a complementary phase, Φ_2. When M_1 goes off, M_2 switches on and creates a channel under its gate. Thus, the electrons necessary to set up the channel of M_2 can balance the injection of electrons from M_1. The threshold and the source (or drain) voltages, as well as the value of the specific gate capacitance, C_{ox}, are matched. Thus, a suitable sizing of M_2 achieves the expected compensation. Since the storing

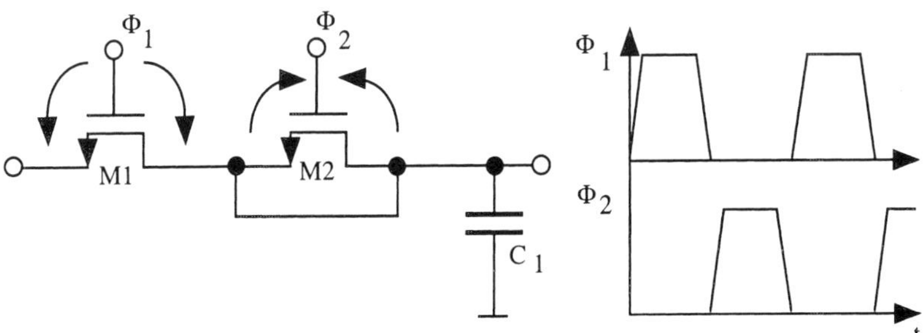

Fig. 2.21 - Use of a dummy switch to compensate the charge injection

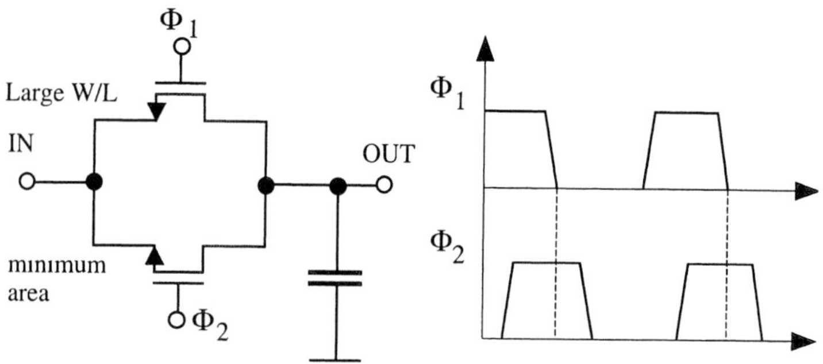

Fig. 2.22 - Use of parallel transistors with different aspect ratio for alleviating clock injection

capacitor receives only a fraction of the M_1 charge channel, the gate area of M_2 should accordingly be smaller. For simplicity's sake, we assume $\alpha = 1/2$; hence, $(WL)_1 = 2\,(WL)_2$.

> **WARNING**
>
> Be sure that the dummy switch acts after the working switch opens.

Compensating the charge from the overlap capacitance is more difficult because it is made up of two components. What is normally done is to assume that $\beta = 1$, thus neglecting the fraction of charge that escapes toward the input terminal in the ΔT interval (see Fig. 2.20); we obtain $W_1 = 2W_2$.

An important point to remember is that the charge injected to compensate must remain in the high impedance node. This cannot be the case if M_1 is still closed while we are in the process of closing M_2. To avoid this, the clock phase driving M_2 should be slightly delayed compared to the clock controlling M_1. An easy way to achieve a reasonable delay is using a chain of inverters (normally three are enough) to generate Φ_2 starting from Φ_1.

Use of parallel transistors

When the hold capacitor has a large value it is necessary to use transistors with a large aspect ratio. This obtains a low on-resistance that, in turn, keeps the time constant to an acceptable value. However, enlarging the transistor's width to drive a large capacitance increases the charge injected as well, and the resulting voltage offset remains the same. For this specific situation, the use of two n-channel (or p-channel) parallel transistors alleviates the charge injection effect; The technique is shown in Fig. 2.22: one of the transistors has the required large aspect ratio; the second has a minimum area. The phases driving the two elements are slightly delayed. When the structure switches on, the bigger transistor closes first and ensures the proper low on-resistance. When the switch goes off, we open the transistor with the larger *W/L* first, thus

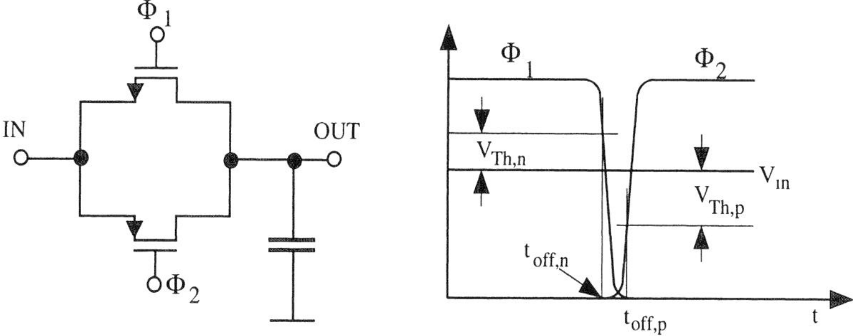

Fig. 2.23 - Compensation of the charge injection with complementary transistors

causing a significant injection of charge in the high impedance node. However, the second switch is still closed and drains this charge (large from the error point of view but quite small compared to the signal) towards the low impedance node. After a given delay, the second switch opens and injects a minimum amount of charge into the big capacitor. The resulting offset voltage is therefore quite small.

Use of complementary transistors.

This compensation technique (Fig. 2.23) exploits the obvious fact that the n-type channel is made of electrons while the p-type one is made of holes. Switching off two complementary transistors leads to a simultaneous injection of negative and positive charges that one hopes will compensate each other. However, to have an exact charge injection cancellation, it is necessary to inject exactly the same amount of positive and negative charge, and this is not easy.

> **(OBVIOUS) NOTE**
>
> N-channel transistors inject electrons; p-channel devices inject holes.

If we look at Fig. 2.23, where the possible phases used to drive the complementary transistors are shown, we can make a number of considerations. Phase Φ_2, used to drive the p-channel transistor, is most likely achieved by inverting Φ_1. We see that it is slightly delayed compared to Φ_1. Moreover, it shows the input voltage which is approximately halfway between V_{DD} and ground. The crossing of Φ_1 with $V_{in} + V_{Th,n}$ provides $t_{off,n}$ while the crossing of Φ_2 with V_{in} - $V_{Th,p}$ provides $t_{off,p}$. We see that the time at which the n-type transistor goes off is anticipated compared to the time at which the p-type transistor opens. Therefore, all of the charge, or any part of it injected by the n-type element, can be driven toward the low impedance node. We can also observe that the overdrive voltages of the two transistors are different. They depend on the input voltage so that, for an increase of V_{in}, $V_{ov,n}$ decreases, while $V_{ov,p}$ increases. Correspondingly, $t_{off,n}$ anticipates while $t_{off,p}$ delays.

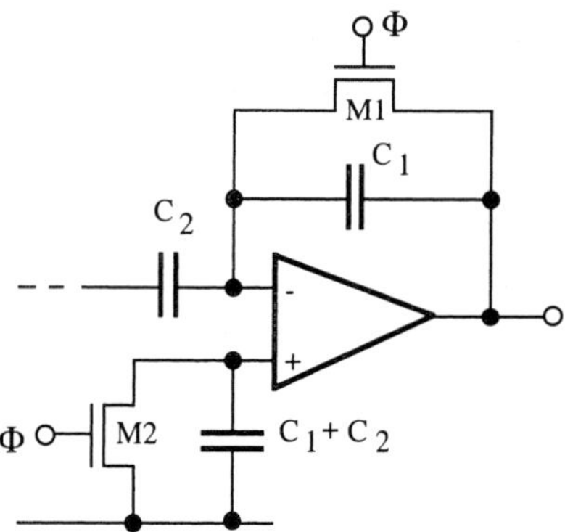

Fig. 2.24 - Charge injection reduction by the use of a compensation network

From the discussion above, we see that many factors affect the total charge injection on the high impedance node. In order to make the compensation effective, it is necessary to establish some parameters and carefully control others. Since the injected charge depends on the input signal, it should be kept at a constant level. This is possible if we connect the switch to the virtual ground of an operational amplifier. Moreover, the relative delay of the two phases Φ_1 and Φ_2 should be properly controlled.

Use of compensation networks

The charge injection can be compensated if we are able to reproduce its effect on a network which is a replica of the circuit concerned. By subtracting the charge injection error and the error achieved with the replica, we obtain a satisfactory cancellation. One possible implementation of this technique is shown in Fig. 2.24. The circuit contains a capacitor C_1 in feedback around an operational amplifier. C_1 integrates the charge injected by C_2 connected between the virtual ground and a low impedance node. Periodically, the switch M_1 resets capacitor C_1. Opening the switch produces a charge injection into the virtual ground and causes an offset error. As we can see, the replica network controls the non-inverting terminal of the op-amp. When M_2 switches off, it injects a charge into the capacitor (C_1+C_2) which is likely to be equal to the one injected by M_1. The charge, integrated on (C_1+C_2), changes the voltage of the non-inverting terminal and, because of the virtual ground effect, changes the inverting input as well. It is easy to verify that the virtual ground voltage modification pulls a charge on the plates of C_1 and C_2 which is equal to what is injected by M_1; thus, we obtain a compensation effect.

The different boundary conditions which are seen by capacitors C_1, C_2 and (C_1+C_2) can cause possible errors affecting the technique. In the latter element, the bottom terminal is connected to ground. For C_1 and C_2 we will most likely have nodes with non-zero equivalent impedance.

Use of fully differential solution

We shall see in a successive chapter that a very convenient method for reducing the noise caused in the digital part is the utilization of fully differential signal processing. We elaborate the signal twice, using two complementary paths. The result is the difference of the two complementary signals achieved. In this way, any common mode variation cancels the other and will not affect the result. The fully differential technique requires fully differential operational amplifiers, which are special blocks capable of providing two complementary outputs.

Fig. 2.25 shows how the fully differential technique can be used to compensate charge injection. The switching off of the two reset transistors across capacitors C_1 produces an injection of charge in the two op-amp inputs. However, this injection corresponds to a common mode signal that is rejected by the circuit operation. Clock coupling is then cancelled within the limits of the matching accuracy between the two processing paths. In the case of minimum area transistors, the matching in the gate area (and, consequently, the charge stored in the channel) is not particularly high. Normally, we assume that compensation is effective with a 90% cancellation of the error.

> **NOTE**
>
> Clock-feedthrough cancellation is difficult to analyse with computer circuit simulators. We compensate the effect by balancing charges.
>
> Many circuit simulators claim charge conservation features but, often, their achieved accuracy is not enough for our purposes.

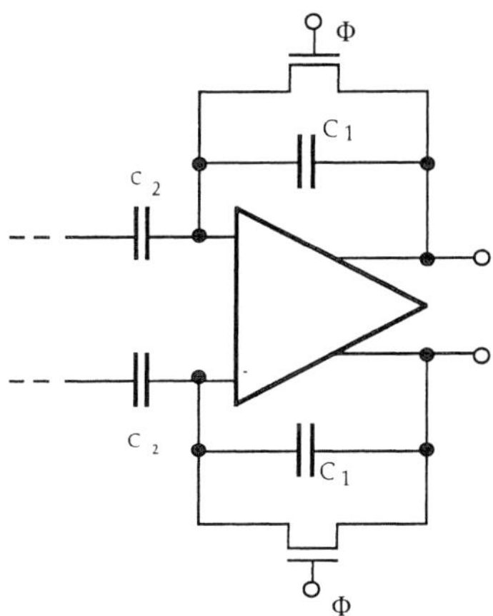

Fig. 2.25 - Compensation of the charge injection with a fully differential structure

2.4 LAYOUT OF SWITCHES

A switch is a critical element in analog processing systems. It is typically used to transfer packets of charge from one analog node to another, but it is also controlled by a digital clock applied to the gate of MOS transistors. This kind of operation: analog function with a digital control, is on the border between the analog world and the digital world and can become a source of possibly undesired coupling that should be avoided.

We can distinguish between the use of only one transistor and the use of complementary elements. In the former case, the only coupling between the analog path and the digital control is established by the crossing of the gate with an active area.In the latter case, it is topologically impossible to avoid an additional crossing between analog lines and digital control.

Fig. 2.26 shows the typical layout of a single transistor switch and a transistor switch made by a complementary element. Transistors have the minimum size. We see that in the former case, we have a transistor directly achieved in the silicon substrate (the n-channel for an n-well technology). The digital signal runs along the top and the analog function affects the bottom of the layout; thus, avoiding crossing interference. In the latter case, we have one transistor in the substrate and the other in the well. Again the digital controls run along the top of the layout. In addition, this strategy used is to place the transistor in the well at the analog side. In this manner, we have some self-protection against the spur coming from the digital side. Possible noise is intercepted by the well and, hopefully, driven towards a low impedance point that biases the well. A similar strategy is used for the single transistor switch. Here the noise intercept is achieved by a substrate bias used to separate the analog and the digital world. The layouts in Fig. 2.26 have the switches on the digital side. We often apply a second protection between switches and the digital bus using a substrate biasing or a well.

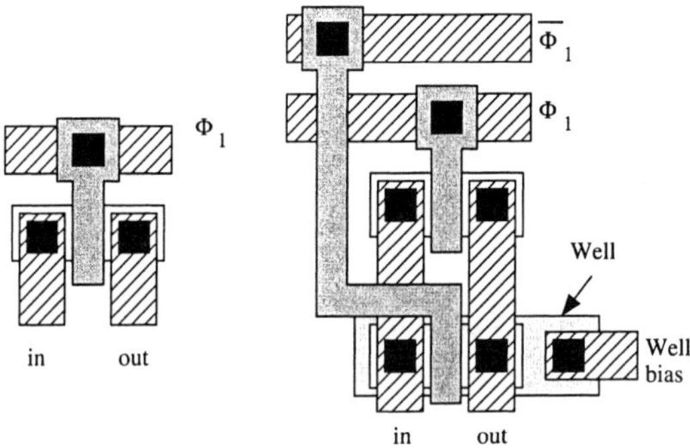

Fig. 2.26 - Layout of a transistor and a complementary transistor switch. Note that the protections that avoid spur injection from the digital lines are not indicated.

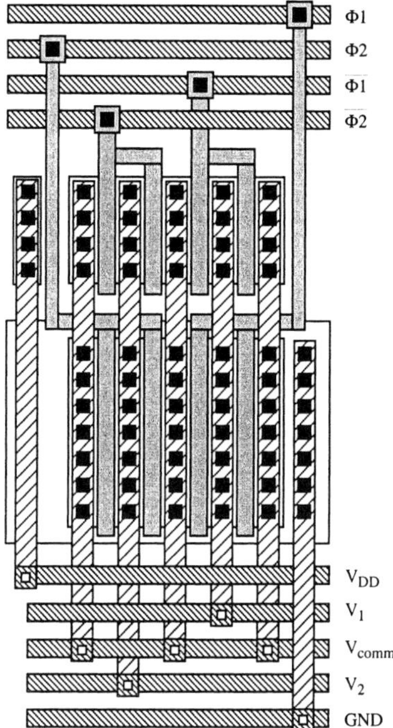

Fig. 2.27 - Toggle switch realized by the parallel connection of complementary transistors

Fig. 2.27 shows the layout of a toggle switch. It connects the common terminal to the voltage V1 during phase 1 and to V2 during phase 2. The complementary transistors of the switch have a large aspect ratio. In order to obtain a reasonable shape the transistors are splitted into the parallel connection of two elements. Fig. 2.27 includes the substrate bias and the well bias connected respectively to VDD and ground.

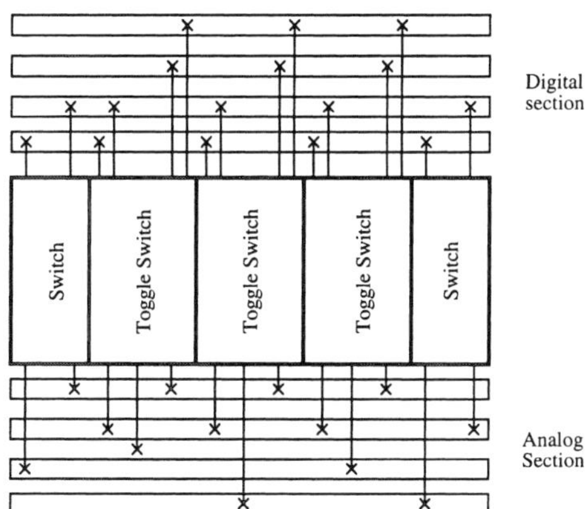

Fig. 2.28 - Floor plan of switches.

When we have a number of switches controlled by the same phases, it is a good procedure to place all the switches side by side forming a stack. The analog world is on one side of the stack and the digital bus carrying all the required phases is on the other (Fig. 2.28). Proper shielding obtained with substrate biasing and additional well rings allows us to intercept possible noise injected from the digital side to the analog side, thus limiting spur effects.

2.5 REFERENCES

- M. Ismail, T. Fiez, *Analog VLSI. Signal and Information Processing*, McGraw-Hill Publishing Company, New York, N.Y. 1994
- R. S. Soin, F. Maloberti, J. Franca, *Analogue-Digital ASICs*, Peter Peregrinus Ltd., London, UK, 1991
- D. J. Alstot, W. C. Black, *Technological design considerations for monolithic MOS switched capacitor filtering systems*, Proc. of IEEE, vol. 71, pp. 967-986, 1983
- M. J. M. Pelgrom, A. C. J. Duinmaiger, and A. P. G. Welbers, *Matching properties of MOS transistors for precision analog design*, IEEE J. Solid-State Circuits, vol. 24, pp. 1433-1439, 1989
- B. J. Sheu, J. H. Shich, M. Patil, *Modeling Charge Injection in MOS Analog Switched*, IEEE Transactions on CAS, Vol. CAS 34, pp. 214-216, 1987

-B.J. Sheu and C. Hu, *Switch-induced error voltage on a switched capacitor*, IEEE Journal ofSolid-State Circuits, Vol, SC-19, 1984, pp. 519-525.

- V. Colonna, F. Maloberti and G. Torelli, *Clock feedthrough compensation with phase slope control in SC circuits*, Electronics Letters, vol. 32, pp. 864-865, 9 May, 1996

2.6 PROBLEMS

2.1 Design the layout of a 8 kΩ resistor made by n^+ diffusion (0.3 μm depth). Try to achieve an almost square design. The sheet resistance is 42 Ω/□ and the lateral diffusion is 80% of the junction depth. Assume that each square corner counts half a square.

2.2 A 400 square resistance is made (3 μm width); the minimum distance between resistive elements is 1 μm and a corner counts half a square; we want dummy elements all around the resistor. What is the minimum silicon area to achieve the resistance?

2.3 An integrated capacitor is defined by a second poly plate with a size of 50 x 35 μm. Oxide thickness is 18 nm and ε_r=4.1. Find the capacitance. The undercut is 0.2 μm and its standard deviation is 0.02 μm. Find the actual capacitance and the accuracy achieved.

2.4 Two nominally equal capacitors (25 μm x 25 μm) are 100 μm apart. The undercuts in the two positions are 0.12 μm and 0.16 μm respectively. Find the relative mismatch.

2.5 Draw the layout of two capacitors having the ratio 3.6. The two structures must have the same perimeter to area ratio. Draw again the layout assuming that the poly-2 contact is achieved using a donut-like structure. The area used for the contact is 3 x 3 μm.

2.6 Find the on-resistance of an n-channel switch whose aspect ratio W/L is 12. k_n=97 μA/V^2, V_{ov}=0.45 V. Find the same result using SPICE and the BSIM models. Use different length of the transistor.

2.7 Consider a sample and hold using a 4 pF storing capacitor. We want to achieve an accuracy better than 0.01% in 100 nsec. Find the maximum allowed on-resistance.

2.8 Estimate the charge stored in the channel of an n-channel and a p-channel transistor in the on state. The effective gate sizes are W= 3μm L= 1μm. Assume C_{ox}=2.3 fF/μ^2, $V_{G,n}$=3.3 V, $V_{G,p}$=0 V, $V_{th,n}$ = $V_{th,p}$ = 0.75, V_{in} =1.9 V

2.9 Repeat the previous problem with the same aspect ratio but with a designed L=0.35μm. Assume that the overlap gate-diffusion is 0.05 μm. Find the charge injected in the two nodes when the gate voltage switch from 3.3 V to 0 V (and Vice-versa). Assume α and β equal to 1/2.

2.10 Simulate with SPICE the circuit given in Fig. 2.20. Design the transistor sizing such that the 2 pF capacitor approaches within 0.1% the 1.3 V input in 100 nsec. Assume V_G=3.3 V. Find the charge injected when the switch goes off. Use the BSIM models given in Appendix A.

2.11 We have to charge a big capacitor (C = 30 pF). The required time constant is 250 nsec. Design the n-channel switch using the same parameters given in the previous problem. Derive the charge injection effect caused by 50% of the channel charge.

2.12 Design a complementary switch that always keeps its on-resistance below 5 kΩ. Use the process parameters given in Problem 2.8.

2.13 Simulate with SPICE (BSIM Model) the use of dummy switch to compensate the charge injection. The circuit doesn't need speed. Define the transistor sizing that achieve the minimum offset with 100 psec, 400 ps, 1 nsec rising and falling time of the clock phases used.

2.14 Design a parallel transistor switch that permits us to charge at 1.5 V a *100 pF* capacitor with a *10 nsec* time constant ($V_{DD} = 3.3\ V$). Design the small switch in order to settle the output voltage within 0.1 mV after *10 nsec* the big transistor opens. Use the models given in Appendix B.

2.15 Simulate with Spice (models in Appendix B) a complementary transistor switch charging a *2 pF* capacitor. Use for the n-channel transistor $W/L = 1/0.5\ \mu m$. Compensate the charge injection with an accuracy better than *0.5 mV*. Assume that the clock phases have 100 psec rising-falling time and that the two phases cross symmetrically. The input voltage is *1.3 V* and $V_{DD} = 3.3\ V$.

2.16 Sketch the layout of two toggle switches connecting sequentially a common mode to 4 different voltages. How many phases are necessary?

Chapter 3

BASIC BUILDING BLOCKS

An analogue system is the suitable interconnection of basic cells and passive components. In turn, the basic cells, namely operational amplifiers and comparators, may be assumed to be a suitable innterconnection of basic building blocks. These blocks are simple gain stages, differential pairs, differential to single ended convertors, voltage and current references, etc. Using this hierarchical view, it becomes easier to analyse and understand the features and performance of complex basic cells. However, one should also acquire prior knowledge of the operation and limits of the basic building blocks that are used as elements of the cell. In this Chapter their operation will be analysed and their features studied together with their limitations.

3.1 INVERTER WITH ACTIVE LOAD

The simplest form of gain stage is the inverter with active load. It is represented for an n-channel input transistor (a) and for a p-channel input transistor (b) in Fig. 3.1. Unlike digital inverters, the input signal is applied to only one transistor while the gate of the complementary element is biased with a fixed

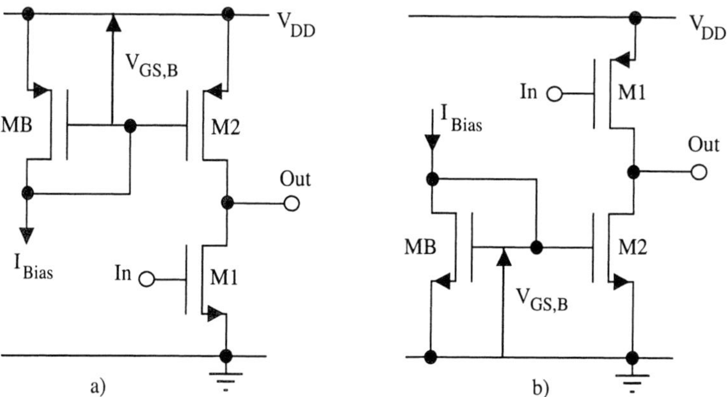

Fig. 3.1 - CMOS inverter with active load; a) n-channel input transistor. b) p-channel input transistor

voltage and operates as an active load. The biasing voltage $V_{GS,B}$ is obtained by transistor M_B connected in the so-called diode configuration (drain and gate connected) and carrying a given current I_{Bias}. The next chapter will show that the configuration used to bias M_2 corresponds to the simplest version of a current mirror.

The gate to source voltage of the active load is constant, therefore its *dc* voltage-current characteristic is univocally defined. If the bias of M_1 is less than its threshold limit, V_{Th}, the current in the circuit will be negligible and the operating point (see Fig. 3.2 a) will be very close to V_{DD}. Under this condition, the input transistor operates in the sub-threshold or in saturation, while M_2 is in the triode region. As V_{in} increases the input transistor exits the sub-threshold and begins to conduct. However, the output voltage remains close to V_{DD} up to the point for which the current in M_1 approaches the saturation current of M_2. At this point, the *dc* transfer characteristic (Fig. 3.2 b) displays a relatively high (negative) slope. In this region, both transistors are in saturation. Finally, when the input voltage is such that the current in M_1 tends to be larger than the saturation current in M_2, the output voltage becomes so low as to push M_1 into triode region. The output voltage drops down close to ground.

Fig. 3.2 b) shows that the slope of the transfer characteristics, which, as known, corresponds to the incremental or small signal gain, is relatively high when both transistors are in saturation. Therefore, to have a satisfactory signal amplification, the inverter should always operate in this region. By contrast, an inverter used for digital applications mainly works with the output voltage close to V_{DD} (logic *1* at the output) or close to ground (logic *0* at the output). In these two states, the circuit must dissipate virtually zero power. This condition, in CMOS, is obtained by driving both complementary transistors with

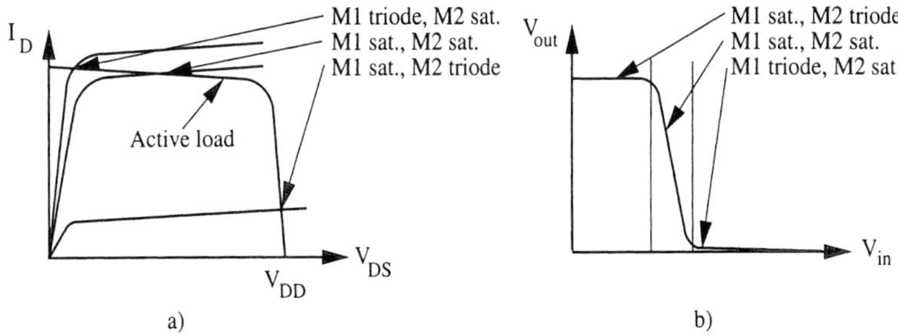

Fig. 3.2 - a) Intersections of the active load curve with the input transistor static characteristics. b) dc transfer characteristics of the inverter with active load.

the input signal. However, the incremental gain achieved in the transition region is quite poor (even if it is enough to guarantee a reasonable digital noise margin). Moreover, the current in the transition region is not well controlled, unlike for the inverter in Fig. 3.1.

3.1.1 Small Signal Analysis

Fig. 3.3 shows the simplified small signal equivalent circuit of the inverter in Fig. 3.1. The two transistors are modelled solely by the transconductance generator and the output resistance $1/g_{ds}$. The model that we studied in Chapter 1 takes into account any frequency-dependent behaviour of MOS transistors by the capacitances C_{gs}, C_{gd}, C_{db} and the two overlapping terms $C_{gs,ov}$, $C_{gd,ov}$. These capacitors will appear in the small signal equivalent circuit. Some of them are in parallel. Fig. 3.3 groups them together and accounts for an output capacitive load, C_L. The capacitances C_1, C_2, C_3 shown, represent the following combination of parasitic elements

$$C_1 = C_{gs1} + C_{gs1,ov} \tag{3.1}$$

$$C_2 = C_{gd1} + C_{gd1,ov}$$

$$C_3 = C_{db1} + C_{db2} + C_{gd2} + C_{gd2,ov} + C_L$$

At low frequency, the small signal gain is given by

$$A_v = \frac{V_{out}}{V_{in}} = \frac{-g_{m1}}{g_{ds1} + g_{ds2}} \tag{3.2}$$

if, as normally happens, the two transistors M_1 and M_2 are in saturation, the *dc* small signal parameters can be approximately expressed by

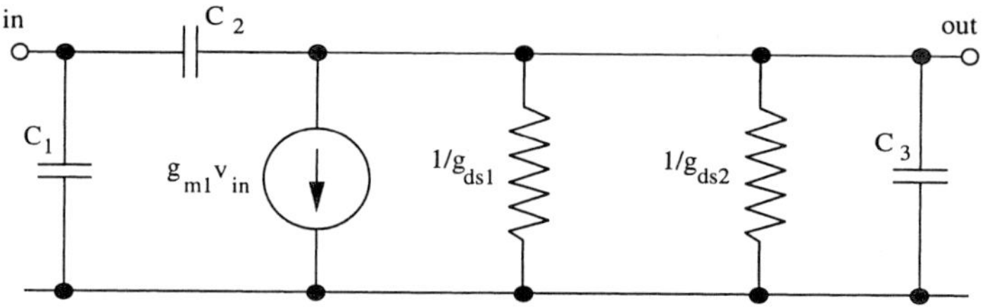

Fig. 3.3 - Small signal equivalent circuit of the inverter with active load.

$$g_m = \sqrt{2\mu C_{ox}\frac{W}{L}I_D} \qquad\qquad g_{ds} = \lambda I_D \tag{3.3}$$

Substituting in (3.2) gives

$$A_v = -\frac{\sqrt{2\mu_1 C_{ox}\left(\frac{W}{L}\right)_1}}{\sqrt{I_D}(\lambda_n + \lambda_p)} \tag{3.4}$$

Thus, the *dc* gain increases as the square root of the bias current decreases. The result, as made evident by (3.3), is due to the dependence of the transconductance on the square root of the drain current and on the dependence of the output conductance on the first power of the drain current. It should be noted that relation (3.4) will hold for as long as equations (3.3) remain valid. If the current is reduced to very low levels, the input transistor enters in the subthreshold region and the first relation in (3.3) is no longer valid. In subthreshold, the transconductance becomes proportional to the first power of the drain

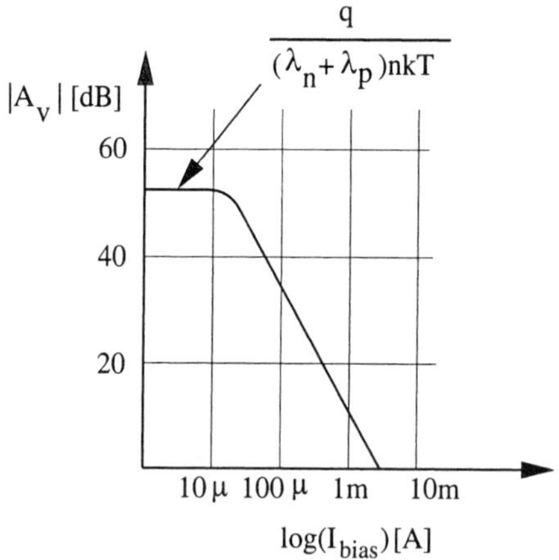

Fig. 3.4 - Typical dependence of the dc gain, A_v, on the bias current.

current

$$g_m = \frac{I_D}{n\frac{kT}{q}} \tag{3.5}$$

The *dc* gain becomes independent of the biasing current, and holds

$$A_v = -\frac{1}{n\frac{kT}{q}(\lambda_n + \lambda p)} \tag{3.6}$$

For a typical CMOS process, λ_n and λ_p are in the order of $0.03 \div 0.1\ V^{-1}$ while n ranges from *1.5* to *2*; thus, the maximum achievable gain of the inverter, at room temperature, is around *50 dB*. Fig. 3.4 shows the above-described dependence of the *dc* gain on the bias current.

Example 3.1

The n-channel input transistor of a CMOS analog inverter has an aspect ratio equal to W/L = 50μm/0.5μm; the λ factor of the n-channel and the p-channel devices are $3\ 10^{-2}\ V^{-1}$ and $6.5\ 10^{-2}\ V^{-1}$ respectively. Moreover, μ_n =460 $cm^2/Vsec$, C_{ox}= 2 f F/μ^2 and n= 1.6. Find the dc gain for I_D= 100μA and calculate its maximum achievable value. Estimate the level of bias current that brings the input transistor to sub-threshold.

Solution:
Assuming that the bias current I_D= 100μA and that a proper voltage applied to the input node brings both transistors to saturation, equation (3.4) can be used to express the dc gain. Remembering that C_{ox}= 2fF/μ^2= 0.2μF/cm^2 the following results

$$A_v = \frac{\sqrt{2\mu_1 C_{ox}\left(\frac{W}{L}\right)_1}}{\sqrt{I_D}(\lambda_n + \lambda_p)} = \frac{\sqrt{2 \cdot 460 \cdot 2 \cdot 10^{-7} \cdot 100}}{\sqrt{10^{-4}} \cdot 9.5 \cdot 10^{-2}} = 142.7$$

The maximum achievable gain (using (3.6)) is calculated in

$$A_v = \frac{1}{n\frac{kT}{q}(\lambda_n + \lambda_p)} = \frac{1}{1.6 \cdot 26 \cdot 10^{-3} \cdot 9.5 \cdot 10^{-2}} = 253$$

The limit of sub-threshold is estimated by using (1.61)

$$\overline{I_D} = \left(n\frac{kT}{q}\right)^2 2\mu C_{ox}\left(\frac{W}{L}\right) = 31.84\mu A$$

which confirms that even with 100 μA transistors still operate in saturation.

The small signal frequency response of the inverter is calculated by considering again the circuit in Fig. 3.3. Capacitance C_2, connected between the input and the output of the gain stage can be decoupled by the use of Miller's theorem. C_2 results split into two elements, one $C_2' = C_2\,(1 - A_v)$ loading the input, the other term $C_2'' = C_2\,(1 - 1/A_v)$ loading the output. If the voltage gain is high enough (and frequency independent), the total output capacitance sums up to $C_T = C_2 + C_3$ and the output resistance holds $1/(g_{ds1} + g_{ds2})$. It immediately follows that the gain transfer function is characterized by a pole located at the angular frequency ω_p

$$\omega_p = \frac{g_{ds1} + g_{ds2}}{C_2 + C_3} = \frac{(\lambda_n + \lambda_p)I_D}{C_2 + C_3} \tag{3.7}$$

Actually, because of the feedback capacitor C_2 we should also account for a zero in addition to the pole. In fact, the node equation at the output node yields

$$(v_o - v_{in})sC_2 + g_{m1}v_{in} + (g_{ds1} + g_{ds2})v_0 + v_o sC_3 = 0 \tag{3.8}$$

That confirms the value of the already estimated pole and locates the zero at the angular frequency $\omega_z = g_m/C_2$. Therefore, the small signal voltage gain is given by

$$A_v(s) = \frac{-g_{m1}}{g_{ds1} + g_{ds2}} \frac{1 - s/j\omega_z}{1 + s/j\omega_p} \tag{3.9}$$

Observe that the location of the zero is in the right half plane at a much higher frequency than the pole: the ratio between the angular frequencies of zero and pole is given by $A_0 \cdot (1 + C_3/C_2)$.

An important parameter of an amplifier is its unity gain frequency f_T (or gain-bandwidth product), which is the frequency at which the gain becomes $0\ dB$. From the equations above, neglecting the effect of the zero, f_T is evaluated in

$$f_T = \frac{1}{2\pi}\omega_p|A_v(0)| = \frac{1}{2\pi}\frac{g_{m1}}{C_2 + C_3} = \frac{1}{2\pi}\frac{\sqrt{2\mu_1 C_{ox}\left(\frac{W}{L}\right)}}{C_2 + C_3}\sqrt{I_D} \tag{3.10}$$

it follows that the unity gain frequency increases as the square root of the bias current increases. This result is exactly the opposite of the previous recommendation derived for increasing the *dc* gain: for fixed transistor sizes, the bias current had to be reduced. Of course, the above rules hold for the approximations used. Computer simulations may show more complex behaviour; nevertheless, we derived the above "rule of thumb" (and we will do the same with the other basic blocks studied in this chapter) with the aim of acquiring a set of "fuzzy rules" that constitute the background knowledge of circuit designers.

A point that is worthwhile underlining is the Miller amplification of the capacitance C_2 at the input terminals. The voltage gain can be as high as a hundred or more. Therefore, even if C_2 is very small *(some fF)*, the Miller amplification transforms it into a not negligible input load, which can result in a problem for the stage driving the inverter.

> **BE AWARE!**
>
> The output dynamic range of an inverter with active load is one of the best that designer can achieve with MOS technology.

Another point to consider concerns the output dynamic range of the inverter. One key requirement for an efficient operation of the circuit is to keep both the input transistor and the active load in saturation. This means that the drain to source voltages must be larger than the respective overdrive voltage (or saturation voltage) $V_{ov} = V_{sat} = V_{GS} - V_{Th}$. Therefore, the output voltage cannot approach V_{DD} or ground at an amount equal to the saturation voltages of the p-channel and n-channel transistor respectively. The output swing is limited by

$$V_{sat,n} \leq V_{out} \leq V_{DD} - V_{sat,p} \tag{3.11}$$

Since the saturation voltage is only a few hundred *mV*, the output dynamic range is normally quite large even for low supply voltages like *3.3 V* or less.

Example 3.2

Simulate an inverter with active load (V_{DD} = 5 V). Assume for the input transistor W_1= 100μm L_1= 1μm and for the active load W_2= 40μm L_2=2μm. The M_{B1} transistor (see Fig. 3.1) has the same sizes as M_2 and carries 100μA. Using the Spice parameters given in Appendix A, find the dc gain and the unity gain frequency.

Note: the reader should repeat the example using the Appendix B models and scaling down widths and lengths by a factor 2. Comparing the two set of results achieved is quite informative.

Solution:

The first step is the estimation of the dc input bias that leads both

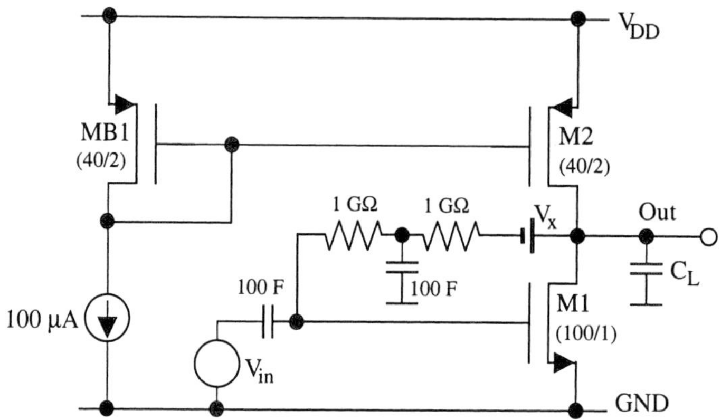

transistors to saturation. Such a voltage can be found by successive trials where the input dc level changes and the output voltage is checked as being around 2.5 V.

Unfortunately this approach is time-consuming; a more convenient method is shown in the figure. The circuit includes a feedback network between input and output; the values of the components used are such that the network sets up a feedback control at a very low frequency; practically, the feedback is open at only a few hertz. Moreover, a big capacitance connects the input generator to the input transistor. We also have the battery V_x setting up a level shift between the input voltage (close to $V_{Th,n}$) and the quiescent output level (close to 2.5 V to optimize output swing).

It is worth noting that the elements of the feedback network have a quite unusual value; this is done to push the operation of the feedback network to a very low frequency. Fortunately, the simulator quietly accepts such large and unrealistic values until they produce problems in the numerical convergency.

The artifices above allow us to solve the problem in one shot by an .AC simulation. The input Spice net list is the following

```
INVERTER WITH ACTIVE LOAD
.OPTIONS NODE NOPAGE
 M1  2 1 0 0 MODN W=100U L=1U AD=100P AS=100P PD=200U PS=200P
 M2  2 4 3 3 MODP W=40U  L=2U AD=40P  AS=40P  PD=80U  PS=80P
 MB1 4 4 3 3 MODP W=40U  L=2U AD=40P  AS=40P  PD=80U  PS=80P
 CL  2 0 0.5P
 R1  1 5 1G
 R2  5 6 1G
 C1  5 0 100
 C2  7 1 100
 VX  2 6 1.575
 IB  4 0 100U
 VDD 3 0 5
```

```
VIN 7 0 0 AC 1
.AC DEC 10 1K 1G
.PRINT AC VDB(1) VP(1) VDB(2) VP(2)
.NODESET V(4)=5
```

The input list includes a possible capacitive load at the output. It is a small (but realistic) capacitance.
Note that, even for this simple circuit, the simulator needs a.NODESET card to favor convergency. This, of course, depends on the simulator used and on the specific circuit. However, it should be noted that the .NODESET card works properly even when it sets a quite different voltage from the real value. From the following list, in fact, the voltage of node 4 is actually 3.932 V.

```
node voltage  node  voltage  node  voltage node voltage
(1) 0.6114 (2)  2.5110  (3) 5.0000  (4) 3.9320
(5) 0.6114 (6)  0.6114 (7) 0.0000
```

The results of the simulation are reported in the following Figure. Observe that the gain achieved is about 47dB; the unity gain frequency is fairly good, being around 500 MHz, and the phase margin is about 87 degrees.

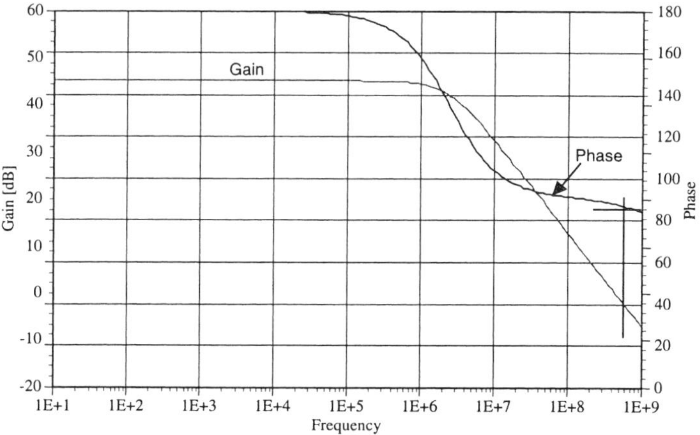

Gain and phase frequency response of the inverter with active load

3.1.2 Noise Analysis

We study the noise performance of the inverter with active load by incorporating the noise sources of transistors used in the small signal equivalent circuit. Fig. 3.5 includes the noise contributions in the schematic of the inverter. Since

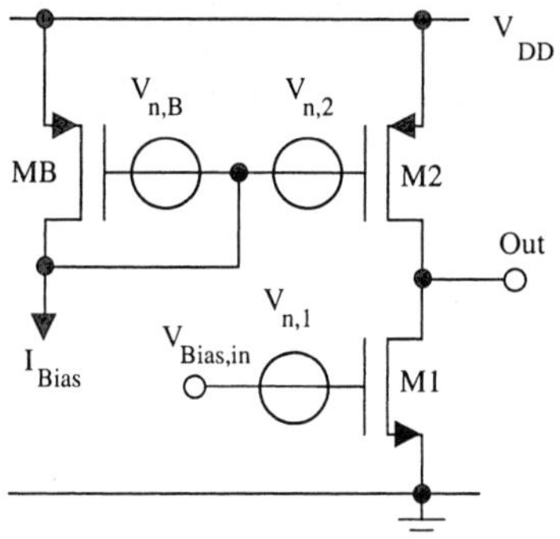

Fig. 3.5 - Circuit schematic for the noise analysis of the inverter with active load

the noise generators are uncorrelated their effect, measured at the output (or referred to the input), are superposed quadratically. Therefore, we calculate the noise considering each noise generator and estimate its small signal effect at the output. The three output voltages achieved are combined quadratically and the resulting output voltage is eventually referred at the input to provide the input referred noise generator.

By inspection of the circuit in Fig. 3.5 one can verify that the amplification of the noise generators at the gates of M_2 and M_B are equal: the transconductance generator of transistor M_2 (or M_B) delivers a current signal proportional to the noise voltage, $g_{m2}v_{n2}$ (or $g_{m2}v_{nB}$). In addition, the noise source of M_1 is multiplied by the transconductance g_{m1}. Therefore, the low frequency output voltage becomes

$$v_{n,out}^2 = \frac{v_{n,1}^2 g_{m1}^2 + (v_{n,2}^2 + v_{n,B}^2) g_{m2}^2}{(g_{ds1} + g_{ds2})^2} \tag{3.12}$$

we can refer the result to the input by dividing it by A_v^2. Assuming the two transistors M_2 and M_B matched, $v_{n,2}^2 = v_{n,B}^2$, we obtain

$$v_{n,in}^2 = v_{n,1}^2 \left(1 + 2 \frac{v_{n,2}^2 \cdot g_{m2}^2}{v_{n,1}^2 \cdot g_{m1}^2} \right) = v_{n,1}^2 (1 + \eta) \tag{3.13}$$

therefore, the input referred noise voltage is the noise of the input transistor multiplied by the amplifying factor, $(1+\eta)^{1/2}$. The amplification depends on the active load and the bias network.

We will see that not only is it advisable to keep the input noise small but also it is recommended that the input noise should be dominated by the input device. Therefore, we have to achieve an η factor which is negligible with respect to *1*. We know that the spectrum of the noise generators in Fig. 3.5 has two components the first is white the second one has a *1/f* spectrum. If we consider these two terms separately and we use equations (1.69) and (1.70), the η factor becomes

$$\eta_{white} = 2 \cdot \frac{g_{m2}}{g_{m1}} \qquad \eta_{1/f} = 2 \cdot \frac{K_{f,2} L_1^2}{K_{f,1} L_2^2} \tag{3.14}$$

Where we have used for g_m the relationship $(2\mu C_{ox}(W/L)I_D)^{1/2}$. From the above equation we can observe that we achieve a minimum η factor in designing the transconductance of the input transistor larger than that of the active load. Moreover, to minimize $\eta_{1/f}$, the type of active load must show the lower flicker noise coefficient and the length of the active load (and the bias transistor load) must be greater than that of the input device.

KEEP IN MIND!

The noise of any gain stage should be dominated by the input device contribution. The noise of the active load must be less than the one of the input device by a factor 2 at least.

The above recommendations hold for a stand-alone inverter with active load. In a complete circuit we have the interconnection of various gain stages. For a given specific circuit architecture the designer should identify the most critical blocks and design those critical blocks following the above recommendations.

3.1.3 Design of Inverters with Active Load

The design of an inverter with active load requires a few parameters to be defined: the type of input transistor (n-channel or p-channel), the transistor sizing and the bias current. Previous sections derived a number of approximate equations for appraising key circuit performance parameters. On the basis of these and the given circuit specifications, we can proceed to properly design the circuit.

The first design decision concerns the type of input transistor. Often this depends on the previous stage used; in a two stage op-amp (that will be studied in full detail in a successive Chapter), an inverter with active load is the second stage of amplification. If the first stage has *n-channel* input transistors, the second one normally utilizes a p-channel at the input and vice versa. In all cases, noise considerations typically drive the choice of the input transistor type. We already mentioned that noise in a gain stage is (and should be) controlled by the noise performance of the input device. Therefore, recalling that an n-channel transistor displays a better white noise but a worse *1/f* term than a p-channel element, the circuit uses an n-channel input device when the white noise is the major concern whereas it utilizes a p-channel input element when the *1/f* noise contribution must be limited.

> **RULE OF THUMB**
>
> Keep the overdrive of transistors in any output stage low: your output swing will be quite large, and the dynamic range will increase accordingly.

Once defined, the type of input transistor we have to determine the transistor sizing. They depend on a number of implications of which output swing is one normally accounted for (especially in low voltage circuits). Conditions given in (3.11) prescribe that the overdrive voltage for both transistors must be lower than the room left available by the desired output swing. The overdrive voltages depend on bias current and the transistor aspect ratio. Therefore, we can write the following relationship

$$\sqrt{\frac{2I_{bias}}{\mu_1 C_{ox}(W/L)_1}} < V_{out} < V_{DD} - \sqrt{\frac{2I_{bias}}{\mu_2 C_{ox}(W/L)_2}} \qquad (3.15)$$

We already mentioned that designers normally want the output swing to be as large as possible, especially in output stages. This because of the need for a maximum signal-to-noise ratio (SNR).

Equation (3.15) provides a relationship between bias current and transistor aspect ratio. To achieve the proper value of the aspect ratio we have to define the bias current. To do this, the gain required, power consumption and speed specifications will all direct the choice of the bias current.

The bias current controls the finite gain and the bandwidth of the stage. Assuming the transistors to be in saturation, the equations that we have to use to properly design the bias current are (3.4), and (3.10). The first defines the finite gain and the second characterizes the unity gain frequency. For the reader's convenience the two equations are reported again below

$$A_v = -\frac{\sqrt{2\mu_1 C_{ox}\left(\frac{W}{L}\right)_1}}{\sqrt{I_D}(\lambda_n + \lambda_p)} \qquad f_T = \frac{1}{2\pi}\frac{\sqrt{2\mu_1 C_{ox}\left(\frac{W}{L}\right)_1}}{C_2 + C_3}\sqrt{I_D} \qquad (3.16)$$

For a given technology, the parameters $\sqrt{2\mu_1 C_{ox}}/(\lambda_n + \lambda_p)$ and $\sqrt{2\mu_1 C_{ox}}$ have a given value. Therefore, equations (3.16) can be solved to find out the value of bias current and aspect ratio. Let us assume, for example, that we use an n-channel input transistor whose process transconductance parameter is $180\ \mu A/V^2$, also we choose the following reasonable values for $\lambda_n = 8.5\ 10^{-2}\ V^{-1}$ and $\lambda_p = 3.5\ 10^{-2}\ V^{-1}$. A typical value of $(C_2 + C_3)$ is $0.75\ pF$. Using the above parameters, and expressing the current in mA, system (3.16) becomes

$$A_v = -5\sqrt{\frac{\left(\frac{W}{L}\right)_1}{I_D}} \qquad f_T = 1.27 \cdot 10^8 \sqrt{\left(\frac{W}{L}\right)_1 I_D} \tag{3.17}$$

This can be solved for given values of finite gain and unity gain frequency. If we want $A_v = -80$ and $f_T = 300\ MHz$, we obtain $(W/L)_1 = 38$ and $I_D = 147\ \mu A$.

> **REMEMBER!**
>
> The output driving capability of the inverter with active load is asymmetrical. A large swing of the input voltage permits a wide control of the device current. By contrast, the current in the active load can't exceed the saturation value.

In addition to the above relationship, we have to ensure a proper large-signal driving of the output node. The load of the inverter is always a capacitance (otherwise, a finite resistance would dissolve the voltage gain) but during transients, the circuit must provide a drain or sink current to charge or discharge that capacitive load. When the current is provided by the input transistor we have no limitations. In contrast, when the current is furnished by the active load, we have an upper limit. It is given by the current generated by the active load itself. This, in turn, sets the slew rate limit

$$\frac{dV_{out}}{dt} \le \frac{I_{bias}}{C_{Load}} \tag{3.18}$$

where C_{Load} is the total output capacitance, $C_2 + C_3$, as shown in the equivalent circuit of Fig. 3.3. If the bias current designed is large enough to comply with the slew-rate specification the designer can proceed to the next design step. Otherwise, he has to increase the bias current, and take a decision for some trade-off between the conflicting requirements of finite gain and speed.

Once the aspect ratio of the input transistor is defined, its length (and accordingly width) must be determined. For this since we have assumed given value of λ_n and λ_p we have to find the lengths that achieve them. In reality, it is sum of the two lambda that matters. Therefore, we have a degree of freedom that we can use to optimize some circuit performance parameters. A proper use of the degree of freedom is for optimizing noise performance. We have seen that the input referred flicker noise contribution is reduced if the length of the active load is larger than that of the input device. Therefore, the designer can decide to have quite a resistive active load and to achieve the required $(\lambda_n + \lambda_p)$ by properly designing the length of the input transistor. Always, normally, the lengths used is rarely larger than a few times the minimum feature allowed by the technology used.

Example 3.3

Design an inverter with active load. Use an n-channel input transistor and the Spice models given in Appendix B. The circuit is loaded by 0.5 pF; it must meet the following specifications:

- *Output swing from 0.2 V to 3.1 V with V_{DD}=3.3V*
- *dc gain better than 40 dB*
- *slew rate better than 20 V/µsec*
- *f_T better than 140 MHz*

Solution:

A good starting point for this design is an understanding of the features of the transistor which operates in saturation. To achieve these features, it is necessary to decide the necessary bias current preliminarily. From the slew rate specifications and the value of capacitive load the following results

$$I_{bias} > \frac{dV_{out}}{dt} \cdot C_L = \frac{20}{10^{-6}} 10^{-12} = 20\ \mu A$$

To ensure some margin, let us assume I_{bias}=30 µA.

The features of transistors to be used are derived from a preliminary simulation where we use a number of transistors in the so-called diode-connection configuration. The features that we have to use are in the Spice print-out. We use the following test circuit

```
DIODE CONNECTED TRANSIATORS
.OPTIONS NODE NOPAGE
M1 1 1 Gnd Gnd MODN L=0.5u W=5u  AD=66p PD=24u AS=66p PS=24u
M2 2 2 Gnd Gnd MODN L=1u   W=10u AD=66p PD=24u AS=66p PS=24u
M3 3 3 Vdd Vdd MODP L=0.5u W=5u  AD=66p PD=24u AS=66p PS=24u
M4 4 4 Vdd Vdd MODP L=1u   W=10u AD=66p PD=24u AS=66p PS=24u

i1 Vdd 1 30uA
i2 Vdd 2 30uA
i3 3 Gnd 30uA
i4 4 Gnd 30uA
vdd Vdd Gnd 3.3
.op
```

that contains two n-channel and two p-channel transistors. The bias current is the above-defined 30 µA; the transistor length are the minimum allowed and two times that minimum. The widths chosen correspond to an aspect ratio equal to 10. The resulting table is

AC SMALL-SIGNAL MODELS

	M1	M2	M3	M4
MODEL	MODN	MODN	MODP	MODP
TYPE	NMOS	NMOS	PMOS	PMOS
ID	3.00e-005	3.00e-005	-3.00e-005	-3.00e-005
VGS	7.60e-001	7.27e-001	-1.08e+000	-1.26e+000
VDS	7.60e-001	7.27e-001	-1.08e+000	-1.26e+000
VTH	5.44e-001	5.08e-001	-7.36e-001	-8.42e-001
VDSAT	1.70e-001	1.67e-001	3.35e-001	3.91e-001
RS	9.84e+000	4.92e+000	9.60e+000	4.80e+000
RD	9.84e+000	4.92e+000	9.60e+000	4.80e+000
GM	2.52e-004	2.47e-004	1.42e-004	1.26e-004
GDS	2.25e-006	8.81e-007	9.24e-006	2.43e-006
GMB	6.21e-005	6.57e-005	1.97e-005	2.39e-005
GBD	0.00e+000	0.00e+000	0.00e+000	0.00e+000
GBS	0.00e+000	0.00e+000	0.00e+000	0.00e+000
CGS	9.09e-015	3.60e-014	3.95e-015	1.65e-014
CGD	1.03e-015	2.10e-015	1.58e-015	3.31e-015
CGB	3.14e-016	8.18e-016	6.42e-016	2.24e-015
CBD	5.50e-014	5.50e-014	3.40e-014	3.29e-014
CBS	8.34e-014	8.54e-014	4.05e-014	4.41e-014

Observe that our target of a less than 0.2 V saturation voltage is marginally achieved for M_1 and M_2 while it is fairly tight for the other test transistors. Since the overdrive voltage decreases with the square root of the aspect ratio, it will be necessary to increase the width of the p-channel element by at least a factor from 2 to 4.

At a first approximation, g_{ds} does not change with the transistor width. Choosing L_n=0.5μm and L_p=1μm, the total output conductance would result 4.68 10^{-6} Ω^{-1}; therefore, the required gain (40 dB) demands a transconductance gain better than 4.68 10^{-4} Ω^{-1}. The table gives only g_m=2.52 10^{-4} $\Omega^{-1}$for M_1. It is therefore necessary to increase the input device aspect ratio by about a factor of 4.

Equation (3.10) allows us to estimate the value of f_T. The specification requires f_T = 140 MHz, therefore, neglecting the parasitic contribution and using only the capacitive load, C_L = 0.5 pF

$$g_m = 2\pi C_L f_T = 6.59 \cdot 10^{-4}$$

which is larger than the value defined by the dc gain specification. This means that the aspect ratio of the input transistor must be further increased to match the bandwidth requirements to at least 90. After some Spice simulations, it is possible to find out $(W/L)_{in,n}$=50μm/0.5μm; $(W/L)_{load,p}$=25μm/1μm, and I_{bias}=30 μA. They lead to the following results:

A_v= 41 dB; f_T = 150 MHz; $V_{sat,\,n}$= 0.068 V; $V_{sat,p}$= 0.290 V

meeting the specification within only a few percent.

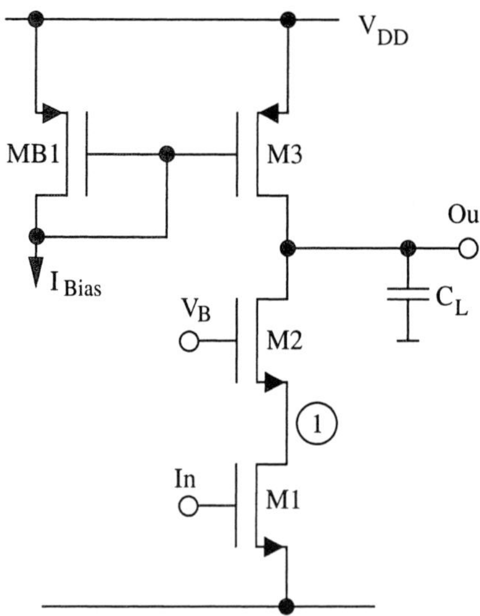

Fig. 3.6 - Cascode with active load

3.2 CASCODE

The previous section pointed out that a potential design problem derives from the Miller amplification of the input-output parasitic capacitance (C_2 in Fig. 3.3). The gain of an inverter with active load is around 100, and such a figure can lead to an amplification of a relatively small capacitance to a level that is not easily driven by the previous stage. This drawback is significantly attenuated by the cascode configuration shown (for n-channel input) in Fig. 3.6. The input transistor is replaced by a cascode arrangement: it is the cascade connection of a common source and a common gate stage. The additional element M_2 yields an extra node (*1*) whose effect is to decouple input from output. Similarly to the simple inverter, a parasitic capacitance ($C_2 = C_{gd1} + C_{gd1,ov}$) between input and node *1* is effective. By using the Miller theorem, this is multiplied by one minus the gain established between the input and the node *1*, A_1. The design target is therefore to maintain A_1 as small as possible, thus keeping the Miller amplification low. Moreover, the gain from input to output must be kept pretty high.

For the circuit to operate correctly, all the transistors should function in the saturation region. The condition defines a given range for the bias voltage, V_B, applied to the gate of M_2. In particular, using a voltage which is too low should be discouraged because it risks pushing M_1 into the triode region. The lower limit is

$$V_B > V_{sat,1} + V_{GS2} = V_{sat,1} + V_{Th,n} + V_{sat,2} = \tag{3.19}$$

$$= V_{Th,n} + \sqrt{\frac{I_1}{2\mu_n C_{ox}\left(\frac{W}{L}\right)_1}} + \sqrt{\frac{I_1}{2\mu_n C_{ox}\left(\frac{W}{L}\right)_2}}$$

For a conventional CMOS technology, the threshold voltage ranges from *0.4* to *1 V*, while the saturation voltage is a few hundred *mV*. Therefore, the minimum value of V_B can go from *0.8 V* to *1.6 V*. In addition to this limit, the value of V_B (or better the approach used to generate it) should take into account the variation of the threshold voltage due to the technological changes.

Besides the constraint we have just discussed for the lower level of V_B (with an n-channel input device), there is also an upper limit: when V_B increases, the voltage applied to the drain of M_1 increases and drives M_1 into better operational conditions. However, a large V_B worsens the minimum permitted value of the negative output swing; in fact, V_B sets the voltage of node *1* at one V_{GS} below V_B itself. The output voltage cannot approach node *1* more than a saturation voltage, it being it crucial to keeping M_2 in saturation. Therefore, for a given minimum desired output voltage, $V_{out,min}$, the following constraint follows

KEEP NOTE!

The cost that designer must pay for the additional transistor used in the cascode configuration is an output dynamic range reduction.

$$V_B < V_{out,min} - V_{sat,2} + V_{GS2} = V_{out,min} + V_{Th,n} \tag{3.20}$$

of course, this condition must be fulfilled together with equation (3.19). The combination of (3.19) and (3.20) allows us to design the proper value of V_B.

3.2.1. Small Signal Analysis

The simplified small signal equivalent circuit of a cascode is shown in Fig. 3.7. Before studying its dynamic operation, it is worth finding the *dc* behavior. For this, it is necessary to solve the network in Fig. 3.7 without accounting for the capacitors' effect. This task is quite simple, however, to obtain simplified relationships, the following calculations also neglect the output conductances g_{ds1} and g_{ds2} to obtain

$$g_{m1}v_{in} = -g_{m2}v_1 = -g_{ds3}v_O \tag{3.21}$$

which immediately results into the two *dc* gains

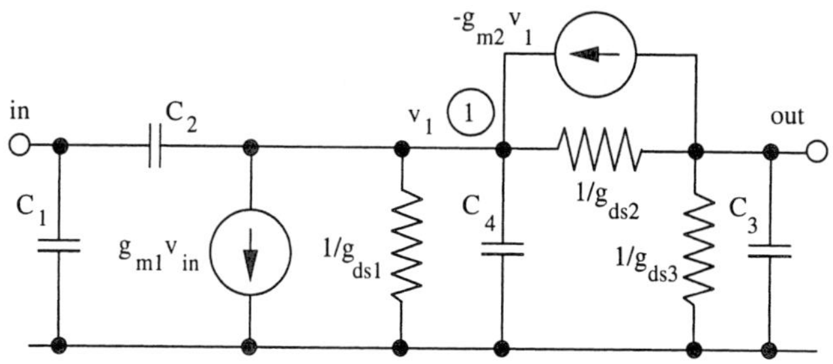

Fig. 3.7 - Small signal equivalent circuit of the cascode

$$A_v = \frac{v_O}{v_{in}} = -\frac{g_{m1}}{g_{ds3}} \tag{3.22}$$

$$A_1 = \frac{v_1}{v_{in}} = -\frac{g_{m1}}{g_{m2}} \tag{3.23}$$

Equation (3.22) and (3.23) hold under the hypothesis that the current generated by the transconductance element $g_{m1}v_{in}$ is fully transferred to the active load, $1/g_{ds3}$. In other terms, we assume that the resistance looking into the drain of M_2 is much larger than $1/g_{ds3}$ and that the resistance looking into the source of M_2 is negligible compared to $1/g_{ds1}$.

Equation (3.22) shows that the *dc* voltage gain is given by the ratio between a transconductance and an output conductance. This is very similar to the result achieved for an inverter with active load. Therefore, as is true for the inverter, the gain is proportional to the inverse of the square root of the bias current

$$A_v = -\frac{\sqrt{2C_{ox}\mu_1\left(\frac{W}{L}\right)_1}}{\sqrt{I_D}\lambda_p} \tag{3.24}$$

Using (3.22) and (3.23) it is possible to note that the gain A_v is surely larger than the one from input to the intermediate node *1*, which is the ratio between two transconductances. Transistors M_1 and M_2 carry the same current, and since they are of the same type, the two transconductance parameters μC_{ox} will match; gain A_1 becomes

$$A_1 = -\sqrt{\frac{W_1}{L_1}}\sqrt{\frac{L_2}{W_2}} \tag{3.25}$$

With transistors M_1 and M_2 having a similar aspect ratio, the Miller amplification of C_2 turns out to be a small number.

Summarizing the small signal *dc* performances, the input-output gain is in the same order as an inverter with active load (around *40 dB*). By contrast, the gain responsible for the Miller amplification of C_2 is fairly low and does not cause any significant problems.

The dynamic performance parameters are calculated considering the small signal circuit in Fig. 3.7 once again. The parasitic capacitances of the transistors have been summed up to give

$$C_1 = C_{gs1} + C_{gs1,ov}$$

$$C_2 = C_{gd1} + C_{gd1,ov}$$

$$C_3 = C_{gd2} + C_{gd2,ov} + C_{gd3} + C_{gd3,ov} + C_{db2} + C_{db3} + C_L$$

$$C_4 = C_{gs2} + C_{gs2,ov} + C_{db1} + C_{sb2} \tag{3.26}$$

The circuit contains three nodes: the input, the output and the decoupling node *1*. The effect on the frequency response of the input node comes from the type of the input generator. If it is an ideal voltage source, we have no frequency limitation. By contrast, with a finite input resistance, R_{in}, an input pole at the angular frequency $\omega_{in} = 1/[R_{in}(C_1 + C_2A_1)]$ would result.

Two capacitances, C_4 (plus C_2 Miller transformed) and C_3, affect the two remaining nodes; because of these the circuit transfer function will have two poles. The frequency of poles is, of course, precisely calculated by solving the network in Fig. 3.7 or a more complete equivalent circuit. However, despite the our efforts, we would obtain a complex expression that makes it difficult to acquire the necessary feeling for the circuit's behavior. We known that the transistor model used in Fig. 3.7 is approximate. Therefore, an exact analysis of a circuit based on it will in any case lead to an approximate result. When more precise results are necessary, the use of a computer simulator like Spice is recommended. At this stage of the study, however, it is probably more profitable to get a feeling of the operation of the circuit and memorize the rules of thumb. Instead of solving the equivalent circuit, it is worthwhile calculating the frequency of the two poles directly, under the assumption that they are decoupled each other, by simply estimate the time constant associated with each node.

Observing the circuit, we see that the resistance of output node is $1/g_{ds3}$ (actually, we also have the resistance looking into the drain of M_2, but it is pretty large and its contribution is negligible). The capacitance of the node is C_3 and the pole results

$$f_{p,out} = \frac{1}{2\pi} \cdot \frac{1}{\tau_{out}} = \frac{1}{2\pi} \cdot \frac{g_{ds3}}{C_3} \tag{3.27}$$

The resistance at node *1* can be approximated by ζ/g_{m2} (g_{ds1} is neglected and ζ is a factor that will be calculated shortly), while the loading capacitance, accounting for the Miller effect over C_2, is $C_4 + C_2(1 + g_{m2}/g_{m1})$. The resulting pole is

$$f_{p,1} = \frac{1}{2\pi} \cdot \frac{1}{\tau_1} = \frac{1}{2\pi} \cdot \frac{g_{m2}^2/\zeta}{g_{m1}(C_2 + C_4) + g_{m2}C_2} \tag{3.28}$$

g_{m1} and g_{m2} are always larger than g_{ds3}. Moreover, C_2 and C_4 are comparable or smaller than C_3, therefore

$$f_{p,out} \ll f_{p,1} \tag{3.29}$$

> **NOTE ON FIG. 3.6**
>
> In addition to the pole related to the output node, it exists another pole, associated to the intermediate node 1. Normally this new pole is at a very high frequency; however, in some particular cases it could affect the circuit' stability.

That just states that the pole associated with the output node is at a much lower frequency than node *1*. However, this is not sufficient: to ensure stability in a system with two or more poles, one of them should dominate over the other. The frequency of the dominant pole must be small enough to ensure that the Bode plot of the gain rolls down with a slope of *20 dB* per decade up and beyond to the *0 dB* level. The non dominant pole can affect the Bode plot but only at frequencies beyond the crossing of the *0 dB* axis, f_T. Therefore, assuming that $f_{p,out}$ is dominant, f_T turns to be

$$f_T = f_{p,dom}|A_v| = \frac{1}{2\pi}\frac{g_{m1}}{C_3} \tag{3.30}$$

and the following condition

$$\frac{g_{m1}}{C_3} < \frac{g_{m2}/\zeta}{(C_2 + C_4) + C_2 g_{m1}/g_{m2}} \tag{3.31}$$

must be fulfilled. This is not particularly difficult since g_{m1} and g_{m2} are similar and because C_3, due to the capacitive contribution of the load, is larger than C_1 and C_2.

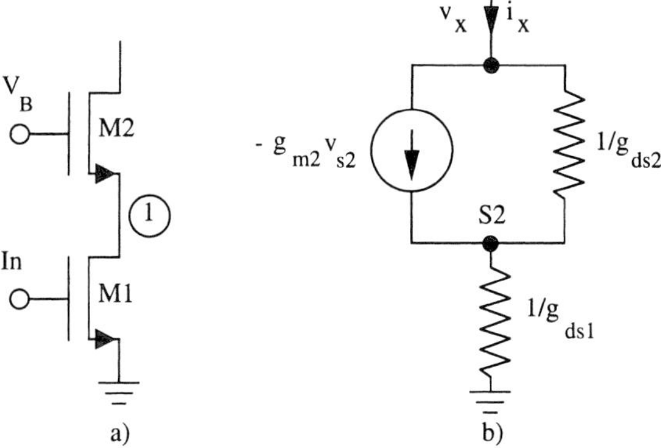

Fig. 3.8 - Schematics used to calculate the output resistance of the cascode

The previous calculation assumes that the impedance of the drain of M_2 is very large. This assumption must verify and we do it now by analyzing the structure in Fig. 3.8 a) and its small signal equivalent circuit in Fig. 3.8 b). The equivalent resistance is calculated, as usual, by injecting a test current i_x and by estimating the resulting voltage v_x. We can write

$$v_x = \frac{i_x}{g_{ds1}} + \frac{i_x + g_{m2} v_{s2}}{g_{ds2}} \tag{3.32}$$

$$v_{s2} = \frac{i_x}{g_{ds1}} \tag{3.33}$$

which yields

$$r_{d2} = \frac{v_x}{i_x} = r_{ds1} + r_{ds2}\left(1 + \frac{g_{m2}}{g_{ds1}}\right) \cong r_{ds1} g_{m2} r_{ds2} \tag{3.34}$$

As a mnemonic rule, the output resistance of the cascode is given by the output resistance of transistor M_1 (the one with the source connected to ground) amplified by the gain of M_2 ($g_{m2}\, r_{ds2}$).

Example 3.4

Verify, using Spice, that the output resistance of a cascode arrangement is as large as predicted by equation (3.34).

Solution:

The figure below shows a network suitable for answering the given request. A cascode configuration contains two equal n-channel

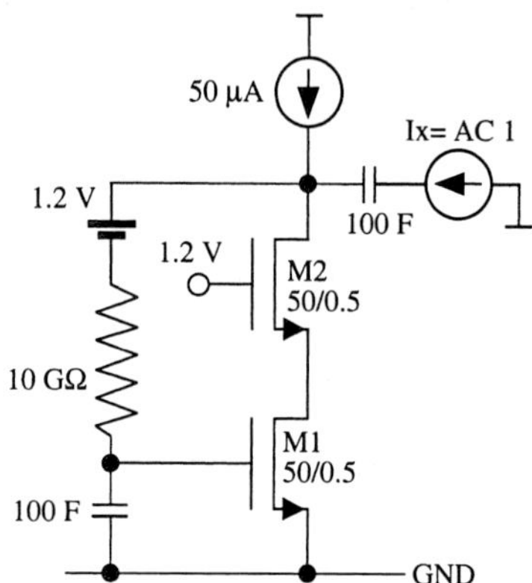

transistors biased with 50 µA. The 10 GΩ resistor and the 1.2 V battery establish a feedback path that self-biases the gate of M_1. The test current enters into the drain of M_2 through a 100 F capacitor (!). Moreover, for the small signals, the input of M_1 is shorted by the 100 F capacitor.
The .AC Spice simulation computes the voltage at output node; as the AC current is 1A, its value corresponds to the numeric value of the output resistance. The result achieved by Spice is: 39.8 MΩ.
The simulation provides the following transistor parameters:

	M1	M2
MODEL	MODN	MODN
TYPE	NMOS	NMOS
ID	5.00e-005	5.00e-005
VGS	6.27e-001	7.01e-001
VDS	4.99e-001	1.33e+000
VBS	0.00e+000	-4.99e-001
VTH	5.46e-001	6.84e-001
VDSAT	8.16e-002	5.71e-002
RS	9.84e-001	9.84e-001
RD	9.84e-001	9.84e-001
GM	1.02e-003	6.90e-004
GDS	6.59e-006	3.58e-006
GMB	3.84e-004	2.37e-004
GBD	0.00e+000	0.00e+000
GBS	0.00e+000	0.00e+000
CGS	8.77e-014	5.28e-014
CGD	1.04e-014	1.04e-014
CGB	3.47e-015	1.16e-014
CBD	5.84e-014	4.66e-014
CBS	9.12e-014	7.27e-014

From these we verify that both transistors are in saturation.Moreover, from the simulated value of g_m and g_{ds}, we find

$$r_{d2} = r_{ds1} g_{m2} r_{ds2} = \frac{6.9 \cdot 10^{-4}}{6.59 \cdot 10^{-6} \cdot 3.58 \cdot 10^{-6}} = 29.2 \mathrm{M}\Omega$$

This is smaller than the value obtained by Spice, which probably accounts for second order contributions. However, the calculated value of r_{d2} is a fairly large value not so different from the Spice result.

Another point that we should verify concerns the assumption that the impedance at node *1* is dominated by the drain to source resistance of M_1. We do that with the help of the circuit in Fig. 3.9. It represents the small signal equivalent circuit looking from the source of M_2 and comprises the equivalent circuit of M_2 and the small signal resistance of the active load r_{ds3}. We calculate the equivalent resistance by injecting a test current, i_x, and then evaluating the voltage produced; it is calculated as

$$v_x = r_{ds3} i_x + r_{ds2}(i_x - g_{m2} v_x) \tag{3.35}$$

leading to the equivalent resistance from the source of M_2

$$r_{s2} = \frac{1}{g_{m2}}\left(1 + \frac{r_{ds3}}{r_{ds2}}\right) = \frac{\zeta}{g_{m2}} \tag{3.36}$$

The same equation allows us to find the expression of the coefficient $\zeta=(1+r_{ds3}/r_{ds2})$ reported previously. Since the resistance of the active load is similar to r_{ds2}, ζ is a small number. Therefore, r_{s2} is only a few times the inverse of the transconductance of M_2 and surely does not exceed r_{ds1}, thus confirming the assumed approximation.

3.3 CASCODE WITH CASCODE LOAD

The gain stages analysed in the previous sections achieve their voltage gain because the signal current, $g_m v_{in}$, induced by the transconductance generator of the input transistor flows into a relatively high resistance (one r_{ds} or the parallel connection of two r_{ds}). This kind of operation is referred to as transconductance

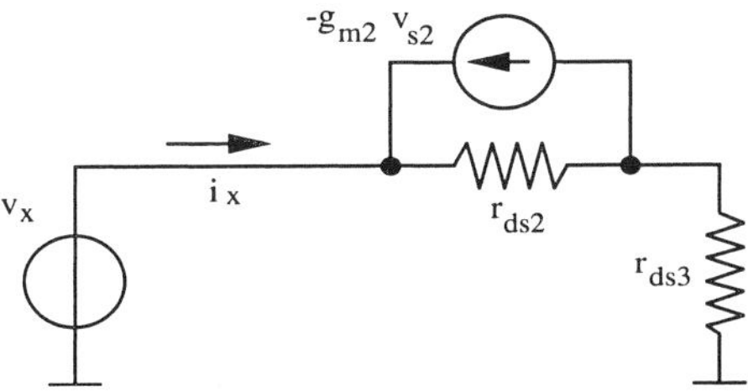

Fig. 3.9 - Equivalent circuit for the calculation of the impedance seen from the M_2 source.

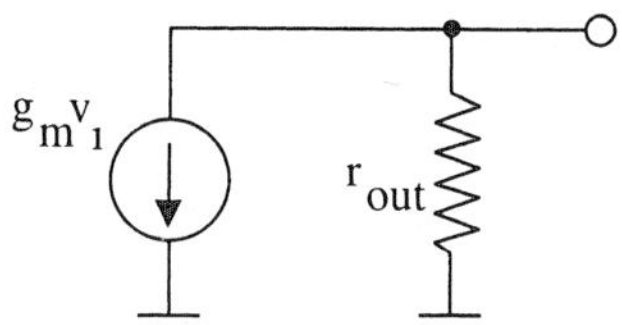

Fig. 3.10 - Equivalent circuit of a transconductance stage

gain amplification (Fig. 3.10) because the active function of the gain component used is transconductor like. With a conventional design, the achieved gain $g_m\, r_{ds}$ ranges around *100*; but, for many practical applications, such a gain level can not be sufficient. Therefore, to increase the gain while using a single stage amplifier, it would be necessary to enhance the transconductance or to augment the output resistance.

> ***OBSERVATION***
>
> The *dc* gain of a cascode with cascode load is approximately the square of the gain of an inverter with active load.

From equation (3.3) we learn that in saturation, an increase of the bias current or an enlargement of the aspect ratio magnifies the transistor's transconductance. However, augmenting the current depresses the output resistance by an amount that is larger than the benefit achieved. Therefore, the only viable way to raise the gain in a simple gain stage is to employ a very high aspect ratio in the input transistor. However, there are practical limits to this; normally the maximum aspect ratio used is not extremely high. A designer rarely uses several hundreds for this ratio, and one or two thousand are used only in very special cases. Hence, a limited increase is expected in the gain by merely pushing up the aspect ratio.

In contrast, increasing the output resistance can lead to better results. The previous section pointed out that the resistance seen from the output drain of a cascode is larger than the output resistance of a simple transistor by a pretty large factor: $g_m\, r_{ds}$. This result suggests that we should use a cascode arrangement even on the active load side, as shown in Fig. 3.11. The resistance of the output node becomes much higher, as it is the parallel connection of two cascode structures. Since the output voltage is given by the product of the signal current and the output resistance, by using the approximate relation (3.34) to express the output resistance of a cascode, a rough expression of the *dc* voltage gain becomes

$$A_v = -g_{m1}\frac{(r_{ds1}g_{m2}r_{ds2})(r_{ds4}g_{m3}r_{ds3})}{r_{ds1}g_{m2}r_{ds2}+r_{ds4}g_{m3}r_{ds3}} \tag{3.37}$$

It is worth noting that the gain is now proportional to the square of the

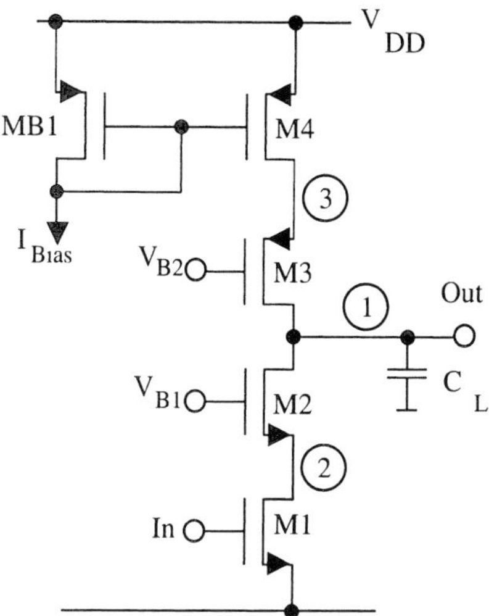

Fig. 3.11 - Cascode with cascode load

product of a transconductance and a transistor output resistance. Therefore, the gain is almost the square of what was achieved by a simple inverter with active load. If the transistors are in saturation, the gain of the simple inverter is inversely proportional to the square root of the bias current. Therefore, for the cascode with cascode load we the gain is inversely proportional to the bias current.

As for the inverter, the gain increases when the current decreases down until the point at which the transistors enters the sub-threshold region: the transconductance becomes proportional to the bias current and the *dc* gain reaches its maximum, typically *80÷100 dB*.

> **NOTE**
>
> The maximum achievable dynamic range in a cascode with cascode load corresponds to 2 V_{sat} distance from the supply voltages. Moreover, the bias voltages used must track the threshold fluctuations.

An important result is that even if the cascode with cascode load has a simple configuration (designers normally say that the circuit is a single stage amplifier, assuming that the gain comes from one transconductance amplification only), it achieves the same gain as the cascade of two inverters with active load. This advantage is counterbalanced by a disadvantage: a reduced output swing. We have already discussed this issue for the case of a simple cascode with active load. In that case, the optimization output dynamic range was associated with the design of one bias voltage. Here, we have to design two bias voltages, one for the gates of M_2 and the other for the gate of M_3. To choose these, we have to

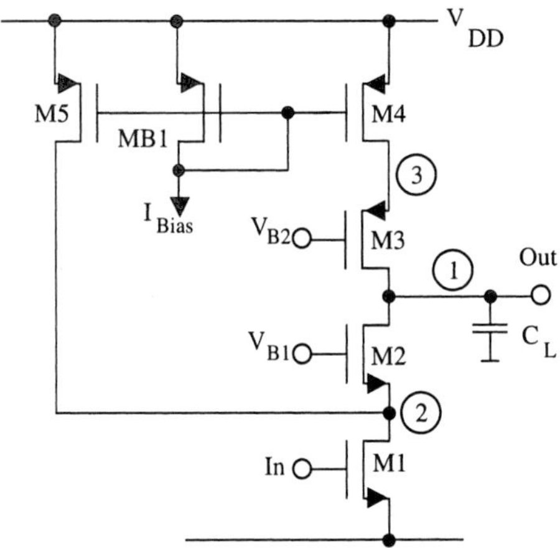

Fig. 3.12 - Gain enhanced cascode load. The additional current is injected into the input transistor to improve the dc gain

remember that the values of V_{B1} and V_{B2} should allow a suitable margin to keep transistors M_1 and M_4 in saturation, but, at the same time, they should minimize the reduction of the output swing. Therefore, for optimum output dynamic range, we should use the following design guidelines for V_{B1} and V_{B2} (which are similar to the ones expressed by (3.19)

$$V_{B1} \sim V_{sat,1} + V_{GS2} + \Delta = V_{Th,n} + 2V_{sat} + \Delta \tag{3.38}$$

$$V_{B2} \sim V_{DD} - V_{sat,4} - V_{GS3} - \Delta = V_{DD} - V_{Th,p} - 2V_{sat} - \Delta \tag{3.39}$$

where Δ is a suitable margin necessary for accommodating the possible mismatch of the threshold voltages and taking the variations of the bias current into account.

The cascode with cascode load discussed above adequately achieves high *dc* gains. Nevertheless, the circuit can be improved upon and the schematic can be re-arranged to better meet specific needs. Below, we consider two modified versions of the cascode with cascode load: the gain-enhanced version and the folded version. The first enhances the *dc* gain. For this, it is worth observing from equation (3.34) that the output resistance, that is the effect of the parallel connection of two cascodes, is dominated by the smaller of the two. It may happen that the smaller one is the active load. In this case, we can enhance the input transconductance by increasing the input transistor bias current. This design choice will lower the input cascode resistance, but this effect is irrelevant until the input cascode resistance becomes similar to the resistance of the active load. The corresponding circuit solution is shown in Fig.

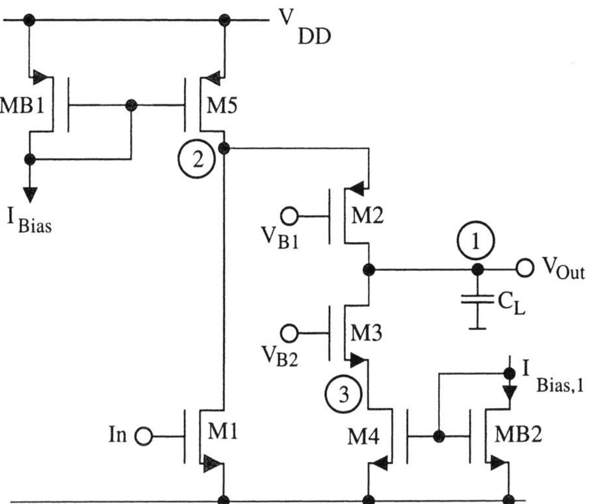

Fig. 3.13 - Folded cascode with cascode load

3.12. Transistor M_5 injects an additional current, I_5, in M_1, thus increasing its transconductance to

$$g_{m1}(I_4 + I_5) = g_{m1}(I_4)\sqrt{\frac{I_4 + I_5}{I_4}} \tag{3.40}$$

The gain increases by the same factor, because the signal current is completely transferred to the output node. As anticipated, the resistance of the output branch through M_2 is lower than in the original circuit. This derives from two effects: the increased current in M_1 reduces r_{ds1} and the additional branch, M_5, adds another term, r_{ds5}, in parallel to r_{ds1}.

Fig. 3.13 shows another possible modification of the basic cascode with cascode load. It is a folded cascode version obtained by complementing all transistors above node *2* of Fig. 3.11 and then bending them down to ground. Branch M_1 and the folded stack M_2-M_3-M_4 are biased by M_5. Since the current in M_4 is a replica of the current in the bias transistor M_{B2}, the current in M_1 and the current in the folded stack are controlled separately through the two bias reference currents I_{Bias} and $I_{Bias,1}$. Therefore, the same points discussed for the circuit in Fig. 3.12 regarding a possible *dc* gain improvement hold true for the folded structure. Furthermore, the folded version allows us to comfortably bias the drain of the transistor M_1. To optimize the output swing, the voltage V_{B1} keeps node *2* rather close to V_{DD}. Instead, in the circuit of Fig. 3.11 and Fig. 3.12, a bias V_{B1}, beneficial for the optimization of the output dynamic range, pushes transistor M_1 close to the limit of the saturation region possibly affecting the transconductance gain, g_{m1}. In the folded version the negative swing is possibly optimized by controlling the drain voltage of M_4 with V_{B2}. This action is not particularly critical being the output resistance of one branch of the cascode structure and not the input transconductance possibly influenced.

Example 3.5

The transistors in the folded cascode shown in Fig. 3.13 have the following sizes (W and L in μm): M_2=150/2; M_3=50/1; M_4=50/1; M_5=150/2. The current in M_5 is 150 μA and the current in M_4 is 30 μA; V_{DD}=5 V. Using the models of Appendix B, find the voltages V_{B1} and V_{B2} that optimize the output swing. Allow a 20% margin above the minimum V_{DS}.

Solution:

The first step is to estimate the saturation voltages of M_4 and M_5. This figure is estimated by a Spice simulation where the two transistors are diode connected and carry the bias currents. The result is:

```
AC SMALL-SIGNAL MODELS
                 M4             M5
MODEL            MODN           MODP
TYPE             NMOS           PMOS
ID               8.06e-005     -1.92e-004
VGS              6.35e-001     -1.16e+000
VDS              5.00e+000     -5.00e+000
VBS              0.00e+000      0.00e+000
VTH              5.41e-001     -8.12e-001
VDSAT            8.97e-002      3.38e-001
RS               1.97e+000      6.40e-001
RD               1.97e+000      6.40e-001
GM               1.25e-003      9.54e-004
GDS              4.96e-005      1.03e-005
GMB              4.20e-004      1.70e-004
GBD              0.00e+000      0.00e+000
GBS              0.00e+000      0.00e+000
 ..........
```

The saturation voltage of the n-channel element is less than 89.7 mV and the saturation voltage of the p-channel is 338 mV. Allow-

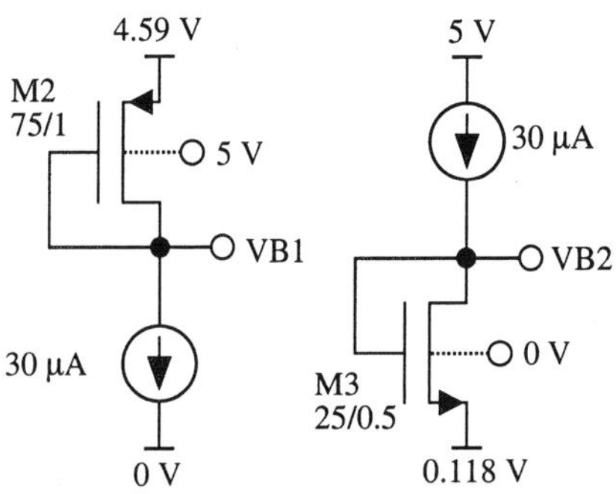

ing a 20% margin, appropriate source voltages for M_2 and M_3 are 4.59 V and 0.118V, respectively.
The value of the bias voltages V_{B1} and V_{B2} are calculated by simulation using the test circuit given in the figure.
It should be noted that the substrates of the two transistors are biased at 0 V and 5 V.
The simulation provides the following figures: V_{B1}= 3.56V; V_{B2}=754mV. Moreover, the saturation voltage of M_2 is 141 mV and the one of M_3 is 76 mV. Therefore, the maximum output swing that keeps transistors in saturation is: (0.118 +0.076) = 0.194V < V_{out} < 4.45 V = (4.59-0.14).

3.3.1 Small Signal Analysis of Cascode Gain Stages

A conventional method of studying the small signal behavior of the circuits discussed above is by analysing the small signal equivalent circuit. The small signal model of the MOS transistors must replace each active element while all of the voltage biases should be shorted to ground. The resulting linear network should be then solved using well-known circuit analysis laws. We followed this approach in the case of the simple stages studied previously. Here, instead of going through the same flow, we try to derive the small signal performances by simply looking directly at the circuit schematic. This should enhance the designer's ability to "understand the circuit at a glance".

The *dc* gain of the simple cascode version (the one in Fig. 3.11) has already been estimated: it determined equation (3.37). For the other two schemes, the difference is that they contain an extra branch: transistor M_5. In the small signal equivalent circuit, M_5 corresponds to an additional resistance, r_{ds5}, connected in parallel to r_{ds1}. Therefore, the *dc* gain is simply obtained by substituting r_{ds1} with the parallel connection of r_{ds1} and $r_{ds5,}$ in the relationship (3.37).

Let us now consider the frequency behavior: the schematics in Fig. 3.11, Fig. 3.12 and Fig. 3.13, shown again in Fig. 3.14, have three nodes: the output node (1) and the intermediate points of the two cascode structures (node 2 and 3). Therefore, at a first approximation, the frequency response of the circuit will display three poles. Assuming that these poles are decoupled from each other, their frequencies simply result from the time constant associated with each node. The output resistance of the output node is given by the contribution of the upper and the lower cascode arrangements

$$r_{out} = \frac{r_{ds2}g_{m2}}{g_{ds1}+g_{ds5}} // \frac{r_{ds3}g_{m3}}{g_{ds4}} \tag{3.41}$$

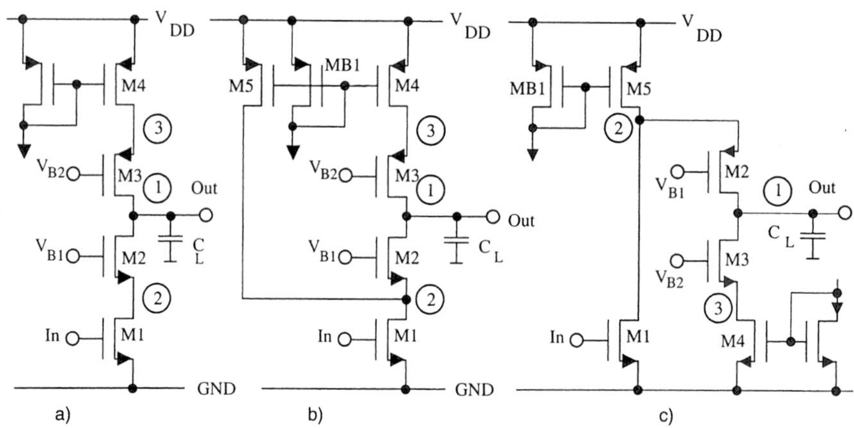

Fig. 3.14 - The three considered cascode with cascode load architecture

where || means parallel connection; the simple version doesn't include the transistor M_5; hence, of course, g_{ds5} must be set equal to *0*. By inspection of the circuit the capacitance of the output node is estimated by

$$C_{out} = C_{gd2} + C_{gd2,ov} + C_{db2} + C_{gd3} + C_{gd3,ov} + C_{db3} + C_L \tag{3.42}$$

that is the sum of all the parasitic affecting the transistors connected to the output node (M_2 and M_3) plus a possible load capacitance, C_L. The associated pole is at the frequency

$$f_{p,out} = \frac{1}{2\pi}\frac{1}{r_{out}C_{out}} \tag{3.43}$$

The capacitances affecting node *2* and node *3* are, respectively,

$$C_2 = C_{gd1} + C_{gd1,ov} + C_{db1} + C_{gs2} + C_{gs2,ov} + C_{sb2} + [C_{gd5} + C_{gd5,ov} + C_{db5}]$$

$$C_3 = C_{gs3} + C_{gs3,ov} + C_{gd4} + C_{gd4,ov} + C_{sb3} + C_{db4} \tag{3.44}$$

BE AWARE!

The calculation of the time constants associated to various nodes of a circuit provides only a rough estimation of poles position.

where the parasitic contribution of M_5 (in brackets) must be accounted for when transistor M_5 is used. Equations (3.44) show that both C_2 and C_3 are made of parasitic contributions only. Therefore, their value is similar to or smaller than C_{out}.

The direct evaluation of the equivalent resistance at nodes *2* and *3* just looking at the circuit schematic seems to be problematic. Let's discuss the issue. In the circuit in Fig. 3.14 we have two paths toward the small signal ground for the nodes *3* and for the node *2* in Fig. 3.14 a); for the other nodes in we have three paths. The resistance of paths through drains is the r_{ds} of the crossed transistor. The resistance of paths through sources depends on the transconductance of the crossed transistor and the load that it has at the drain. We studied a similar problem for the cascode with an active load (cfr. Fig. 3.9). Here, unlike that case, the load at the drain of M_2 (or M_3) is not a simple resistance, but a cascode structure with the load capacitance, C_L, in parallel. Therefore, to estimate the impedance from the source of M_2 (or M_3) in equation (3.36), we should use a proper impedance: $z_{load,2}$ (or $z_{load,3}$), instead of r_{ds3}. It results that

$$z_{s2} = \frac{1}{g_{m2}}\left(1 + \frac{z_{load,2}}{r_{ds2}}\right) = \frac{\zeta_1}{g_{m2}} \qquad z_{s3} = \frac{1}{g_{m3}}\left(1 + \frac{z_{load,3}}{r_{ds3}}\right) = \frac{\zeta_2}{g_{m3}} \tag{3.45}$$

We can't calculate equivalent time constants associated with nodes *2* and *3* any more because we have no simple *RC* networks, but instead have complex equivalent circuits. However, if we assume that the dominant pole of the circuit is assured by the output node, at frequencies for which nodes *2* and *3* produce some effect, $z_{load,2}$ and $z_{load,3}$ will be dominated by C_L. Most likely, at those frequencies the reactance of C_L is much smaller than r_{ds} and the term z_{load}/r_{ds} in the two ζ becomes negligible. According to this approximation, at nodes *2* and *3* we have resistances, namely r_2 and r_3, which are estimated as

$$r_2 = \frac{1}{g_{m2}} \| r_{ds1} \| r_{ds5} \cong \frac{1}{g_{m2}} \qquad r_3 = \frac{1}{g_{m3}} \| r_{ds4} \cong \frac{1}{g_{m3}} \tag{3.46}$$

Where, for the circuit in Fig. 3.14, $r_{ds5} = \infty$. Now, since capacitances C_2 and C_3 are smaller or comparable to the capacitance of the output node, and resistances r_2 and r_3 are much smaller than the output resistance, the frequency of the pole associated with the output node will occur at a much smaller frequency than the poles of nodes *2* and *3*. Moreover, the output node pole is dominant if

$$f_{p,out}A_v < f_{p,2} \qquad f_{p,out}A_v < f_{p,3} \tag{3.47}$$

$$\frac{g_{m1}}{C_{out}} < \frac{g_{m2}}{C_2} \; ; \; \frac{g_{m1}}{C_{out}} < \frac{g_{m3}}{C_3} \tag{3.48}$$

Normally, one designs the circuit such that the transconductance of the input transistor is not much larger than the one of the common gate transistors. In addition, the output capacitance is set larger than the parasitic terms C_2 and C_3. Thus, relationships (3.48) are normally fulfilled. In addition, when necessary, a proper additional capacitance loading the output permits us to further reduce the first terms in (3.49), thus enforcing the dominant pole condition.

Example 3.6

Consider the folded cascode amplifier shown in the following figure. Use the Spice models in Appendix B and find the gain and phase shift from input to output and from input to node 2.

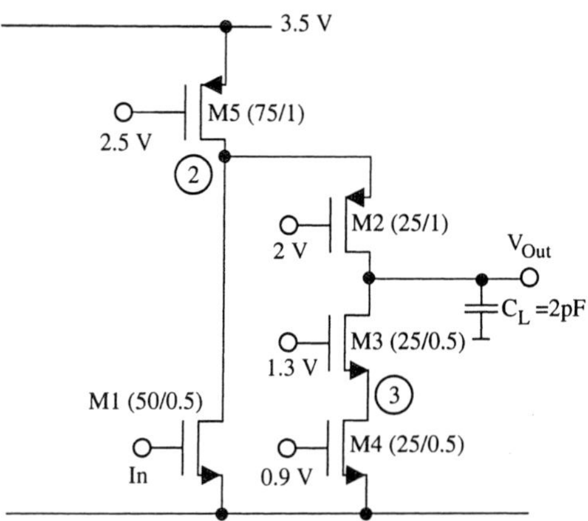

The use of a network like the one employed in Example 3.2 allows us to self-bias the input transistor. Since the rail voltage is 3.5V, a convenient output quiescent voltage is 1.75 V. Therefore, in the above mentioned feedback network, a battery of 0.8 V is adequate. Use the following input list:

```
M1  2 1   Gnd Gnd MODN L=0.5u W=50u AD=100p PD=24u  AS=100p PS=24u
M2  out   6 2 Vdd MODP L=1u   W=75u AD=66p  PD=24u  AS=66p  PS=24u
M3  out   7 3 Gnd MODN L=0.5u W=25u AD=66p  PD=24u  AS=66p  PS=24u
M4  3 9   Gnd Gnd MODN L=0.5u W=25u AD=66p  PD=24u  AS=66p  PS=24u
M5  2 4   Vdd Vdd MODP L=1u   W=25u AD=66p  PD=24u  AS=66p  PS=24u
MB1 9 9   Gnd Gnd MODN L=0.5u W=25u AD=66p  PD=24u  AS=66p  PS=24u
MB2 4 4   Vdd Vdd MODP L=1u   W=25u AD=66p  PD=24u  AS=66p  PS=24u

R1 1   10   1G  TC=0.0, 0.0
R2 10  N7   1G  TC=0.0, 0.0
CL out Gnd  2pF
C1 10  Gnd  100
C2 11  1    100
i3 Vdd 9    25uA
i4 4   Gnd  75uA
v6 7   Gnd  1.3
v7 6   Gnd  2
v8 out N7   0.8
vdd Vdd Gnd 3.5
v5  11 Gnd  0.0 AC 1.0 0.0
........
.print ac vdb(out) vp(out)  vdb(2) vp(2)
```

Solution: *After performing the Spice simulation, we obtain the Bode plots depicted in the given diagram.*

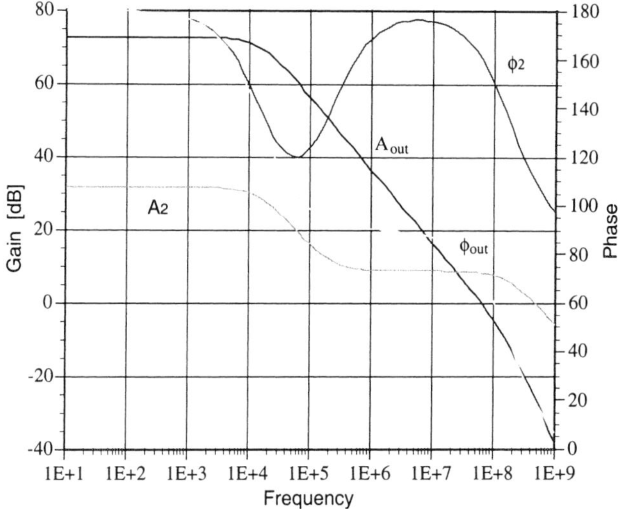

We observe that the gain and phase plots of the output gain show a 20 dB roll-off with a good phase margin (it is close to 70 degrees). The low frequency gain is 73 dB and the unity gain frequency is around 62 MHz. The behavior of the gain from input to node 2 is interesting: at frequencies higher than the dominant pole, it holds 9 dB, 3 dB more than the expected value: g_{m1}/g_{m2}. *Instead, at a low frequency,* A_2 *climbs to 32 dB. This happens because at low frequency, the factor* ζ *is no longer negligible, as discussed in Section 3.3.1: the resistance at node 2 becomes in the order of* r_{ds}. *In the region where* A_2 *goes from high to low gain, there is also a phase shift.*

3.3.2 Gain Enhancement Techniques

The cascode with cascode load provides a fairly high voltage gain. We have seen that it is as large as the cascade of two inverters with active load. Such a gain (around *80 dB*) is sufficient for the majority of analog systems. However, for applications where high gain is crucial, the approach can be extended to further enhance the cascoding effect. Fig. 3.15 a) shows a cascode with a double cascode load. It includes two transistors in addition to the transistors of a cascode with cascode load: M_3 and M_4 operating as further common gate stages. As for the previously studied circuits, the enhanced output resistance multiplied by the transconductance of the input transistor, g_{m1}, obtains the voltage gain. By inspection of the circuit, R_{out} is

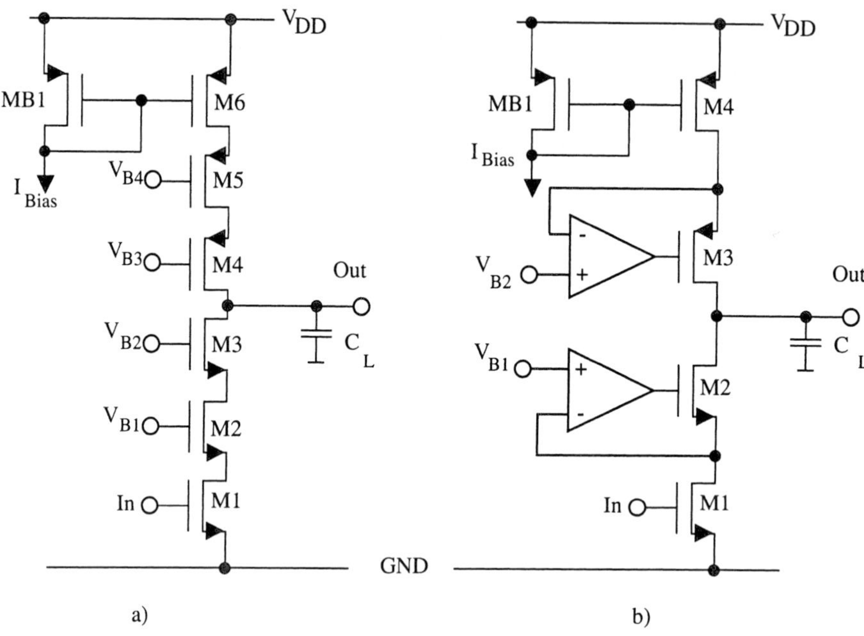

Fig. 3.15 - a) Double cascode with double cascode load. b) Regulated cascode amplifier

$$R_{out} = \frac{[r_{d1}(g_{m2}r_{d2})(g_{m3}r_{d3})][r_{d6}(g_{m5}r_{d5})(g_{m4}r_{d4})]}{[r_{d1}(g_{m2}r_{d2})(g_{m3}r_{d4})] + [r_{d6}(g_{m5}r_{d5})(g_{m4}r_{d4})]} \tag{3.49}$$

Therefore, the voltage gain becomes proportional to the cube power of $g_m r_d$.

Of course, the two added transistors, M_3 and M_4, restrict the output dynamic range by at least two V_{sat}. Moreover, the four necessary voltage references must be designed carefully: they should allow transistors to operate in saturation in all possible conditions, including the variation due to process changes. in addition they should ensure the widest output dynamic range. Therefore, after accounting for proper security margins and using the supply biasing imposed by technologies that can be (as low as *3.3 V*), the output dynamic range goes down to very low values (*1 V* or less).

The circuit in Fig. 3.15 b) enhances the output resistance and, at the same time, ensures a better output swing. It is named regulated cascode amplifier since the active load is a regulated cascode as well (we will analyse the regulated cascode configuration shortly, when the current mirrors will be studied). The regulated cascode utilizes a local feedback to keep the source voltage of the common gate transistor constant. If the source of M_2 (or M_3) tries to change its value in regard to the voltage defined by the reference voltage (V_{B1} or V_{B2}), the amplified difference is applied to the gate of *M2* (or *M3*) and the feedback loop stabilizes it.

The calculation of the output resistance for a regulated cascode is performed with the help of the equivalent circuit in Fig. 3.8 b). It refers to a simple cascode structure. The difference with respect to the present circuit is that we must use a transconductance generator g_{m2} amplified by the gain of the additional gain stage used in Fig. 3.15. Thus, the cascode resistance increases by the gain of the amplifier employed. It results that

$$R_{out} = \frac{A_1 A_2 [r_{d1}(g_{m2} r_{d2})][r_{d4}(g_{m3} r_{d3})]}{A_1 [r_{d1}(g_{m2} r_{d2})] + A_2 [r_{d4}(g_{m3} r_{d3})]} \quad (3.50)$$

Therefore, assuming that A_1 and A_2 are similar, the output resistance and the gains are enhanced by factor A_1 compared to the cascode structure. The output swing of the regulated cascode is equivalent to the output swing of the cascode with cascode active load. Accordingly, the voltage difference between the output node and the rail lines can be as low as the saturation voltage of two transistors. This result is beneficial; however, the circuit requires two additional gain amplifiers which increase power consumption, increase silicon area and, more importantly, can lead to stability problems in the local loops. Moreover, the two inputs of the auxiliary gain stages appears to work close to V_{DD} or to ground.

3.4 DIFFERENTIAL STAGE

A differential pair is widely used as the input stage of the operational amplifiers. Fig. 3.16 shows its CMOS configuration. It is made of two transistors with their source in common, fed by a current source. The transistors may either be n-channel (as shown in the figure) or p-channel, and they are matched to each other. If the two transistors are in the saturation region, we can write

$$I_1 = \frac{\mu C_{ox}}{2}\left(\frac{W}{L}\right)_1 (V_{GS1} - V_{Th})^2 \quad (3.51)$$

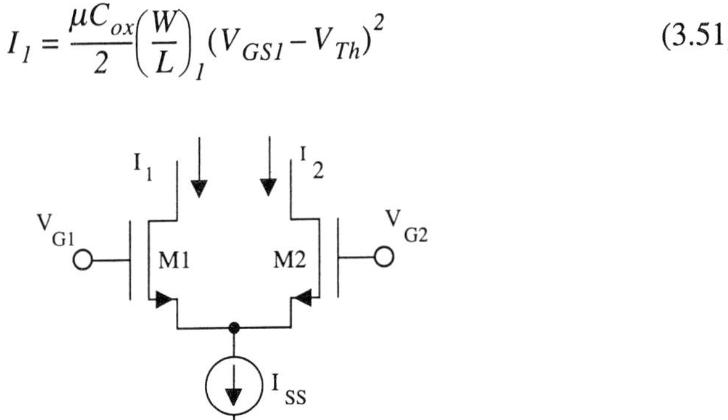

Fig. 3.16 - MOS differential stage. The output is the differential current

$$I_2 = \frac{\mu C_{ox}}{2}\left(\frac{W}{L}\right)_2 (V_{GS2} - V_{Th})^2 \qquad (3.52)$$

where $(W/L)_1$ and $(W/L)_2$ are nominally equal, the transistors being matched. Moreover, in the above equations the output conductance has been neglected.

The input signals can be expressed as

$$V_{GS1} = V_{GSO} + \frac{V_{in}}{2}; \quad V_{GS2} = V_{GSO} - \frac{V_{in}}{2} \qquad (3.53)$$

where V_{GSO} is the common node component and V_{in} is a differential signal.

Assuming the differential current, ΔI, as the output variable of the circuit, we have

$$\Delta I = I_1 - I_2 = \mu C_{ox}\left(\frac{W}{L}\right)_1 V_{in}(V_{GSO} - V_{Th}) \qquad (3.54)$$

Since the bias current can be expressed as

$$I_{SS} = I_1 + I_2 = \mu C_{ox}\left(\frac{W}{L}\right)_1 (V_{GSO} - V_{Th})^2 \qquad (3.55)$$

the differential current becomes

$$\Delta I = V_{in}\sqrt{\mu C_{ox}\left(\frac{W}{L}\right)_1 I_{SS}} = V_{in} g_m \qquad (3.56)$$

which states that the output of the differential stage is proportional to the first power of the input voltage and to the square root both of the aspect ratio and the bias current. Therefore, in the differential stage, like in the case of the inverter with active load, the transconductance gain increases with the square root of the bias current.

KEEP NOTE!

The relationship between the large signal differential input voltage and the differential output current of an MOS differential pair is linear (if transistors are in saturation).

It is important to remark the linear relationship between the large signal differential input voltage and the large signal differential current. This is the result of the specific quadratic current-voltage relationship describing *MOS* transistors in saturation. In calculating ΔI, the quadratic terms cancel each other out and only the double products remain. Reader can verify, as a useful exercise, that the achieved result equals the approximate expression that holds true for the small signals.

Example 3.7

Using Spice, verify equation (3.56). Consider an n-channel differential pair using: (W/L)= 50μm/0.5 μm and I_{SS}=100μA.

Solution:
The input list suitable for the required simulation is the following:

```
DIFFERENTIAL STAGE
.OPTIONS NODE NODEPAGE

M1 1 2 3 3 MODN L=0.5u W=50u AD=100p PD=24u AS=100p PS=24u
M2 4 5 3 3 MODN L=0.5u W=50u AD=100p PD=24u AS=100p PS=24u
R1 Vdd 1 1 TC=0.0, 0.0
R2 Vdd 4 1 TC=0.0, 0.0
iss 3 Gnd 100uA
vdd Vdd Gnd 3
v3 7 Gnd 1.5
esign 5 Gnd 7 Gnd 1
vsign 2 7 0
.dc vsign -300M 300M 4M
.include modn.md
.include modp.md
.print dc V(1,4)
```

The list above produces the diagram given above (the vertical axis shows the normalised ΔI). Actually, the transconductance transfer function is fairly linear over a wide range of the input signal. It starts to saturate only when the input signal approaches the overdrive voltage of the differential pair (around 150 mV).

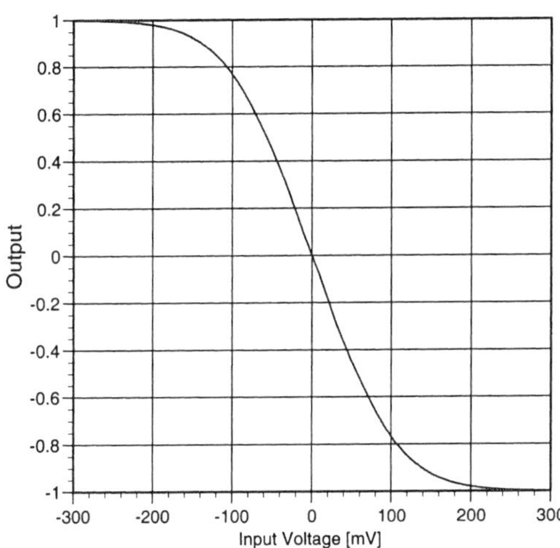

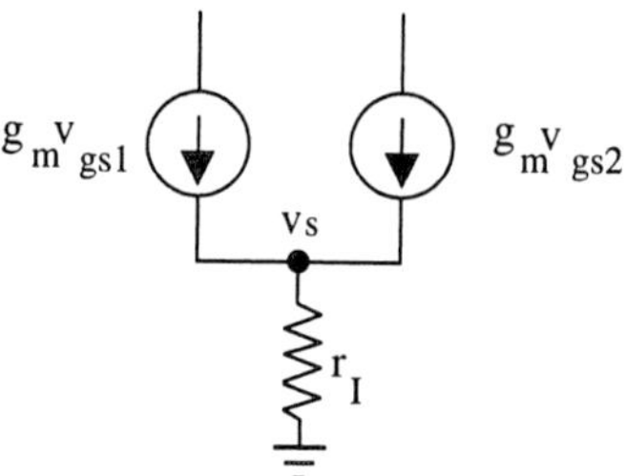

Fig. 3.17 - Simplified equivalent circuit of the MOS differential pair

The main function of the differential stage is to amplify the differential input signal, $\pm V_{in}$, and, if possible, reject any common-mode component, V_{GS0}. Therefore, it is essential to estimate the circuit's response to a common mode input. We can do this by using the simplified small signal equivalent circuit in Fig. 3.17. It contains only the circuit elements relevant to our purpose: the transconductance generators, describing the active effect of the MOS transistors, and the resistance r_I, representing the non-ideality of the current source. If a common mode signal, v_{CM}, is applied to the two inputs, by observing the circuit, we can write

$$2g_m(v_{CM} - v_s) = v_s / r_I \tag{3.57}$$

from which the signal current flowing in the two output nodes can be evaluated

$$i_{CM} = \frac{g_m v_{CM}}{1 + 2g_m r_I} \approx \frac{v_{in}}{2} \tag{3.58}$$

The common mode rejection, defined as the ratio between the differential current and the common mode current, is given by

$$CMMR = \frac{i_d}{i_{CM}} \cong 2g_m r_I \tag{3.59}$$

Thus, a good *dc CMMR* can be achieved by increasing either the transconductance of the differential pair or the equivalent resistance of the current source.

3.5 SOURCE FOLLOWER

The previous sections discussed the main features of gain stages. An inverter with active load or a cascode arrangement produces a convenient volt-

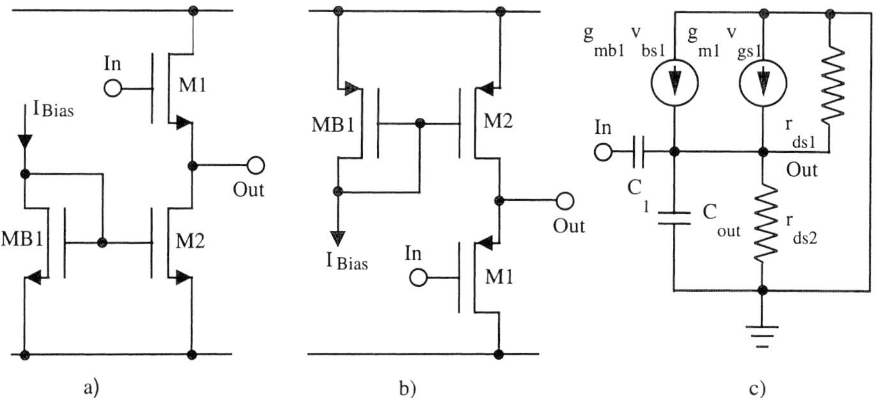

Fig. 3.18 - Source follower. a) with n-channel input transistor. b) with n-channel input transistor. c) small signal equivalent circuit.

age gain. However, the output impedance is rather high and is not suitable for driving low resistive or large capacitive loads. Therefore, an output stage becomes necessary at times. A source follower is the basic block used to achieve low output impedance and, as we will see, a level shifter. The source follower is shown in Fig. 3.18 for an n-channel input transistor and for a p-channel input transistor. The signal is applied to the gate of transistor M_1 and, as we will find, it is replicated almost identically at the output node. Observing the circuit, we see that the *dc* difference between input and output is given by V_{GS1}. This difference is sometimes used as a voltage-level shifting, which is downward for an n-channel input transistor and upward for a p-channel input transistor. The value of the voltage shift is around *1 V* or less for modern technologies.

Since all of the transistors of the circuit are assumed to be operating in saturation, the output voltage is not allowed to approach ground (or V_{DD}) more than the saturation voltage of M_2. Moreover, since the input voltage cannot exceed V_{DD} for an n-channel input transistor (or, for a p-channel input transistor, cannot become smaller than ground), the output swing will be also limited to one V_{GS} below V_{DD} (or one V_{GS} above ground, for p-channel input). To sum up, the output dynamic range (for n-channel input scheme) is limited by

$$V_{out,max} = V_{DD} - V_{GS1} = V_{DD} - V_{Th,n} - V_{sat,1} \tag{3.60}$$

$$V_{out,min} = V_{sat,2} \tag{3.61}$$

for maximum and minimum levels, respectively (n-channel input).

As already mentioned the voltage level shift from the input to the output is given by V_{GS1}. If transistor M_2 is in saturation, its current, and hence the current in the stage, is substantially controlled by its bias. Therefore, assuming $I_{M1} = I_{M2}$, V_{GS1} is expressed by

$$V_{GS1} = V_{Th,1} + \sqrt{\frac{2I_2}{\mu C_{ox}\left(\frac{W}{L}\right)_1}} \tag{3.62}$$

It is independent of the input and output voltages only if the threshold voltage remains constant. We have seen in Chapter 1 that the threshold voltage depends on the substrate bias V_{SB} through the body effect coefficient γ (see equation 1.31). For large signals the threshold changes in a non-linear fashion. Therefore, when the source follower is used as a level shifter, the input transistor M_1 is recommended as being of the same type that is integrated in the well, and should have the bias of the well connected to the source.

FOLLOWERS FOLLOW IF

The transistors substrate should be connected to the source to ensure a true "follower" operation (gain close to one) and, more important, to achieve a proper source follower linearity.

The small signal equivalent circuit of the source follower is shown in Fig. 3.18 c). It is a bit more complex than the circuits used so far since it includes the transconductance element controlled by v_{sb} for the small signal representation of M_1. Assuming the substrate connected to ground, at low frequency, we find

$$(g_{ds1} + g_{ds2})v_{out} + g_{mb1}v_{out} - g_{m1}v_{gs1} = 0 \tag{3.63}$$

which leads to

$$A_v = \frac{v_{out}}{v_{in}} = \frac{g_{m1}}{g_{m1} + g_{ds1} + g_{ds2} + g_{mb1}} \tag{3.64}$$

If the transconductance g_{m1} is large compared to the other terms in the denominator as it happen when we consider output conductances only, the voltage gain is close to one. However, when we have to include the effect of g_{mb1} the gain is lower than one since g_{mb1} is not negligible with respect to the transconductance g_{m1}. For highly doped substrates, it can be as high as 10% of the transconductance g_{m1} itself.

Example 3.8

The input transistor of a source follower is n-type with W/L =50μm/0.5 μm. Its source is connected to an ideal current genera-

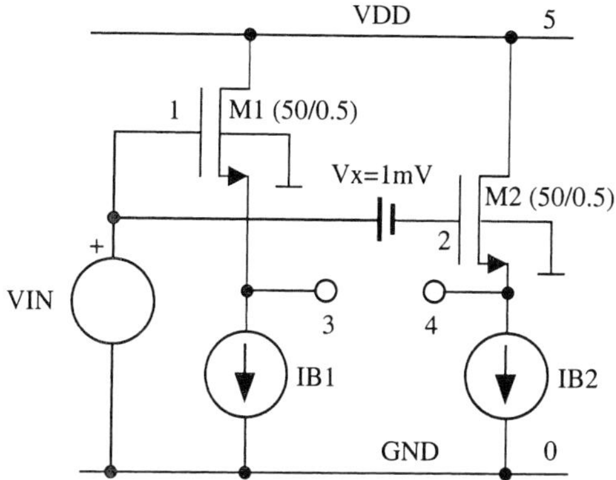

tor $I_B = 0.1mA$. The technology used is n-well and the Spice models are the ones given in Appendix B. $V_{DD} = 3.3$ V. Simulate the large signal behavior and derive the dc small signal voltage gain.

Solution:

A suitable circuit for the required Spice simulation is shown in the figure. It includes two identical source followers whose gates are driven by slightly different voltages (V_x=1 mV). The voltage at node 3 provides the large signal response; the drop voltage between node 3 and 4 furnishes the dc voltage gain (multiplied by 1 mV).

```
M1  Vdd 1 3 Gnd MODN L=0.5u W=50u AD=66p PD=24u AS=66p PS=24u
M2  Vdd 2 4 Gnd MODN L=0.5u W=50u AD=66p PD=24u AS=66p PS=24u
vdd Vdd Gnd 3.3
vin  1 Gnd 3.3
vx   2 1 1M
ib1  3 Gnd 100uA
ib2  4 Gnd 100uA
.include modn.md
```

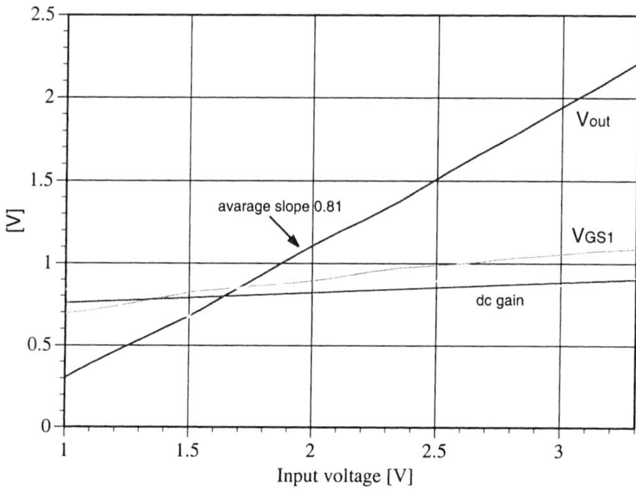

```
.dc vin 3.3 1 50M
.print dc v(3) V(1,3) V(3,4)
```

The plots summarise the results of the simulation. The output voltage practically follows the input shifted by V_{GS}. However, due to the body effect, the value of V_{GS} is not constant. It rises from 706 mV to 1.09 V in a non slight linear fashion. Therefore, the slope of the input-output characteristics is not 1, but 0.81, on average. The figure also shows the dc gain (V(3,4)/1 mV): its value ranges from 0.79 to 0.91. These figures match quite well with the values predicted by equation (3.64). In fact, for V_{in}=1V we have: $g_m=1.35\cdot10^{-3}$ A/V and $g_{mb}=2.47\cdot10^{-4}$ A/V (gain=0.83) while for V_{in}=3.3 V we have: $g_m=1.97\cdot10^{-3}$ A/V and $g_{mb}=2.91\cdot10^{-4}$ A/V (gain=0.87).

The output capacitance and the feedforward capacitance of the source follower small equivalent circuit (Fig. 3.18 c) leads to the frequency response. C_{out} groups up the capacitive load and parasitic components

$$C_{out} = C_L + C_{gd2} + C_{gd2,ov} + C_{db2} + C_{sb1} \tag{3.65}$$

the feedforward capacitance C_1 is given by

$$C_1 = C_{gs1} + C_{gs1,ov} \tag{3.66}$$

The output capacitance determines a pole in the circuit transfer function while the feedforward capacitor produces a zero. The node equation at the output node gives

$$(v_{in} - v_{out})sC_1 + g_{m1}(v_{in} - v_{out}) - v_{out}G - v_{out}sC_{out} = 0 \tag{3.67}$$

where $G = g_{ds1} + g_{ds2} + g_{mb1}$

From this, the voltage gain results in

$$A_v(s) = \frac{g_{m1}}{g_{m1} + G} \cdot \frac{1 + sC_1/g_{m1}}{1 + s(C_1 + C_{out})/(g_{m1} + G)} \tag{3.68}$$

The output capacitance is normally bigger than the feedforward capacitance. Hence, the angular frequency of the pole [$\omega_p=(g_{m1}+ g_{ds1} + g_{ds2} + g_{mb1})/ C_1+C_{out})$] occurs at a frequency which is much lower than the frequency of the zero. Observe that at a very high frequency, the transistor does not contribute at all, and only the capacitive attenuator C_1 and C_{out} produces the output signal.

An important parameter of the source follower is its output impedance. It is obtained by applying a test voltage v_x to the output node and measuring the resulting current i_x. Doing this in the equivalent circuit in Fig. 3.18 it results that

$$i_x = (g_{ds1} + g_{ds2} + g_{mb1} + g_{m1})v_x \tag{3.69}$$

from which we get the output resistance that is given by

$$R_{out} = \frac{1}{g_{ds1} + g_{ds2} + g_{mb1} + g_{m1}} \cong \frac{1}{g_{m1}} \tag{3.70}$$

The availability of a stage with a relatively low output resistance can be very helpful. However, for a *MOS* circuit, the achievable transconductances may not be sufficient to produce a very low output resistance. The transconductance is expressed by $g_m = 2I_D/(V_{GS} - V_{Th})$; in a common situation the overdrive voltage $(V_{GS} - V_{Th})$ is a few hundred mV and the drain current is no larger than a few hundred μA, leading to output resistances in the $k\Omega$ range (if $(V_{GS} - V_{Th}) = 260\ mV$, with a bias current equal to $I_{MOS} = 260\ \mu A$, $1/g_{m,MOS} = 500\Omega$). This figure is larger than its bipolar counterpart driving the same current by a substantial factor (with a *BJT*, in fact $g_{m,BJT} = I_C/(kT/q)$, $kT/q = 26\ mV$; with a bias current equal to $I_c = 260\ \mu A$, $1/g_{m,BJT} = 100\Omega$).

3.6 THRESHOLD INDEPENDENT LEVEL-SHIFT

The source follower studied in the previous section furnishes at the output a replica of the input shifted up or down by the extent given by equation (3.62). The threshold dependency and the relatively high shift achieved are not suitable for many applications: with a typical technology used, the source follower shift is in the order of *1 V* and, because of technological variations, the inaccuracy can be as high as some hundred of *mV*.

In many cases the designer needs a small voltage shift, threshold independent. So, the source follower can not be used. The circuits in Fig. 3.19 is a possible answer to those needs. Complementary configurations achieve the same function. The circuit in Fig. 3.19 a) has a diode connected transistor (M_1) at the input which shifts the input voltage upwards than a source follower M_2 shifts the result downwards. Instead, the circuit in Fig. 3.19 b) has a source follower M_2 before and the diode connected transistor (M_1) after. The small signal behavior results from a direct inspection of the circuits. One can verify that small signal resitance between input and output is approximated by

$$r_{eq} = 1/g_{m1} + 1/g_{m2} \tag{3.71}$$

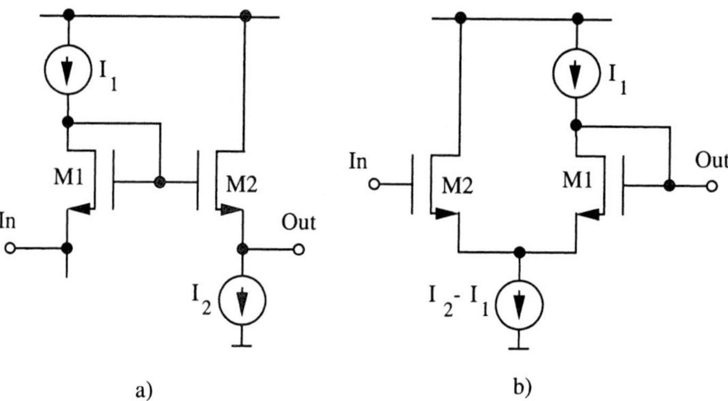

Fig. 3.19 - Threshold independent level-shifters

therefore, for the small signals, the circuits in Fig. 3.19 act like a the source follower with an additional output series resistance equal to $1/g_{m1}$.

Both solutions achieve the voltage shift through the series of a shift-up and a shift down. By inspection of the two circuits we have

$$\Delta V = V_{Th,1} + \sqrt{\frac{2I_{D1}}{\mu C_{ox}\left(\frac{W}{L}\right)_1}} - V_{Th,2} - \sqrt{\frac{2I_{D2}}{\mu C_{ox}\left(\frac{W}{L}\right)_2}} \tag{3.72}$$

since the two shifts are due to the same kind transistors, the thresholds will match and the result is just the difference between two overdrive

$$\Delta V = \frac{2}{\mu C_{ox}} \cdot \left\{ \sqrt{I_{D1}\left(\frac{W}{L}\right)_1} - \sqrt{I_{D2}\left(\frac{W}{L}\right)_2} \right\}. \tag{3.73}$$

Observe that, in addition to a threshold independency, the circuits achieve the level shift value by a proper the choice of transistor aspect ratios and bias currents. Therefore, designer can achieve positive or negative and even pretty small voltage shifts.

3.7 IMPROVED OUTPUT STAGES

We have seen that the simple source follower provides output resistances in the $k\Omega$ range. This is normally enough for the charge or the discharge of low-medium capacitive loads. However, the performances achieved become poor when the buffer is required to drive resistive loads or large capacitors at high

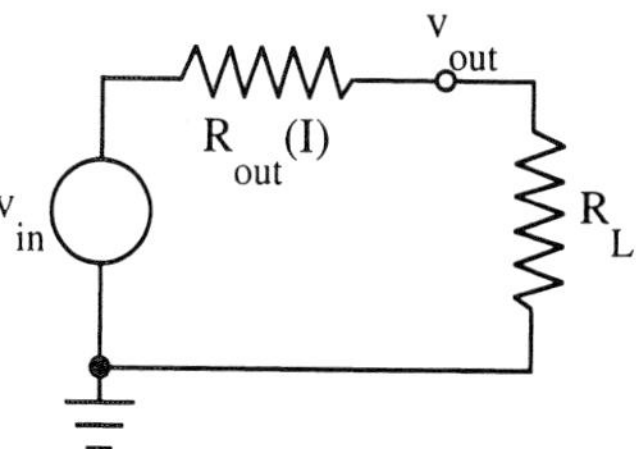

Fig. 3.20 - Equivalent circuit used to discuss the non-linear behaviour of an output stage

frequency. For this design situations the output stage should provide an output resistance significantly lower than the load impedance that it drives. Moreover, to keep the harmonic distortion under control, the variation of output resistance induced by current swings must be only a very small fraction of the load. For a better understanding of this point, let us consider Fig. 3.20. It represents the equivalent circuit of an output stage driving a resistive load. The output voltage is, of course

$$V_{out} = V_{in}\frac{R_L}{R_{out} + R_L} \tag{3.74}$$

which, assuming $R_{out} = R_{out,0}\{1 + \alpha(I)\}$ and α a suitable function expressing the non-linear behavior, can be represented as

$$V_{out} \approx V_{in}\frac{R_L}{R_{out,0} + R_L}\left\{1 - \frac{R_{out,0}\alpha(I)}{R_{out,0} + R_L}\right\} \tag{3.75}$$

The output voltage is, therefore, an attenuated replica of the input (and this is acceptable). But, in addition we have the term $R_{out,0}\ \alpha(I)/(R_{out,0} + R_L)$ which is responsible for a non-linear response. Many applications can tolerate an attenuation of the signal. Instead, non-linearities are not admissible beyond a given limit since they produce harmonic distortion. In order to prevent this sort of problem, the designer can follow two possible strategies: to desentityze the non-linearity effect, by decreasing $R_{out,0}$ so that it becomes much smaller than R_L or to improve the output resistance linearity. Both techniques are discussed for a specific circuit solution. The reader can find in the references other possible methods achieving the same result.

3.7.1 Source Follower with Local Feedback

A typical way to reduce the output resistance is to use feedback. This technique can be conveniently applied to the source follower, as shown in Fig. 3.21

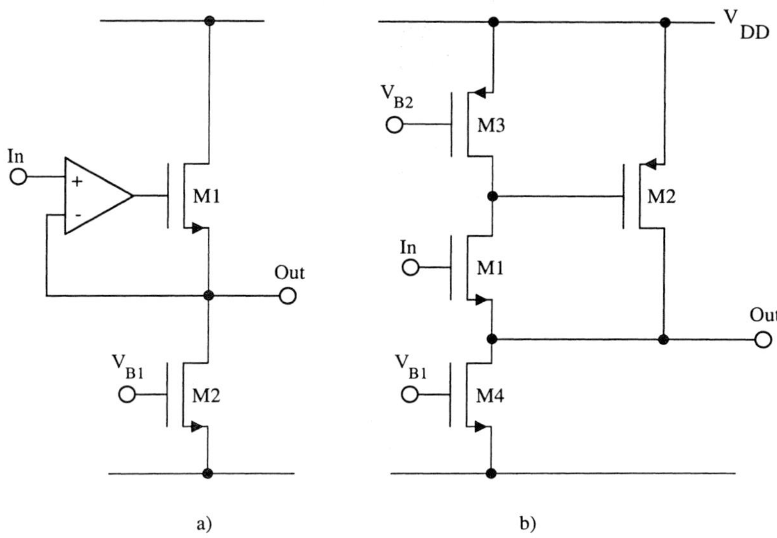

Fig. 3.21 - Source follower with local feedback

a). The gain stage amplifies the difference between the input and output of the source follower, thus reducing the output resistance by the amplification of the gain stage itself. The gain stage can be a simple scheme or a more complex structure like the operational amplifiers that will be studied in a next Chapter.

The output resistance and its linearity can also be improved by using local feedback. Since non-linearities possibly result from the current variation in the MOS transistor achieving the follower behaviour. Feedback can be used to keep constant the current as shown by the circuit in Fig. 3.21 b). Just like the conventional source follower, the input signal is replicated at the source of M_1, but it is also amplified at the drain of the same transistor. The amplified signal is fed to the input of M_2 which operates as the input element of a second gain stage. A feedback loop is thus established through the M_1 - M_2 path. Note that any change of the current in M_1 is caught by the feedback loop that works to keep I_{M1} constant. Thus, g_{m1} doesn't change significantly as required to improve the linearity. However, since the feedback loop gain influences the output resistance, the output resistance itself can show non-linearity because of the loop gain variation with the large signal output current.

The small signal equivalent circuit is shown in Fig. 3.22. We can write

$$i_x = (g_{m1} + g_{ds2} + g_{ds4})v_x + g_{m4}v_2 ; \; v_2 = g_{m1}r_{ds3}v_x \tag{3.76}$$

which enables us to obtain

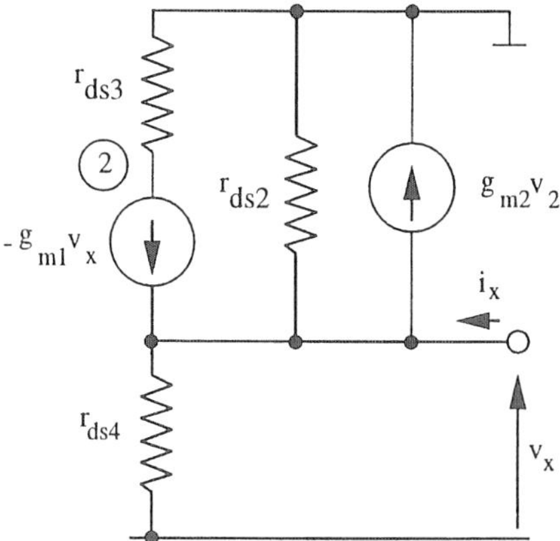

Fig. 3.22 - Small signal equivalent circuit of the source follower with local feedback.

$$R_{out} = \frac{1}{g_{m1}(1 + g_{m4} r_{ds3}) + g_{ds2} + g_{ds4}} \tag{3.77}$$

The term $g_{m1} g_{m4}\, r_{ds3}$ is dominant in the denominator of (3.77).Therefore, the output resistance is reduced by the factor $g_{m4}\, r_{ds3}$ (the feedback loop gain) with respect to the $1/g_{m1}$. Observe that the benefit of the local feedback vanishes when the current in M_2 goes to zero. This happens if the output node sinks a current larger than the quiescent current of M_2 itself. Under those conditions the feedback loop is no more working. Therefore, the non-linearity is minimised within given limits of the current to the output node.

More in general, the circuit in Fig. 3.21 b) (which is also common to the simple source follower) has an asymmetrical and limited output-current driving capability. It can be easily verified that if the input voltage rises, transistors M_2 is driven-on and the output node can drain a current that is limited only by transistor size and its maximum overdrive. However, if the input voltage decreases and the stage is required to sink a large current, M_2 will be turned off. A further decrease of the input voltage will reduce the current in M_1 until M_1 itself turns off. Therefore, the maximum sink current is limited by the saturation current of transistor M_4.

As an important final remark we have to observe that, since the circuit contains a feedback loop with a cascade of two gain stages, the stability of the circuit should be analysed carefully. Depending on the specific load that the circuit drives, it might be necessary to provide a compensation network. The same remark holds for any implementations based on the scheme of Fig. 3.21 a): it incorporates a feedback loop: therefore, stability must be considered. Since the loop contains a buffer its phase shift must be accounted for. It possibly can determine a phase margine degradation.

3.7.2 Push-Pull Output Stage

Another circuit that keeps under control the output resitance non-linearity is the push-pull output stage shown in Fig. 3.23. Two complementary transistors, M_1 and M_2, are driven by the same (shifted) signal. The gate of M_2 is connected directly to the input node, while the gate of M_1 is fed by the shifting down of the input signal. The transistors M_3, M_4 and the bias current generator M_5 achieve the shifter. Analysis of the circuit provides

$$V_{12} = V_{GS3} + V_{GS4} = V_{Th,n} + V_{Th,p} + \sqrt{I_5}\left(\sqrt{\frac{2L_3}{\mu_n W_3 C_{ox}}} + \sqrt{\frac{2L_4}{\mu_p W_4 C_{ox}}}\right) \quad (3.78)$$

For transistor sizes, we set the following matched relationship

$$\left(\frac{W}{L}\right)_1 = k\left(\frac{W}{L}\right)_4 \qquad \left(\frac{W}{L}\right)_2 = k\left(\frac{W}{L}\right)_3 \quad (3.79)$$

If all the transistors are in saturation, because of the symmetry, the current in M_5 determines the current in the output transistors in the zero output current case. Since the voltage difference V_{12} is also applied to the gates of the transistors in the output branch, we can write

$$V_{12} = V_{Th,n} + V_{Th,p} + \sqrt{I_2}\left(\sqrt{\frac{2L_2}{\mu_n W_2 C_{ox}}} + \sqrt{\frac{2L_1}{\mu_p W_1 C_{ox}}}\right) \quad (3.80)$$

combining (3.78) and (3.80) and using (3.79), it turns out that

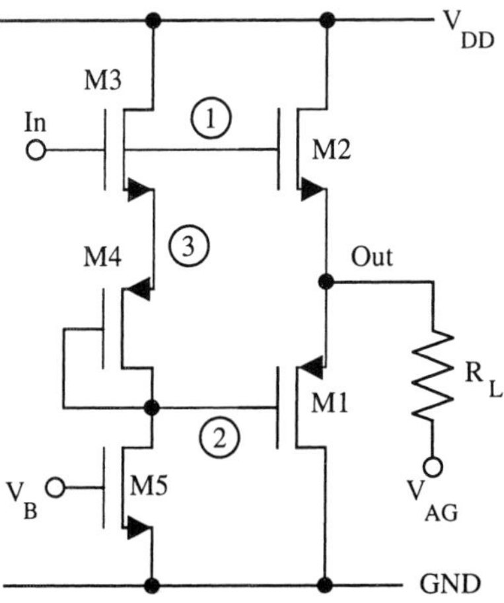

Fig. 3.23 - Class AB push-pull output stage

$$I_2 = kI_5 \tag{3.81}$$

The output small signal conductance of the circuit is determined by the parallel connection of the two paths through M_1 and M_2. It results in

$$g_{out} = \frac{1}{R_{out}} = g_{m1} + g_{m2} \tag{3.82}$$

With a zero output current ($R_L = \infty$, $I_{out} = 0$), and assuming a perfect match between the threshold voltages and between the geometrical dimensions, the output of the circuit will equal the voltage of node *3*, since $V_{GS1} = V_{GS3}$. Moreover, a proper input voltage, $V_{in,q}$, will lead to the condition $V_3 = V_{out} = V_{AG}$. For a finite resistive load connected between the output node and the analog ground and for an input voltage different to the quiescent value $V_{in,q}$, a current, either positive or negative, flows through the output node. It is given by the difference between the currents in the two transistors of the output stage; therefore it is given by

$$I_{out} = I_2 - I_1 \tag{3.83}$$

A difference between I_1 and I_2 can result only if the overdrive of M_1 and M_3 (as well as the ones of M_2 and M_4) are no longer matched. This means that the output voltage becomes different to the voltage of node *3* by, say, ΔV. Such a difference is the expression of a finite output resistance: output voltage drops when an output current must be delivered. This is, in a certain sense, inevitable (and acceptable for a limited extent) in real output stages. However, as we have already discussed, what is important is that the output resistance is as linear as possible; that is, the drop voltage ΔV should be proportional to the output current. Since the drop voltage ΔV increases the overdrive voltage of M_2 and reduces the overdrive voltage of M_1 by the same amount, we can write

$$I_2 = \frac{\mu_n C_{ox}}{2}\left(\frac{W}{L}\right)_2 (V'_{ov,n} + \Delta V)^2 \tag{3.84}$$

$$I_1 = \frac{\mu_p C_{ox}}{2}\left(\frac{W}{L}\right)_1 (V'_{ov,p} - \Delta V)^2 \tag{3.85}$$

where $V'_{ov,n}$ and $V'_{ov,p}$ are the overdrive voltages for $\Delta V = 0$. For them we get

$$\mu_n C_{ox}\left(\frac{W}{L}\right)_2 V'^2_{ov,n} = \mu_p C_{ox}\left(\frac{W}{L}\right)_1 V'^2_{ov,p} \tag{3.86}$$

Substituting (3.84) and (3.86) in (3.83) results in

$$I_{out} = \mu_n C_{ox}\left(\frac{W}{L}\right)_2 V'^{2}_{ov,n}\left\{\Delta V\left(\frac{1}{V'_{ov,p}} + \frac{1}{V'_{ov,n}}\right) + \frac{1}{2}\Delta V^2\left[\frac{1}{V'^{2}_{ov,p}} - \frac{1}{V'^{2}_{ov,n}}\right]\right\} \quad (3.87)$$

Thus, if we want to cancel the non-linear term, we should fulfil the condition

$$V'_{ov,n} = V'_{ov,p} \quad (3.88)$$

which, in turn, leads to

$$\mu_n\left(\frac{W}{L}\right)_2 = \mu_p\left(\frac{W}{L}\right)_1 \quad (3.89)$$

stating that the ratio between aspect ratios of the two output transistors should be equal to the inverse of the ratio of the corresponding mobilities. Unfortunately, the above condition is difficult to achieve since the surface mobility of electron and holes are not well controlled by the technology.

REMEMBER

A push-pull output stage shows a linear response if the aspect ratios of the output transistors are inversely proportional to the corresponding surface mobilities.

When the current in one output transistor goes to zero the stages becomes non-linear.

The output resistance lasts being linear until one of the two output transistors is turned off. After this point, the output current is provided by a transistor in saturation alone. The relationship current-overdrive voltage rises quadratically with a resulting non-linear output-resistance. As already mentioned this behavior is responsible for harmonic distortion.

Example 3.9

Determine, with a Spice simulation, the large signal output resistance as a function of the output current of the push-pull output stage in Fig. 3.23. Use the following transistor sizing: $(W/L)_1=(W/L)_3=50/0.5$ μm, $(W/L)_2=(W/L)_4=125/0.5$ μm. In addition, replace transistor M_5 with an ideal current source $I_{bias}=100\mu A$. The Spice models are given in Appendix B. From the given figures, we can observe that the mobility ratio $\mu_n/\mu_p=(460\ /181)$ (mobilities expressed in $cm^2/Vsec$) is equal to 2.54
Assume that $V_{DD}=3.3$ V and $V_{in}=2.3$ V.

Solution:
The input file suitable for the required simulation is the following:

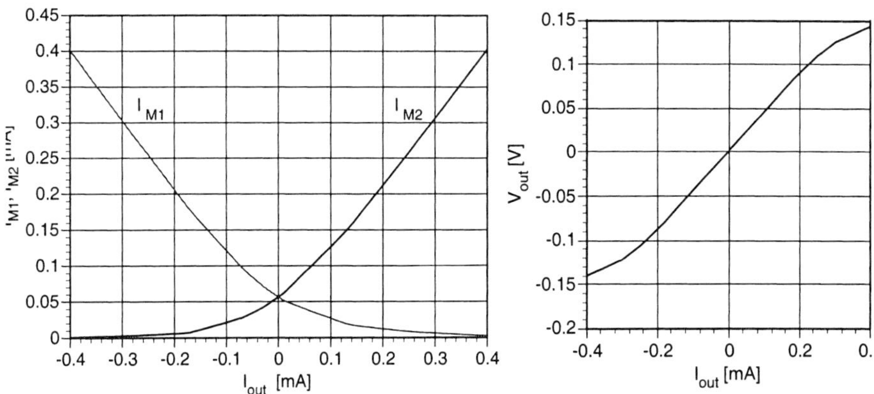


```
PUSH-PULL OUTPUT BUFFER
M1 6 1 4 And    MODN L=0.5u W=50u AD=66p PD=24u AS=66p PS=24u
M2 7 2 4 Add    MODP L=0.5u W=125u AD=66p PD=24u AS=66p PS=24u
M3 Add 1 3 And MODN L=0.5u W=50u AD=66p PD=24u AS=66p PS=24u
M4 2 2 3 Add    MODP L=0.5u W=127u AD=66p PD=24u AS=66p PS=24u
add Add And 3.3
in 1 And 2.3
vt1 Add 6 0
vt2 7 And 0
.dc i2 -400u 400u 10u
i1 2 And 50uA
i2 4 And 0uA
.op
.print dc i(vt1) i(vt2) v(3,4)
```

The output resistance is calculated by a test current source, i_2, connected to the output node. Two test voltage sources, v_{t1} and v_{t2}, allow us to measure the current in the output transistors of the push-pull. The figures show the results of the simulation. The middle of the diagram demonstrates the push-pull operation: the current in one transistor increases while decreasing in the other. Correspondingly, the output voltage changes almost linearly.

The output resistance, calculated by making the ratio V_{out}/I_{out}, ranges around 450 Ω and is almost constant for output currents lower than 150 μA.

The class AB stage in Fig. 3.23 requires a level shifting necessary to generate the two split versions of the input signal. This result can be achieved together with a gain function, as shown in Fig. 3.24. In this way, designers save the power consumption required to achieve some voltage gain. The *dc* voltages of nodes *1* and *2* are shifted by the gate-to-source voltages of transistors M_3 and M_4; also, the signal developed on the two nodes is the amplified version of the input signal. Actually, the gains from the input to nodes *1* and *2* are not exactly the same since there is the small signal resistance between them given by the

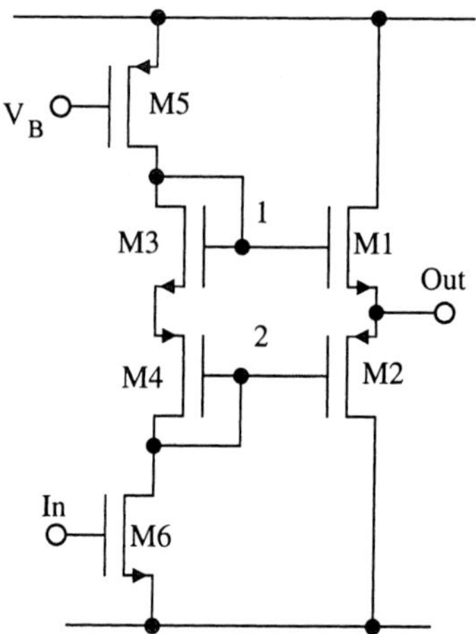

Fig. 3.24 - Push-pull output stage with combined gain stage

two diode-connected elements M_3 and M_4: $(1/g_{m3} + 1/g_{m4})$. However, the difference between the two gains is negligible if

$$\frac{1}{g_{m3}} + \frac{1}{g_{m4}} \ll r_{d5} \tag{3.90}$$

It is worth noting the output swing limits of the push-pull cascode. By inspection of the schematics, we observe that the output transistors connect their source to the output node: thus, the driving of output elements requires more than an upwards or downwards threshold voltage. In addition, the node driving output transistors needs at least a saturation voltage from the bias rails. Therefore, we can write the following limiting conditions

$$V_{out,max} = V_{DD} - V_{sat,p} - V_{sat,n} - V_{th,n} \tag{3.91}$$

$$V_{out,min} = V_{sat,n} + V_{sat,p} + V_{th,p} \tag{3.92}$$

For present technologies, the decrease in output swing with respect to the voltage supply used can be as much as *2 V*, resulting in a significant limitation for circuits with a low voltage supply.

3.8 REFERENCES

- R. Gregorian, *Introduction to CMOS OP-AMPS and Comparators*, J. Wiley and Sons, New York, NY, 1999.
- K.R. Laker, W. M. C. Sansen, *Design of Integrated Circuits and Systems*, McGraw-Hill, New York, NY, 1894
- R. Gregorian, G. C. Temes, *Analog MOS Integrated Circuits,* J. Wiley & Sons, New York, NY, 1986
- D. A. Johns, K. Martin, *Analog Integrated Circuit Design,* J. Wiley & Sons, New York, NY, 1997
- P.E. Allen, D. R. Holberg, CMOS Analog Circuti Design, Holt, Rinehart, and Winston, New York, NY, 1887

3.9 PROBLEMS

3.1 Consider the CMOS inverter in Fig. 3.1. The substrate doping of M_1 is $6\ 10^{-14}\ cm^{-3}$ and the substrate doping of M_2 is $4\ 10^{-14} cm^{-3}$. Also, $\mu_n = 500\ cm^2/V\text{-}sec$ and $C_{ox} = 2.5\, fF/\mu^2$. Find the *dc* gain for $(W/L)_1 = 10, 100, 1000, 10000$.

3.2 The output swing of the inverter in Fig. 3.1 must approach the ground and the positive rail (*3.3 V*) by just *0.2V*. Determine the aspect ratio of M_2 for a $100\mu A$ bias current in the inverter. Use Spice and the models in Appendix B.

3.3 Simulate a CMOS inverter and extract from the Spice results the parameters of the small signal equivalent circuit given in Fig. 3.3. Use the following aspect ratios: $(W/L)_1 = 50/0.7\mu m$; $(W/L)_2 = 30/1\mu m$ and a bias current equal to $10\mu A$. Use Spice and the model in Appendix B

3.4 Determine the *dc* gain of an inverter with active load as a function of the bias current. Use the design parameters: $(W/L)_1 = 100/0.7\mu m$; $(W/L)_2 = 50/1\mu m$ and a bias current ranging from $1\mu A$ to $100\mu A$. Use Spice and the model in Appendix B. Estimate from the simulation results the sub-threshold limit and compare that figure with the one predicted using (1.61).

3.5 Determine the unity gain frequency for the inverter with active load used in Problem 3.4. Use the following bias currents: $50\mu A$, $100\mu A$, and $200\mu A$. Assume that $0.2\ pF$ loads the output node.

3.6 Calculate the input referred noise (white term) for the inverter with active load used in Problem 3.3.

3.7 Repeat Example 3.3 using the following more challenging specifications: output swing from 0.12 to $3.15\ V$ and f_T better than $150\ MHz$. Negotiate, if necessary, the value of the capacitive load.

3.8 Determine with simulations the input capacitance in the cascode with cascode load in Fig. 3.6 with a suitable test circuit. Use the following design parameters: $(W/L)_1 = 50\ \mu m/0.5\mu m$; $(W/L)_2 = 100\ \mu m/1\mu m$; $(W/L)_3 = (W/L)_{MB1} = 50\ \mu m/2\mu m$; $I_{bias} = 130\ \mu$A; $V_B = 2\ V$; $V_{DD} = 3.3\ V$. Use the Spice model in Appendix B.

3.9 Using Spice and the model in Appendix B, determine the impedance seen from node *1* of the cascode with cascode load in Fig. 3.6. Use the same transistor sizing, current and voltages of Problem 3.8. Determine how the impedance changes with $(W/L)_3 = (W/L)_{MB1} = 100\ \mu m/4\ \mu m$.

3.10 Using a *DC* Spice simulation and the model in Appendix B, find the maximum output swing for the circuit used in Problem 3.7. Ensure a margin of $0.1\ V$ in the transistors' saturation.

3.11 Solve the small signal equivalent circuit in Fig. 3.7 and find the exact expressions of A_v and A_I. Compare the results obtained with the simplified relationships (3.22) and (3.23) and estimate the errors produced by the approximations used.

3.12 Using Spice and the model in Appendix B, determine the dc gains A_v and A_I in a cascode with active load. Use the following design parameters: $(W/L)_1 = 100\ \mu m/0.5\mu m$; $(W/L)_2 = 150\ \mu m/1\mu m$; $(W/L)_3 = (W/L)_{MB1} = 50\ \mu m/2\ \mu m$; $I_{bias} = 40\ \mu$A; $V_B = 2\ V$; $V_{DD} = 3.3\ V$.

3.13 Using Spice and the model in Appendix B, estimate the value of the parasitic capacitances C_1, C_2, C_3 and C_4 (Fig. 3.7) for the cascode circuit considered in Example 3.4.

3.14 The transconductance of the n-channel input transistor of a cascode amplifier is 100 μA/V. We want to achieve $A_v = 250$. Find the length of the active load. Assume $n_D = 5\ 10^{-14}\ cm^{-3}$.

3.15 Design a cascode with cascode load capable of achieving $A_v = 75\ dB$. Use the Spice models in Appendix B. Find the bias voltages that maximize the output swing. Find the unity gain frequency and the phase margin for $C_L = 4\ pF$. Estimate the output load that brings the phase margin to $60°$.

3.16 Using Spice and the model in Appendix B, optimize a gain enhanced cascode with cascode load (Fig. 3.12). Use the following design parameters: $(W/L)_1 = 100\mu m/1\ \mu m$; $(W/L)_2 = 150\mu m/1\mu m$; $(W/L)_3 = 100\mu m/1.5\ \mu m$; $(W/L)_4 = (W/L)_{MB1} = 50\mu m/2\ \mu m$; $I_{bias} = 40\ \mu A$; $V_B = 2\ V$; $V_{DD} = 3.3\ V$. Determine the M_5 sizing that maximizes the dc gain.

3.17 Repeat Example 3.6 with an additional capacitor of *0.25 pF* connected between the node *2* and ground. Justify the result achieved. Estimate the value of the high-frequency poles associated to nodes *2* and *3*. Identify the poles in the output gain and phase plots.

3.18 Using Spice and the model in Appendix B, simulate a differential pair made by *n-channel* transistors. The transistors of the pair have $(W/L) = 75/0.5\mu m$. The current source, $I_{SS} = 80\ \mu A$, is generated by transistor with $(W/L) = 100\mu m/1\mu m$. Determine the *CMRR*.

3.19 Repeat Example 3.8 using *p-channel* input transistors. Utilize same transistor sizing voltage and currents. Assume the substrate connected to V_{DD} or tight to the source. Compare the results.

3.20 Design, at the transistor level with the model in Appendix B, the regulated cascode amplifier shown in Fig. 3.15 *b*. Use $V_{DD} = 3.3\ V$. The bias current is $200\ \mu A$, V_{B1} and V_{B2} are $0.4\ V$ and $2.8\ V$ respectively. It is required to ensure an output swing not smaller than *1 to 2.3 V*. The two gain stages are simple differential stages with active load and, if necessary, level shift.s.

3.21 Design a threshold independent level shift with a *0.3 V* upward shift. Use Spice and the models in Appendix B. The current available is $50\ \mu A$. $V_{DD} = 3.3\ V$. Use real current sources. Determine the input-output relationship for input voltages that swing from *1.5* to *2.9 V*.

3.22 Using Spice and the model in Appendix B, simulate the source follower with local feedback shown in Fig. 3.21 b. Use the following design parameters: $(W/L)_1 = 100\mu m/0.5\ \mu m$; $(W/L)_2 = (W/L)_3 = 50\mu m/1.5\ \mu m$; $(W/L)_4 = 50\mu m/1\ \mu m$; $I_{M3} = 40\ \mu A$; $I_{M4} = 80\ \mu A$; $V_{DD} = 3.3\ V$. Determine the input output characteristics and estimate the output resistance.

3.23 Repeat the Problem 3.22 using a complementary scheme. Connect the substrate of the input transistor M_1 to its source. Compare the results with the ones of Problem 3.22 and remark the limitation produced by the substrate transconductance.

3.24 Design a push-pull output stage like the one shown in Fig. 3.23. The supply voltage is $5\ V$. The small signal output resistance should be equal to $2\ K\Omega$. Moreover, the output node should be able to drain or sink $1\ mA$ with minimum harmonic distortion. Determine the voltage-current relationship and derive a polynomial fitting equation. Assume that a resistive load of $2\ K\Omega$ connects the output with a node at $2.5\ V$. Estimate the harmonic distortion in the output voltage when a $1\ V$ peak sinewave is applied to the input.

3.25 Using Spice and the model in Appendix B, simulate the push-pull stage shown in Fig. 3.24. Use the following design parameters: $(W/L)_1 = 1000\mu m/0.5\ \mu m$; $(W/L)_2 = 3000\ \mu m/1\ \mu m$; $(W/L)_3 = 100\ \mu m/0.5\ \mu m$; $(W/L)_4 = 300\ \mu m/1\ \mu m$; $(W/L)_5 = 100\ \mu m/1\ \mu m$; $(W/L)_6 = 100\mu m/0.5\ \mu m$; $I_{M5} = 100\ \mu A$; $V_{DD} = 3.3\ V$. Determine the small signal gain A_0. Connect a resistive load of $2\ K\Omega$ between output and a node at $1.65\ V$. Apply, superposed to a suitable input bias, a sinewave with amplitude $0.5/A_0\ V$. Determine the harmonic distortion of the output voltage. (Hint: use the trick suggested in Example 3.2 and, eventually, use a dummy replica of the push-pull stage for biasing purposes).

Chapter 4

CURRENT AND VOLTAGE SOURCES

The basic building blocks studied in the previous Chapter require, as essential elements for their operation, current generators and voltage biases. We know that without effective biases and references no circuit can operate properly; moreover, the performance of current and voltage sources affect power consumption, speed, dynamic range and noise performance. It therefore becomes important to study how to design these "auxiliary" blocks appropriately, to recognize their functional limits, and to estimate costs and benefits for the best design decision.

4.1 CURRENT MIRRORS

As the name itself suggests a current mirror is used to generate a replica (if necessary, it may be attenuated or amplified) of a given reference current. If we look at the electric function of the circuit, a current mirror is a current controlled current source (*CCCS*). Fig. 4.1 shows its ideal representation. It consists of a branch that measures the reference current and the *CCCS*. In real circuits, as we will see shortly, current mirrors are not able to accomplish the function of a *CCCS* exactly. The current gain factor can only be positive while the output

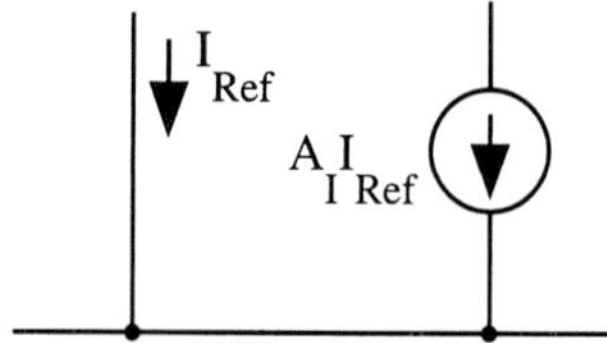

Fig. 4.1 - Equivalent circuit of an ideal current mirror.

impedance, the dynamic range and the speed are finite. Moreover, the current to be copied is not measured ideally as it would be necessary to show a short circuit. Instead, to measure the reference current, a diode connected MOS transistor is normally used.

We can design a number of circuits which accomplish the current mirror function. The ones mostly used (and studied below) are:

- the simple current mirror
- the Wilson current mirror
- the improved Wilson current mirror
- the cascode current mirror
- the modified cascode current mirror
- the high compliance current mirror
- the regulated cascode current mirror

4.1.1 Simple Current Mirror

The implementation of the current mirror shown in Fig. 4.2 is the simplest form of the required function: it is composed of two transistors, of which one, M_1, is diode-connected. M_1 receives the reference current I_{Ref} and measures it by developing at its gate the voltage V_{GS1}; this voltage biases the gate of M_2. We assume that both transistors operate in the saturation region; therefore, the currents are

$$I_{Ref} = I_1 = \frac{\mu C_{ox}}{2}\left(\frac{W}{L}\right)_1 (V_{GS1} - V_{Th})^2 (1 + \lambda V_{DS1}) \tag{4.1}$$

$$I_{out} = \frac{\mu C_{ox}}{2}\left(\frac{W}{L}\right)_2 (V_{GS1} - V_{Th})^2 (1 + \lambda V_{DS2}) \tag{4.2}$$

which enable us to express the output current, I_{out}, as a function of I_{Ref}, V_{out} and V_{DS2}. To simplify algebraic calculations, we assume the term $\lambda V_{DS1} = 0$. Thus, V_{GS1} is directly derived from the former equation and after substitution

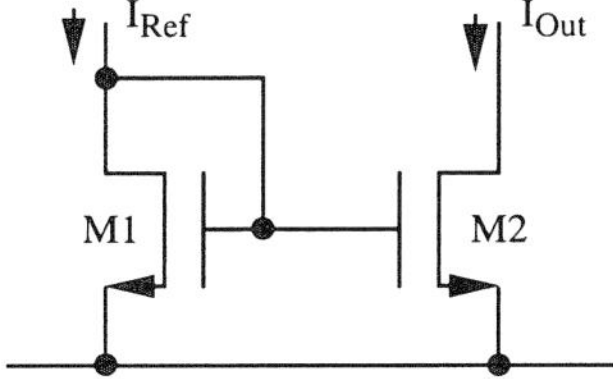

Fig. 4.2 - Simple current mirror (Widlar).

in the latter, gives

$$I_{out} = I_{Ref}\frac{(W/L)_2}{(W/L)_1}(1 + \lambda V_{out}) \tag{4.3}$$

Apart from the term $(1 + \lambda V_{out})$, the output current is a replica of I_{Ref} multiplied by the aspect ratios, W/L, of the transistors M_2 and M_1. The term $(1 + \lambda V_{out})$ takes into account the finite output resistance of M_2, that, for small signal, is

$$r_{out} = \frac{1}{\lambda I_{out}} \tag{4.4}$$

Unfortunately, the value of output resistance which can be achieved with the technologies and medium value currents used, is not large enough for a number of applications. Assuming $\lambda = 1/30\ V^{-1}$, r_{out} is as low as $300\ K\Omega$ for $I_{out} = 100\ \mu A$. Thus, as we will study shortly, other solutions must be used when output resistance is a key design issue. However, for low voltage applications the simple scheme in Fig. 4.2 may be preferred because of its excellent output dynamic range: the output node (the drain of M_2) can swing down to the saturation voltage of M_2, whose value, for common designs, is as low as a few hundred mV. More complex solutions allow us to increase the output impedance but we normally have to pay for this benefit with a reduced dynamic range.

Both the simple current mirror and the other schemes that we are going to study may deviate from ideal behaviour. This is due to:

- imperfect geometrical matching
- technological parameter mismatch
- parasitic resistances

We will discuss all these limits in the present sub-section, the extension to more complex current mirrors being straightforward.

Let us again consider equation (4.1) and (4.2). We assumed equal values of technological parameters and properly ratioed geometrical dimensions of transistors to obtain (4.3). This is not attained in a real circuit and certain mismatches in the geometrical dimension and the technological parameters will always exist. Any mismatches will produce an error in the generated current. If all the parameters in equations (4.1) and (4.2) are statistically independent we obtain

$$\left(\frac{\delta I_{out}}{I_{out}}\right)^2 = \left(\frac{\delta W}{W}\right)^2 + \left(\frac{\delta L}{L}\right)^2 + \left(\frac{\delta C_{ox}}{C_{ox}}\right)^2 + \left(\frac{\delta \mu}{\mu}\right)^2 + \quad (4.5)$$
$$+ 2\left(\frac{\delta V_{Th}}{V_{GS} - V_{Th}}\right)^2 + 2\left(\frac{\delta V_{GS}}{V_{GS} - V_{Th}}\right)^2$$

Hence, current incorrectness derives from the quadratic superposing of relative geometrical and technological mismatches. Inaccuracy in geometrical dimensions comes from photolithographic processes and etching. To limit these, the layout should take into account undercut and boundary dependent effects.

NOTE

The simple current mirror is superior to all the other architecture studied here when the output dynamic range is the key target.

Errors due to mobility and oxide thickness mainly come from unavoidable technology gradients along the surface of the chip. We can reduce this effect by using an inter-digitized or a common centroid structures which minimise the distance between the transistors. Inter-digitized arrangements have already been discussed for passive components; similar dispositions can be achieved by splitting transistors into a given number of equal parts to be connected in parallel. Fig. 4.3 shows a current mirror with a unity mirroring factor. Transistors M_1 and M_2 are split into four equal parts. Two parts of M_1 are interleaved with two parts of M_2 thus realizing an inter-digitized configuration. It should be noted that the inner drain and source connecting areas have a gate on each of their two sides. Therefore, the area and parasitic capacitance of the substrate junctions are minimized. Moreover, the metal makes contact with the underlying area at more than one point: in this way series resistance due to the doped layer is reduced. In the layout we have one part of M_1 on one side of the structure and one part of M_2 on the other. This compensates the boundary-dependent etching effect at the two endings. However, very demanding circuits use two dummy transistors at the terminations of the layout to eliminate the ending effect. Finally, half of the current in both transistors flows from right to left and the other half from left to right. This cancels errors produced by anysotropic behaviours.

Going back to equation (4.5) we can observe that the contribution to error

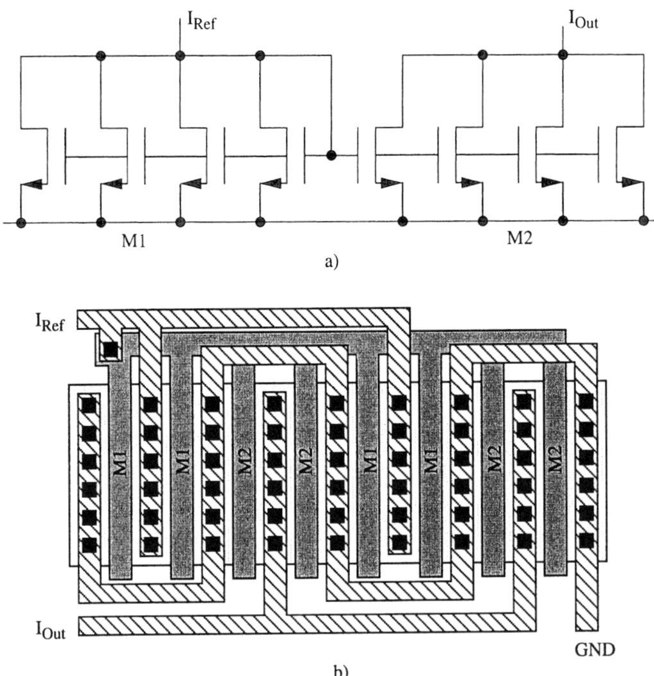

Fig. 4.3 - a) Current mirror with transistors split into four equal parts. b) Its layout.

caused by threshold and source voltage mismatch is proportional to the inverse of the overdrive voltage: we thus secure a good current mirror accuracy when we use a large overdrive. Moreover, we have to acknowledge that the threshold voltage of *MOS* transistors placed in close proximity are matches within few *mV* or less. In contrast, transistors which are hundreds of microns apart have threshold voltages that can differ by a many *mV*. For this reason, when matched currents are required in sections that are quite apart in the same integrated circuit, it is not advisable to generate the voltage bias using a diode connected transistor and to use that voltage throughout the chip to drive the all the mirroring elements. Instead, when the current sources are not one close to each other, it is much better to distribute matched currents and to mirror them locally.

The last term in equation (4.5) refers to a mismatch in V_{GS}. It can derive from resistive drop voltage. The gate connection does not carry current; therefore the voltage of the gates will be the same. Instead, the source connections carry current. Therefore, since metal lines have a specific

WARNING

A possible drop voltage across the metal line connecting the source of the reference transistor to the source of the mirroring one can be responsible for not negligible current mismatch.

resistance ranging from *20* to *50* $m\Omega/\square$, a connection only *20* squares long becomes a *0.4 -1*Ω series resistance. With a *4 mA* current it generates a V_{GS} error as large as *1.6 - 4 mV*, a value which is similar to or larger than the short distance threshold voltage mismatch. Currents as large as many *mA* are not common in *CMOS* integrated circuits; nevertheless, the above observation is important as a means of rethinking a possible problem.

Another limit that we have to account for is the one due to parasitic resistances. Parasitic resistances come from the reversal biased diodes connected between each source and drain and the substrate (or the well). The equivalent resistances of parasitic diodes may affect the output resistance of the current mirror and, at a very low generated current, the leakage current flowing through the diodes may modify the mirror factor.

We should also further discuss the issue of output dynamic range limitation. For n-channel transistors (as shown in Fig. 4.2) the upper limit of the output voltage is only determined by positive rail constraints. In contrast, the lower limit is given by the condition that M_2 should be in saturation

$$V_{out,min} = V_{sat,2} = V_{GS2} - V_{Th2} \tag{4.6}$$

We can again affirm that this result is quite good. We shall see shortly that the more complex mirrors studied below require a higher minimum value of the output voltage.

4.1.2 Wilson Current Mirror

The relatively low value of the output resistance of the simple current mirror can be improved with the Wilson scheme shown in Fig. 4.4. The gate-to-source voltage of M_1 and M_2 are equal, therefore ensuring similar operation to the circuit in Fig. 4.2. However, as we shall see, the addition of M_3 and the established local feedback allows us to increase the output resistance. The small signal equivalent circuit of the stage is shown in Fig. 4.5. Resistance R_L

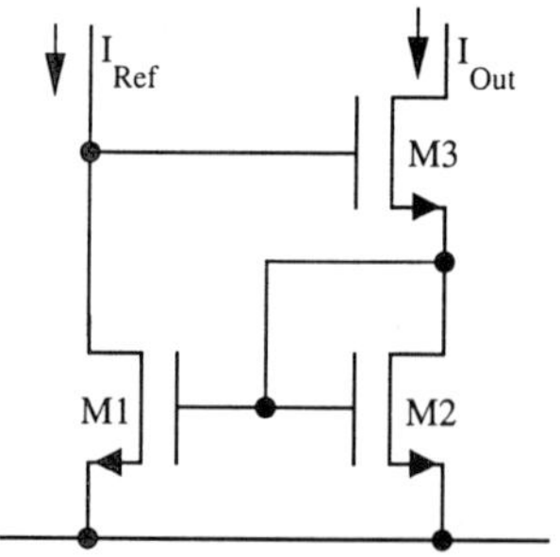

Fig. 4.4 - Wilson current mirror.

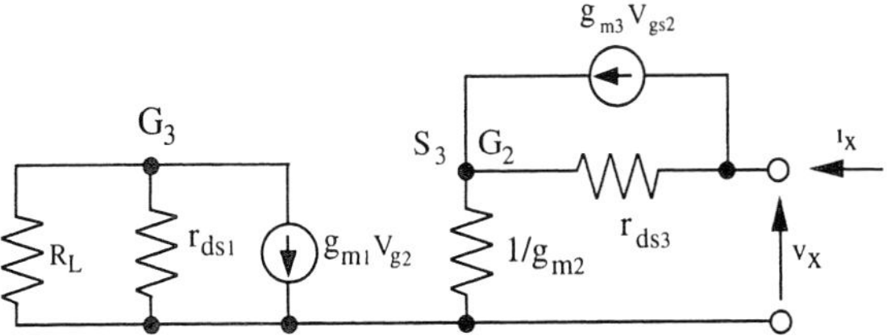

Fig. 4.5 - Small signal equivalent circuit of the Wilson current mirror.

represents the external load seen from the reference current connection. Analysis of the circuit provides

$$v_{g2} = v_{s3} = i_x / g_{m2} \tag{4.7}$$

$$v_{g3} = -g_{m1}v_{g2}r_T \tag{4.8}$$

$$v_x = \frac{i_x}{g_{m2}} + (i_x - g_{m3}v_{gs3})r_{ds3} \tag{4.9}$$

where r_T denotes the parallel connection of R_L with r_{ds1}.

> **KEEP IN MIND!**
>
> The Wilson current mirror earns a high output resistance only if the reference current comes from a high impedance source which small signal resistance must be higher or comparable to r_{ds}.

From (4.7), (4.8) and (4.9) the output resistance is given by

$$r_{out} = \frac{1}{g_{m2}} + r_{ds3}\left[1 + \frac{g_{m3}}{g_{m2}}(1 + g_{m1}r_T)\right]. \tag{4.10}$$

Transconductances g_{m2} and g_{m3} have same order of magnitude because M_2 and M_3 are carrying the same current. Therefore, the output resistance of the circuit is approximately determined by the output resistance of M_3 (r_{ds3}) amplified by the factor $r_T g_{m1}$. It is large if r_T is large and, in turn, if R_L is sufficiently large. This condition is naturally met when the reference current comes from a current source.

The circuit in Fig. 4.4 suffers from a shortcoming. The drain-to-source voltage of M_1 and M_2 are systematically different, in fact

$$V_{DS1} = V_{GS3} + V_{DS2}\,. \tag{4.11}$$

This, because of the channel length modulation effect (or the finite small-signal output conductance), reveals a systematic mismatch between reference current and output current. Approximately, the reference current is larger than the generated one by the amount V_{GS3}/r_{ds1}. Remembering that $1/r_{ds1}=\lambda_1 I_{D1}$ the current difference becomes $\lambda_1 I_{D1} V_{GS3}$.

Example 4.1

Simulate using Spice the Wilson current mirror shown in Fig. 4.4. Use the following transistor sizing (in μ): M_1: 25/0.5, M_2: 25/0.5 M_3:50/0.5.
Use the Spice model of Appendix B. Find the difference between the reference and generated current as a function of I_{ref} in the range 0-200 μA.

Solution:
Since the technology used is n-well, the substrate of M_3 must be connected to ground. A possible larger aspect ratio permits us to reduce the overdrive. Moreover, we increase g_{m3} and, in turn, we increase the output resistance.
The following input list leads to the result shown in the diagram below.

```
WILSON CURRENT MIRROR
M1 1 2 Gnd Gnd MODN L=0.5u W=25u AD=66p PD=24u AS=66p PS=24u
M2 2 2 Gnd Gnd MODN L=0.5u W=25u AD=66p PD=24u AS=66p PS=24u
M3 3 1 2   Gnd MODN L=0.5u W=50u AD=66p PD=24u AS=66p PS=24u
IREF Vdd 1 50uA
vdd Vdd Gnd 2.5
VI Vdd 3 0
.include modn.md
.dc IREF 1u 300u 10u
.print dc I(VI) I(IREF,VI)
```

As expected, the figure shows an output current which is slightly smaller than the reference current. Moreover, the difference is almost linear with the reference current and is around 5% the reference current. This figure can be approximately confirmed by

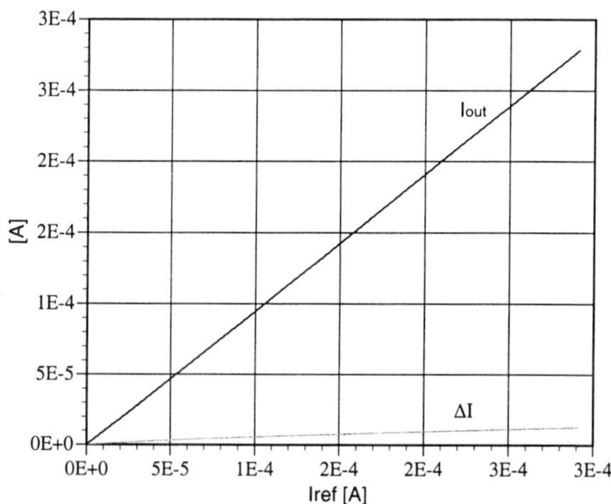

hand calculations; let us compare the point of the curve: I_{ref}=100μA ΔI=5.2μA with the result obtained using the parameters provided by Spice. Since $g_{ds1}=5.95\cdot10^{-6}$A/V and V_{GS3}=0.831 V, we calculateΔI=4.95 μA that is very closed to the above data.

4.1.3 Improved Wilson Current Mirror

The systematic current mismatch in the Wilson current mirror is compensated by the improved solution shown in Fig. 4.6. One additional transistor is used, M_4, which shifts down the voltage of the gate of transistor M_3. Therefore, the drain voltage of M_1 is given by

$$V_{DS1} = V_{GS3} + V_{DS2} - V_{GS4} \tag{4.12}$$

If the gate-to-source voltage of M_3 and M_4 are equal (as will be reasonable if they are matched), the V_{DS} voltage of M_1 and M_2 will result as equal.

The addition of transistor M_4 slightly changes the output resistance small signal analysis. The transistor M_4, diode-connected, adds a resistance $1/g_{m4}$ in series with the resistance, R_L, of the reference current (see Fig. 4.7).

The small signal voltage at the gate of M_3 is expressed by

$$v_{g3} = -g_{m1} v_{g2} r_T' \frac{R_L g_{m4}}{1 + R_L g_{m4}} \tag{4.13}$$

where

$$r_T' = r_{ds1} \| (R_L + 1/g_{m4}) \tag{4.14}$$

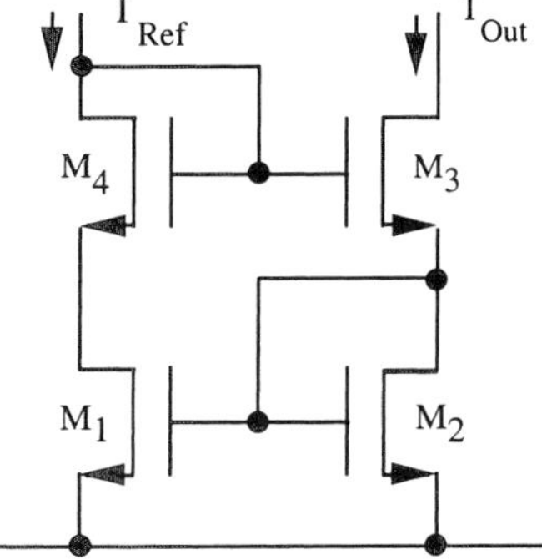

Fig. 4.6 - Improved Wilson current mirror.

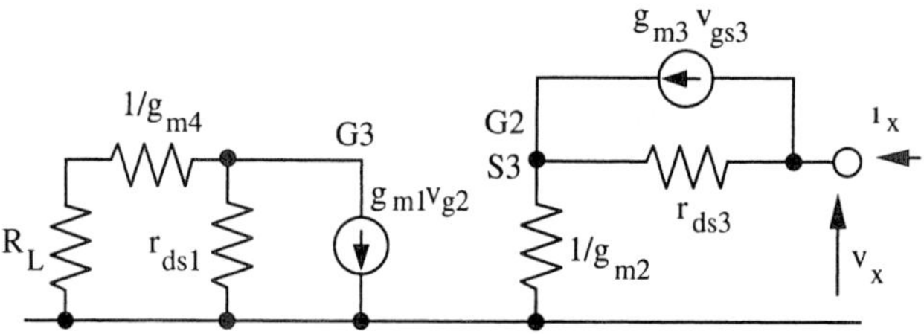

Fig. 4.7 - Small signal equivalent circuit of the improved Wilson current mirror.

while the other equations (4.7) and (4.9) are still valid. Therefore, the output resistance changes according to

$$r_{out} \cong r_{ds3}\frac{g_{m3}}{g_{m2}}g_{m1}r_T\frac{R_L g_{m4}}{1 + R_L g_{m4}} \tag{4.15}$$

this result is substantially unchanged with respect to the plain Wilson configuration if $R_L >> 1/g_{m4}$.

NOTE ON FIG. 4.6

A balanced and symmetrical design optimizes circuit performance. The improved Wilson current mirror prevents a possible systematic offset by balancing the drain voltages of M_1 and M_2.

For both the Wilson current mirror and its improved version, the increase in the output resistance is paid for by a reduction in the output dynamic range. The source voltage of transistor M_3 is V_{GS2} and, to keep M_3 in saturation, the output voltage must be higher than this value by the saturation voltage of M_3. The lower limit of V_{out} is therefore given by the following relationship

$$V_{out, min} = V_{GS1} + V_{sat, 3} = V_{Th, n} + V_{sat, 1} + V_{sat, 3} \tag{4.16}$$

its value can be as high as *0.8 -1.5 V*, which, even for a *5 V* supply, is a significant fraction of the available supply range. When the output voltage drops below its minimum, expressed by (4.16), the feedback network pulls up the gate voltage of M_3 and suddenly the transistor goes into the triode region and induces a drop of the gain $g_{m3}\, r_{ds3}$ of M_3 itself. Therefore, the output resistance falls by the same amount.

Example 4.2

Simulate using Spice and the models of Appendix B the behaviour of the improved Wilson current mirror shown in Fig. 4.6. Use the following transistor sizing: M_1 = M_2: 10μ/0.4μ; M_3=M_4:30μ/

0.4μ. Set V_{DD} = 4V and use a resistance R_L= 20 KΩ connected from the drain of M_4 and V_{DD}. Compare the output resistance achieved for the case where R_L= 40 KΩ and the reference current equals the previous case.

Solution:
The following input list describes the improved Wilson current mirror.

```
M1 5 4 Gnd Gnd MODN L=0.4u W=10u
M2 4 4 Gnd Gnd MODN L=0.4u W=10u
M3 2 1 4   Gnd MODN L=0.4u W=30u
M4 1 1 5   Gnd MODN L=0.4u W=30u
.op
.ac dec 10 10 1G
.print ac idb(vx)
R1 Vdd 1 20K
v1 Vdd 3 0.0 AC 1.0 0.0
vx 3 2 0
vdd Vdd Gnd 4.0
```

The computer simulation leads to r_{out} = 1.41 MΩ. If we use the small signal parameters provided by Spice (g_{ds1} = 1.08 $10^{-5}\Omega^{-1}$, g_{m4} =1.87 10^{-3} Ω^{-1}, g_{ds3} = 2.6 $10^{-5}\Omega^{-1}$, g_{m3}=2.15 10^{-3} Ω^{-1}), we obtain r_{out} = 1.36 MΩ that is in good agreement with the above result (note that the use of a Spice version different from the one used here can lead to slightly different results).
The current in M_1 is 116 μA. To evaluate the output resistance with R_L= 40kΩ we should replace the R_L line with the following

```
R1 Vdd 1 40K
iref Vdd 1 58uA
```

The output resistance becomes 2.24 MΩ. The figure is again in good agreement with hand calculations.

4.1.4 Cascode Current Mirror

An alternative way to increase the output resistance is to use a cascode configuration. Fig. 4.8 shows a possible scheme. The output stage consists of two transistors (M_2-M_3) in the cascode arrangement. Their biases result from two other transistors (M_1-M_4), which are diode-connected. Again, as for the previously studied current mirrors, the V_{GS} voltage of M_1 and M_2 are set equal. Therefore, a replica of the current in M_1 is generated by M_2. The output resistance increases because of the cascode arrangement. Fig. 4.9 shows the small

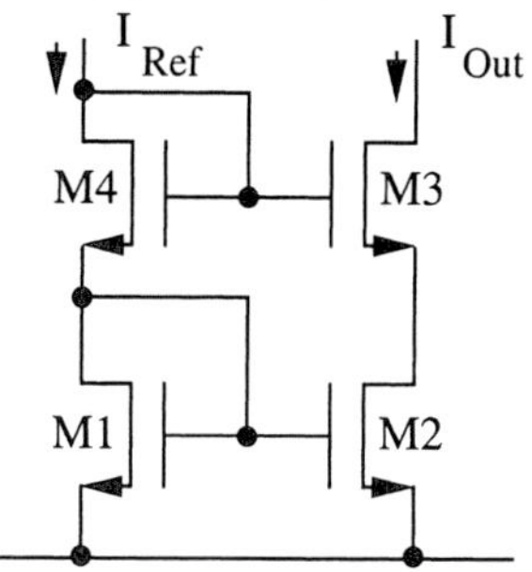

Fig. 4.8 - Cascode current mirror.

signal equivalent circuit. Calculation of the output resistance proceeds as for the already studied cascode configuration. It is obtained from

$$v_x = r_{ds2} i_x + r_{ds3}(1 + g_{m3} r_{ds2}) i_x \tag{4.17}$$

which yields

$$r_{out} \cong r_{ds3} g_{m3} r_{ds2} \tag{4.18}$$

Therefore, as stated by the mnemonic rule given in the previous Chapter, the output resistance is given by the product of the gain of transistor M_3 by the drain resistance of M_2.

Even for the cascode configuration the increased output resistance is paid for by a reduction of the output dynamic range; in fact

$$V_{S3} = V_{GS1} + V_{GS4} - V_{GS3} \tag{4.19}$$

$$V_{out,min} = V_{S3} + V_{sat3} \approx V_{Th} + 2V_{sat} \tag{4.20}$$

thus, the minimum output voltage cannot be less than one threshold plus two saturation voltages.

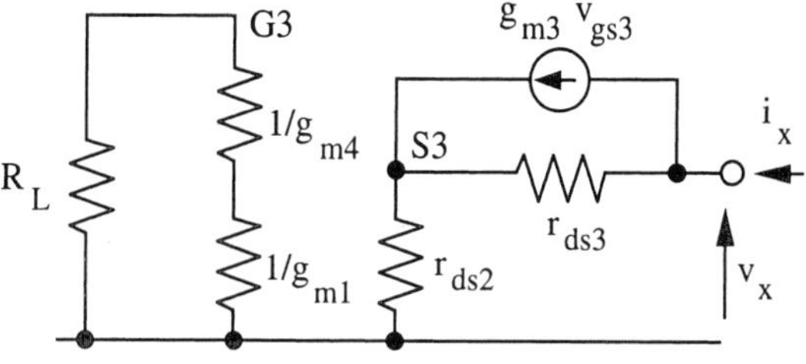

Fig. 4.9 - Small signal equivalent circuit of the cascode current mirror.

4.1.5 Layout of Modified Wilson and Cascode Current Mirrors

The modified Wilson and cascode current mirrors have very similar schemes. The only difference is that in the former architecture transistor M_2 is diode connected, whilst in the latter M_1 is diode connected. As a consequence the layout of the two configurations can be quite similar.

Often the layout of a given basic block is accommodated together with the layout of other basic blocks. Nevertheless, as an exercise, we shall here consider possible layouts of the four transistors used to form the current mirrors. A good design strategy is to choose the same width for all the elements used or, if necessary, to choose widths that are a multiple of the same value. If required, the designer can achieve different aspect ratios by using transistors with different lengths. Fig. 4.10 shows possible layout examples. Note that transistors M_1 and M_2 have a greater length than M_3 and M_4. This allows us to maximize the output resistance while using the same width for all the elements. The four transistors are stacked together with M_3 and M_4 placed at the two sides. As required

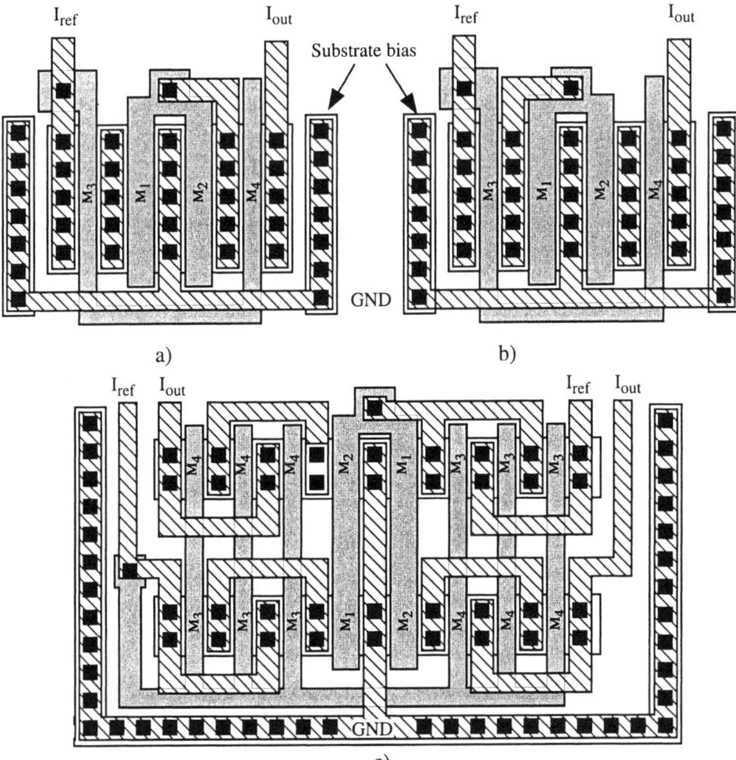

Fig. 4.10 - Simple layout of a modified Wilson (a) and a cascode (b) current mirror. c) Common centroid version of the cascode current mirror.

the gates of M_3 and M_4 are connected to the input terminal. The gates of M_1 and M_2 are connected to the drain of M_2 in the modified Wilson case, a), and to the drain of M_1 for the cascode arrangement, b). we assume the use of an n-well technology. Therefore, we achieve n-channel transistors directly in the substrate and we don't need to design the well. However, it is important to remember to properly and repeatedly bias the substrate.

Neither of the layouts in Fig. 4.10 a) and b) employ inter-digitized or common centroid arrangements. Therefore, possible gradients in the geometrical or technological features may affect the mirroring factor. The layout in Fig. 4.10 c) shows (for different transistor sizing) the common centroid layout of the cascode current mirror. Transistors M_1 and M_2 are split into two parts while M_3 and M_4 are divided into six parts. The cell has two terminals for I_{ref} and two for I_{out}. They should be properly connected to the rest of the circuit outside the cell, possibly using the second metal layer.

Observe that transistors M_3 and M_4 whose matching is less important for mirror factor accuracy are distant from the common centroid, even if they are symmetrically arranged. Moreover, the layout matches possible errors coming from the ending transistors of the two stacks.

4.1.6 Modified Cascode Current Mirror

The cascode current mirror studied in a previous sub-section (the one in Fig. 4.8) biases the gate of M_3 with a diode-connected transistor (M_4) placed on the top of M_1 (also diode connected). If V_{GS3} matches V_{GS4} the drain voltage of M_2 equals the one in M_1. Therefore, the circuit is not affected by any systematic current mismatch. However, this desired feature costs a limitation to the output dynamic range as expressed by equation (4.20).

> **REMEMBER**
>
> If we want to keep transistors in saturation, the minimum voltage at the output of a stack of two transistors should be two saturation voltages at least.

Often, low voltage operation is a key requirement (especially for battery operated circuits). Therefore, current mirrors must ensure the highest output swing together with a high output resistance, even if we have to sacrifice for this result the beneficial systematic match between the reference and generated current. In achieving this compromise solution, we can observe that the output impedance remains quite high so long as the two transistors in the cascode configuration are in saturation. This means the drain to source voltage of M_2 must be at least slightly greater than the overdrive voltage of M_2 itself. Therefore, the minimum value of the gate of M_3 comes to be just one V_{GS} higher than a saturation

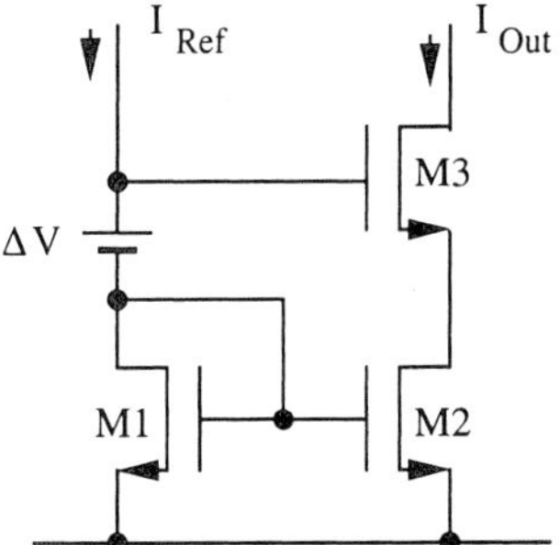

Fig. 4.11 - Conceptual scheme of the modified cascode current mirror.

voltage

$$V_{G3} = V_{sat,2} + V_{GS3} \tag{4.21}$$

It turns out that under these operating conditions the minimum output voltage can become as low as

$$V_{out,min} = V_{sat,2} + V_{sat,3} \tag{4.22}$$

The scheme in Fig. 4.11 depicts the objective. A suitable level shift, ΔV, replaces the role of transistor M_4 used in Fig. 4.8. The value of the level shift ΔV being $V_{GS3} + V_{sat,2} - V_{GS1}$. Observe that the implementation of the conceptual scheme in Fig. 4.11 must account for possible errors affecting the circuit used. In particular, it must ensure that the transistors remain in saturation even if the technology shows variations and the transistors themselves manifest geometrical mismatches. Since transistors M_1 and M_2 are close to each other the geometrical mismatch will not be so important. Moreover, since they are of the same type, a possible threshold alteration will affect both to the same extent. Hence, the level shift ΔV must be threshold independent.

Possible solutions for the threshold-independent level shifter have already been considered in Chapter 3. Those configurations, when applied to the conceptual scheme in Fig. 4.11, lead to the circuits in Fig. 4.12. Both solutions use two additional transistors compared to the conventional cascode structure. In both schemes M_4 is shifting up the gate voltage of M_1 and M_5 is shifting down the result. M_6 is used to control the current in the shifting down element. In the solution of Fig. 4.12 a) the reference current flows into M_4. For the other scheme (Fig. 4.12 b), an additional bias current, $I_{4,5}$ is used to control the current in the pair M_4-M_5 ($I_4 = I_{4,5} - I_6$). Remember that the overdrive of M_5 is controlled by the current in M_6 which mirrors the reference current. Therefore, the level shifting is controlled by transistor size ratios and bias currents and is independent of the threshold of n-channel or p-channel elements (assumed to

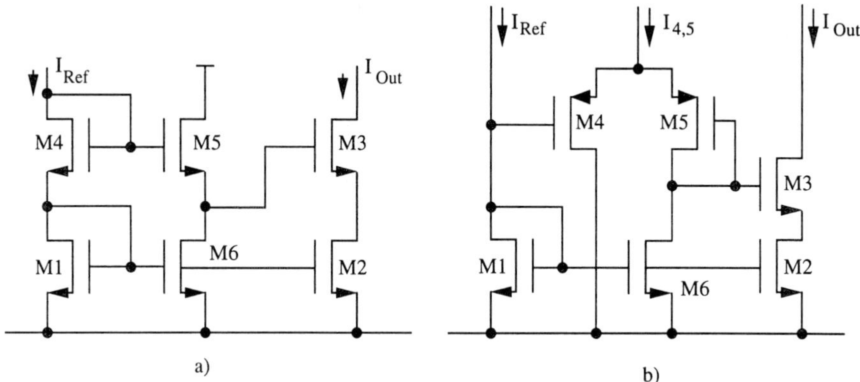

Fig. 4.12 - Modified cascode current mirror with a level shifter provided using: a) n-channel transistors; b) p-channel transistors.

be matched)

$$\Delta V = V_{GS4} - V_{GS5} = \sqrt{\frac{2I_4L_4}{\mu C_{ox}W_4}} - \sqrt{\frac{2I_5L_5}{\mu C_{ox}W_5}} \tag{4.23}$$

Example 4.3

Simulate the output resistance of the circuit in Fig. 4.11 with V_x ranging from 0.1 to 0.6 V. Use for transistors M_1, M_2 and M_3 the following sizing respectively: 20µ/0.5µ, 20µ/0.5µ, 30µ/0.5µ; moreover, I_{ref} =100 µA. Use the Spice models in Appendix B.

Solution:

The Spice description is straightforward. With a voltage at the drain of M_3 equal to 3V it results as:

for Vx=0.6: $g_{m3} = 1.03\ 10^{-3}$; $g_{ds3}=6.43\ 10^{-6}$; $g_{ds2}=9.62\ 10^{-6}$
for Vx=0.4: $g_{m3} = 0.87\ 10^{-3}$; $g_{ds3}=5.76\ 10^{-6}$; $g_{ds2}=1.46\ 10^{-5}$
for Vx=0.2: $g_{m3} = 0.93\ 10^{-3}$; $g_{ds3}=5.86\ 10^{-6}$; $g_{ds2}=5.30\ 10^{-5}$
for Vx=0.1: $g_{m3} = 1.06\ 10^{-3}$; $g_{ds3}=5.39\ 10^{-6}$; $g_{ds2}=2.54\ 10^{-4}$

The above results shows that the small signal parameters of M_3 change a little because the operating point and the current are not constant. However, their variation is within 10%. In contrast the output resistance of M_2 changes significantly: the drain voltage of the transistor goes from 0.54 V to 0.145 V while the saturation voltage ranges around 0.15 V. As a result the output resistance decreases by more than one order of magnitude as shown in the figure.

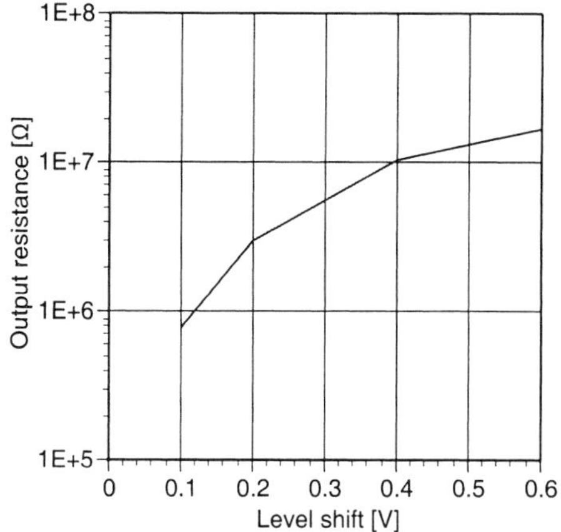

Observe that the value of g_{ds} in normal operating conditions is much lower than the value achieved in Example 4.2. There the designed length of transistors was closer to the minimum. Therefore, the considerable difference of effective length and the short channel effects leads to a significant reduction in output resistance.

4.1.7 High Compliance Current Mirror

The current mismatch inherent to the modified cascode configurations and the additional circuit complexity can be avoided by using the high compliance scheme, shown in Fig. 4.13. By comparing the given circuit with the usual cascode current mirror, we observe that the diode connection of transistor M_1 incorporates transistor M_4. Therefore, the drain-to-source voltage of M_1 is no longer equal to its gate-to-source voltage. Instead, the value of V_{DS1} and that of V_{DS2} are controlled by the gate of transistors M_4 and M_3 respectively. The matching between these two elements ensures identical voltage at the drains of M_1 and M_2, thus leading to a systematic current matching. The gates of M_3 and M_4 should be biased by a voltage that keeps both M_1 and M_2 in saturation and which, at the same time, should avoid M_4 going into the triode region. Therefore

$$V_{bias} - V_{Th,4} - V_{sat,4} > V_{sat,1} \tag{4.24}$$

$$V_{bias} - V_{ds1} - V_{Th,4} < V_{Th,1} + V_{sat,1} - V_{ds1} \tag{4.25}$$

that require keeping V_{bias} between one threshold plus two saturations and two thresholds plus one saturation. This condition can be achieved because V_{th} is normally higher than V_{sat}. The designed value of V_{bias} causes V_{GS1} to split

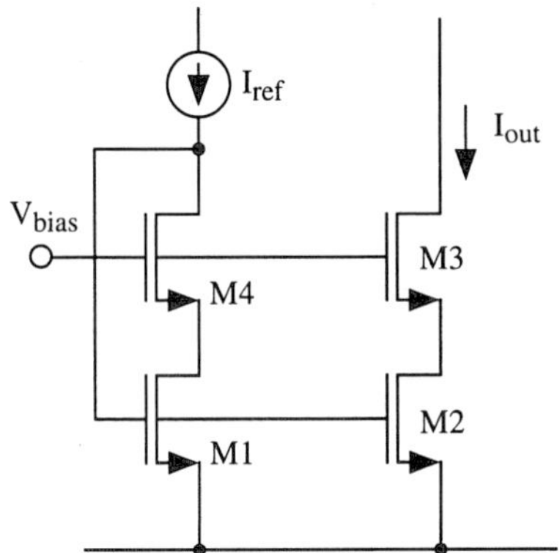

Fig. 4.13 - High Compliance Current Mirror.

between V_{DS1} and V_{DS2}. Typically the chooses value of V_{bias} splits V_{GS1} evenly between the two drain-to-source voltages.

Note that any increase in the threshold voltage caused by technological variations could push the transistors M_1 and M_2 into the triode region. We avoid this by using a bias voltage, V_{bias}, that tracks the threshold changes. A simple example of such a bias generator is given by a diode connected transistor whose overdrive voltage is quite large and equal to the sum of the overdrive voltages of both M_1 and M_4.

Example 4.4

Calculate as a function of the bias voltage V_{bias} the output resistance of the high compliance current mirror shown in Fig. 4.13. Use the following transistor sizing: M_1=M_2: 20μ/0.5μ; M_3=M_4: 30μ/0.5μ. Use the Spice model in Appendix B; moreover, assume I_{ref} = 50 μA and set the output voltage to 1.5 V.

Solution:

After describing the circuit using Spice it is necessary to estimate the value of V_{bias}. It must stay within the range prescribed by the two conditions (4.24) and (4.25): $V_{th}+2V_{sat} < V_{bias} < 2V_{th}+V_{sat}$. The threshold voltage of the NMOS transistors is around 0.5 V and, according to the given current and aspect ratios, the saturation voltage is in the order of 150 mV. Thus, the bias voltage must be higher than 0.8 V and lower than 1.15 V, otherwise M_1 or M_4 would operate in the triode region.

Computer simulation allows us to verify the above limiting conditions. The below figure summarize the simulation results. Actually, the achieved operation range is slightly different than expected.

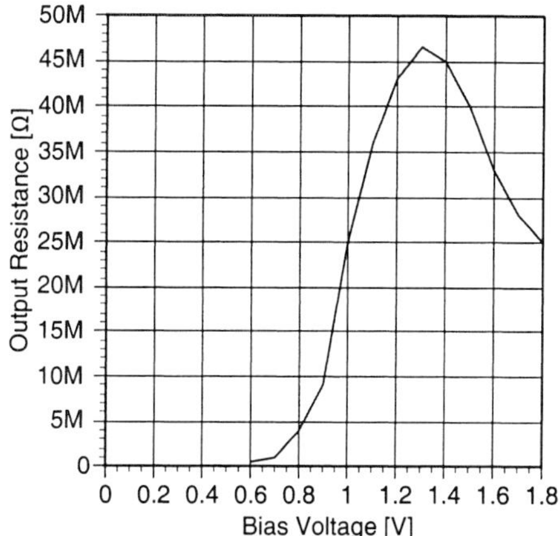

This is because the saturation is slightly higher than 150 mV. Moreover, because of the body effect, the threshold voltage of μ_4 depends on the voltage of the substrate.
Observed that the output resistance has a maximum for V_{bias} = 1.3 V and holds more than 40MΩ. Moreover, since the transistors' sizing is not the same, we obtain a saturation voltage of lower components which is lower compared with the upper ones.

4.1.8 Enhanced Output-Impedance Current Mirror

The output impedance of a current mirror can be improved by using feedback. We understand this by making the following obvious observation. For a given V_{GS} voltage, the drain current of a transistor does not change if we keep its drain-to-source voltage constant. Thus, the differential output resistance goes to infinite. We maintain the voltage variation at the drain of the mirroring transistor M_2 under control with feedback loops like the one shown in Fig. 4.14. The gain stage amplifies the difference between a given bias voltage, V_{bias}, and the drain voltage of transistor M_2. The result controls the gate of M_3; therefore, the action of M_3 is enhanced by gain A. Thus, the term $g_{m3}r_{ds2}i_x$ in (4.17) should be multiplied by A. The output resistance becomes

$$r_{out} \cong A\ r_{ds3} g_{m3} r_{ds2} \tag{4.26}$$

The bias voltage used determines the output swing of the current mirror. Transistor M_3 operates properly until its V_{DS} exceeds the saturation voltage. Therefore, the minimum output voltage is given by

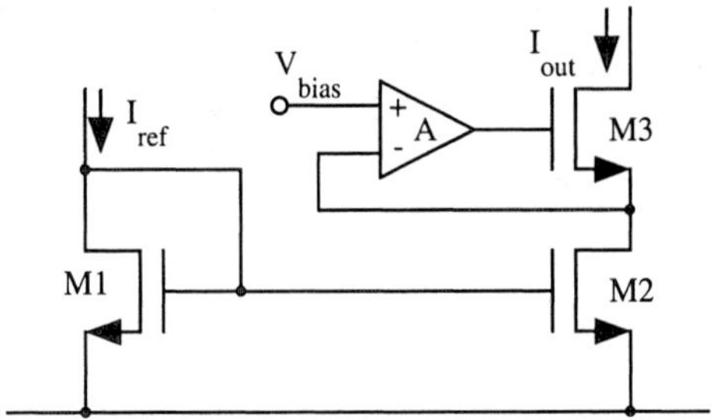

Fig. 4.14 - Enhanced output-impedance current mirror.

$$V_{out,min} = V_{bias} + V_{sat,3} \tag{4.27}$$

where we assume the offset of the gain stage to be negligible with respect to the used V_{bias}.

Any difference between the drain voltages of M_1 and M_2 contributes to a systematic mismatch between the reference and generated current. Therefore, if matching is an important design parameter, the bias voltage, V_{bias}, should be equal to the drain of M_1. In the case shown in Fig. 4.14 the transistor is diode connected. Therefore the bias voltage should be equal to V_{GS1} and the minimum output swing becomes one saturation above that V_{GS}.

If the designer wants to optimize the dynamic range and, at the same time achieve a systematic current matching, the circuit in Fig. 4.14 should be modified as shown in Fig. 4.15. The connection between the gate and drain of M_1 incorporates a replica of the enhancing feedback loop. Note that the feedback loop used around M_4 has no effect on the output resitance. It is only used to achieve the needed matching of the drain voltages of M_1 and M_2.

It should be noted that any enhancing solution encounters an upper limit when improving output impedance. Between the drain of M_3 and the substrate, there is a reversal biased diode. It establishes a parasitic resistance

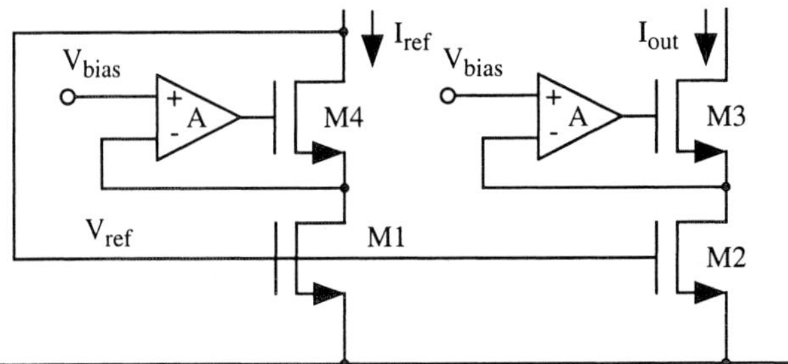

Fig. 4.15 - Current mirror in Fig. 4.14 with optimized dynamic range and systematic matching.

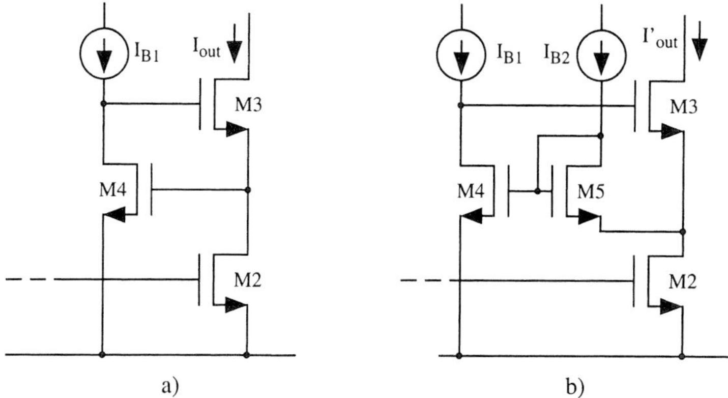

Fig. 4.16 - Possible circuit solutions for the enhanced output-impedance current mirror

between the output node and ground that ultimately limits the value of the output resistance. Therefore, the gain of the amplifier used should be chosen so as to optimize the benefit-cost ratio of the technique studied.

The use of a feedback loop requires some care to avoid possible instability. Moreover, designers must also account for power consumption and frequency limitations. These concerns often lead the designer to use simple schemes to achieve the gain stage whose gain is in the order of *40 dB*. Fig. 4.16 shows two possible solutions. Both schemes use a simple inverter as amplifier. For simplicity, an ideal current source, I_{B1}, represents the active load. The circuit used does not have a non-inverting terminal. However, we know that the inverter operates properly when the input voltage is somewhat larger than the threshold. Therefore, the drain of M_2 of the first scheme will be set somewhere above the threshold of M_4. We thus have a simple circuit whose output dynamic range, however, may not be satisfactory. The second solution in Fig. 4.16 allows us to significantly improve the output dynamic range. Given the level shift on the top of the drain of M_2, the circuit must verify the following condition

$$V_{GS4} = V_{GS5} + V_{DS2} \tag{4.28}$$

that can be achieved with proper transistor sizing and a suitable choice of bias currents. From equation (4.28) it results that the drain voltage of M_2 is given by the difference between the V_{GS} of two n-channel transistors. Therefore, it can be quite low and, also, it is threshold independent.

> **KEEP IN MIND!**
>
> The ultimate upper limit of the current mirror output resistance is given by the conductance of the parasitic diode affecting the output node.

Observe that the output current of the mirror in Fig. 4.16 b) is not a replica of the bias current I_{B1}. This for two reasons: the drain voltage of M_2 and M_4 do not match and part of the current mirrored by M_2 is provided by M_5. Thus, the output current is lower than the one in M_2 approximately by I_{B2}. As it typically happens, a given extent of imprecision is the cost that we have to pay for increasing the dynamic range.

4.1.9 Current Mirrors with Adjustable Mirror Factor

The current mirrors studied so far have a fixed mirror factor determined, at first approximation, by the aspect ratio of two key transistors. In some applications it is desirable to electrically adjust the mirror factor, so that, for example, the offset or other features of a circuit can be controlled.

We can adjust the mirror factor by using source-degenerated current mirrors. Fig. 4.17 a) shows a simple current mirror where the addition of resistors between ground and sources achieves the so called degenerated configuration. Assuming the transistors to be in saturation and neglecting the output resistances, we obtain

$$V_{G1} = R_1 I_{ref} + V_{Th,1} + \sqrt{\frac{2I_{ref}}{\mu C_{ox}}}\sqrt{\frac{L_1}{W_1}} = R_2 I_{out} + V_{Th,2} + \sqrt{\frac{2I_{out}}{\mu C_{ox}}}\sqrt{\frac{L_2}{W_2}} \quad (4.29)$$

Assuming equal thresholds, after simplification, this results in

$$R_1 I_{ref} + \sqrt{\frac{2I_{ref}}{\mu C_{ox}}}\sqrt{\frac{L_1}{W_1}} = R_2 I_{out} + \sqrt{\frac{2I_{out}}{\mu C_{ox}}}\sqrt{\frac{L_2}{W_2}} \quad (4.30)$$

That is a non-linear equation which solution permits to obtain the mirror factor I_{out}/I_{ref}.

In the particular case where the ratio (R_1/R_2) matches the ratio $(W/L)_2/(W/L)_1$ the mirror factor is $(W/L)_2/(W/L)_1$ and the drop voltage across the resistors

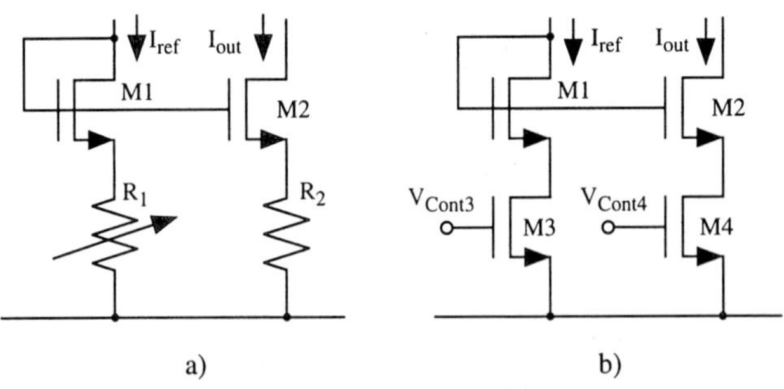

Fig. 4.17 - Current mirror with an adjustable mirror factor.

is the same. More generally, the mirror factor depends both on aspect ratios and resistor values. Moreover, we can possibly change the mirror factor by electrically tuning the resistor used. This can be done with *MOS* transistors in the triode region implement the degeneration resistors. Fig. 4.17 b) shows the circuit solution. The control voltage V_{Cont3} and V_{Cont4} are high enough to bring transistors M_3 and M_4 in triode. Moreover, their aspect ratios match the nominal mirror factor required in the case $V_{cont3} = V_{cont4}$. A possible imbalance of the control voltages changes the equivalent resistors and, in turn, the mirror factor.

Example 4.5

Simulate the circuit in Fig. 4.17 a). Use the transistor model in Appendix B and the following design parameters: $W/L = 20\mu/0.5\mu$; $I_{ref}=100\mu A$; $R_2=1k\Omega$. The nominal value of the resistance R_1 is equal to that of R_2 and can be adjusted by ± 50%. Determine the output current for $R_1 = R_2$ and V_{out} = 3V, 2V, 1V.

Solution:

If the output voltage is 3 V the voltage at the drain of M_1 will be surely lower. Therefore, because of the mismatch we expect an output current slightly higher than the reference. Computer simulation gives I_{out} =106 μA. When the output voltage is reduced to 2 V and 1V the corresponding current diminishes to 103μA and 100 μA respectively. Observe that we have a good matching when the output voltage equal the drain of M_1. The given results permit to estimate the output resistance which is approximately 330 kΩ (= 2V/6μA).

The variation of the resistance R_1 by ±50% while the output voltage is kept equal to 3V leads to the result given in the figure. Surprisingly the current variation is almost linear. However, an accurate analysis of the response leads to the following interpolating equa-

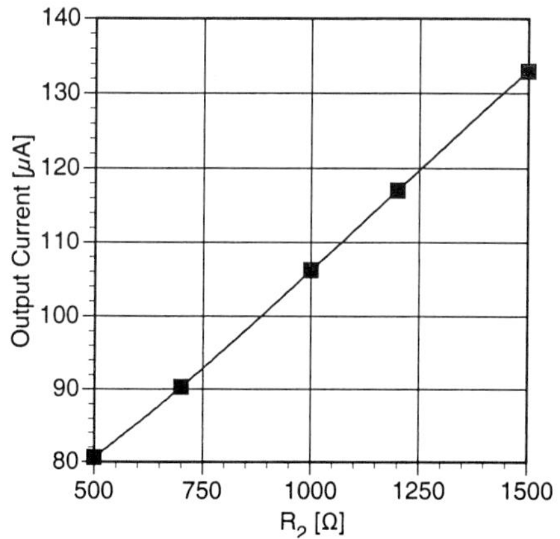

tion:

$$I_{out} = \left[61.5 + 28 \cdot \frac{R_1}{R_{1,nom}} + 23.5 \cdot \left(\frac{R_1}{R_{1,nom}}\right)^2 - 7 \cdot \left(\frac{R_1}{R_{1,nom}}\right)^3\right] \cdot 10^{-3}$$

that show, as expected, the dependence on the square root term in (4.30). It is because of the used transistor sizing and resistors values that the non-linearity is not particularly high here. It is important to underline this point as a warning for the reader: responses that seem linear at first glance, with more close analysis show non-linearity behaviour.

4.2 CURRENT REFERENCES

Most of the basic building blocks use a reference current. We know that the current value controls the transconductance of transistors and, in turn, influences the static and dynamic properties of circuits. Moreover, the currents used establish the power consumption of the design, a parameter which in many cases is very important.

Often, the current in different basic blocks results from the mirroring of one or more references. Therefore, an important designer task is to endow the master currents of the system with the required accuracy and supply voltage independence. This section shall consider the design technique for the generation of current references. Very simple approaches as well as supply independent and micro-current generators will be discussed.

4.2.1 Simple Current Reference

The simplest technique for the generation of a current reference is to use the circuits in Fig. 4.18.The current in the branch, I_{ref}, is given by

$$I_{Ref} = \frac{V_{DD} - V_{GS1}}{R_L} \tag{4.31}$$

which is injected into transistor M_1 whose diode connection provides a useful voltage for current mirroring.

The method used is straightforward: it exploits the Ohm law. However, the accuracy of a current generated in this manner is generally poor. Assuming the

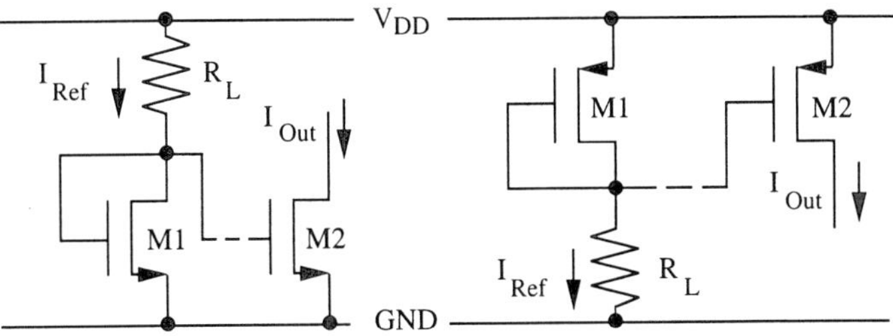

Fig. 4.18 - Simple (supply dependent) current reference generator.

inaccuracies of the voltage and the load resistance are statistically independent we can write

$$\left(\frac{\delta I_{Ref}}{I_{Ref}}\right)^2 = \left[\frac{\delta(V_{DD} - V_{GS1})}{V_{DD} - V_{GS1}}\right]^2 + \left(\frac{\delta R_L}{R_L}\right)^2 \tag{4.32}$$

The V_{DD} voltage is typically supplied to the circuit by a standard voltage regulator whose accuracy is around $\pm 10\%$. Even the accuracy of V_{GS} is not particularly good ($\pm$ *20%*). Thus, if the nominal value of V_{DD} is *3.3 V* and that of V_{Th} is *0.6 V*, the resulting error of the first term in (4.32) is $\pm$ *11.5%*. Moreover, the accuracy of integrated resistors can be $\pm$ *30%* (not considering temperature dependence). It follows that the global inaccuracy is as high as $\pm$ *32%*. The result shows that, in practice, inaccuracies are dominated by the error in the resistance value. Since the resistance value doesn't change in time (or it changes very slowly just because of temperature drifts) the produced error is static. In contrast, all the dynamic variations affecting the supply lines produce spur in the generated current, holding great risks for the dynamic performance of the circuit. To verify that, let us assume that two spur voltages v_{n+} and v_{n-} alter the supply lines' voltage (in the circuit with *n-channel* devices). The effect is studied using the small signal analysis shown in Fig. 4.19. C_1 and C_2 describe the parasitic capacitance in parallel to M_1 and R_L. Moreover, C_3 represents the parasitic coupling between the output node and ground. However, since the coupling expressed by C_3 is normally weak, for the sake of simplicity we will neglect it.

We are interested in the spur voltage across the gate-to-source of the diode connected transistor M_1 being that, $v_{n,out}$, what affects the current in the mirroring element. Circuits analysis (with $C_3 = 0$) leads to

$$v_{n,out} = \frac{[v_{n+} + v_{n-}](1 + sR_L C_2)}{(1 + g_{m1}R_L)\left(1 + s\frac{R_L}{1 + g_{m1}R_L}(C_1 + C_2)\right)}. \tag{4.33}$$

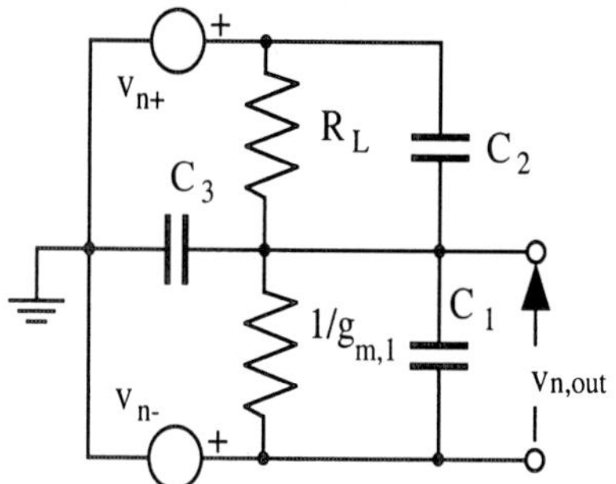

Fig. 4.19 - Small signal circuit used to calculate the noise due to spur on supply lines.

That produces an output noise current

$$i_{n,\,out} = v_{n,\,out}g_{m2} \tag{4.34}$$

Observe that at low frequency the spur voltage signals are given by the spur affecting the supply lines attenuated by the factor $(1+g_{m1}R_L)$. Remembering that, in saturation, $g_m = 2I_D/V_{ov}$, the attenuation factor becomes

$$1 + g_{m1}R_L \cong g_{m1}R_L = \frac{2I_D}{V_{ov}} \cdot \frac{V_R}{I_D} = \frac{2V_R}{V_{ov}} \tag{4.35}$$

That is twice the ratio between the drop voltage on the resistance and the overdrive voltage.

At high frequency the spur signals are transferred to the output node by parasitic coupling and, well above the pole shown in equation (4.33), the attenuating factor becomes

$$\frac{C_2}{C_1 + C_2} \tag{4.36}$$

4.2.2 Self Biased Current Reference

A current reference capable of generating current almost independently of the supply voltage is shown in Fig. 4.20. Transistors M_3 and M_4 form a current mirror developing ratioed currents in the two branches of the circuit. If $(W/L)_3$ and $(W/L)_4$ are identical the two currents match ($I_{Ref} = I_1$). They are sank to ground by transistor M_1 on one side and by resistance R on the other. Thus, a voltage equal to V_{GS1} drops across R. We can write

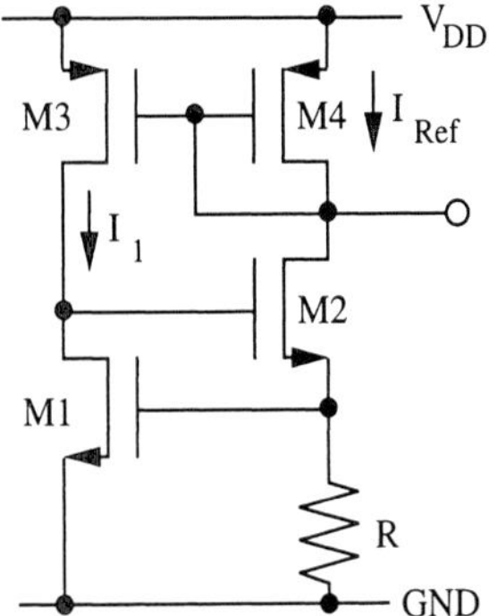

Fig. 4.20 - Self biased current reference.

$$V_{GS1} = RI_1 \tag{4.37}$$

$$V_{GS1} = V_{Th} + \sqrt{\frac{2I_1L_1}{\mu C_{ox}W_1}} \quad for\ V_{GS1} > V_{Th}$$

$$I_1 \cong 0 \qquad for\ V_{GS1} < V_{Th} \tag{4.38}$$

Equations (4.37) and (4.38) form a system of two equations which can be solved to obtain V_{GS1} and I_1. The solution of this non linear system is accomplished graphically in Fig. 4.21. Two results, indicated by the letters *A* and *B* in the figure, emerge. One (*B*) is trivial and corresponds to zero current and zero voltage across the resistance. Depending on its past history, the circuit can stand into one or the other solutions. To ensure that the current reference functions properly, a start-up circuit is required.

Since V_{GS} is around *1 V*, a bias current of *100 μA* is obtained with a resistance in the range of *10 kΩ* which is an achievable value for integrated technologies.

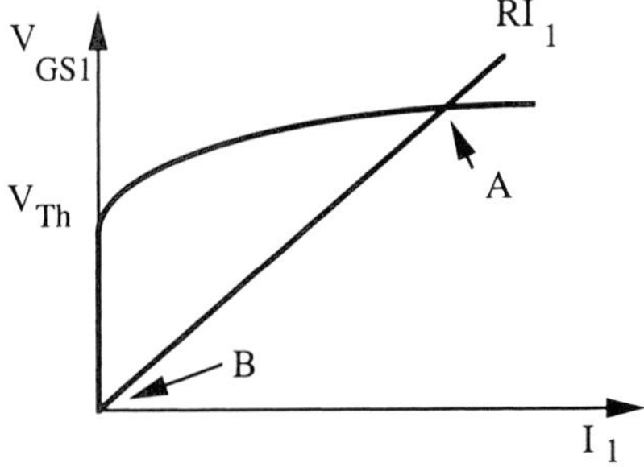

Fig. 4.21 - Graphic solution of the system of equations describing the circuit in Fig. 4.20.

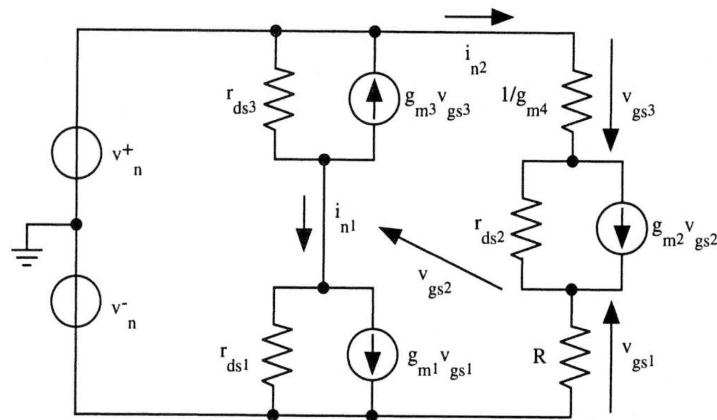

Fig. 4.22 - Small-signal equivalent circuit used to estimate the spur rejection in the self biased reference current.

The value of the generated current weakly depends on the supply voltage because there is a high impedance element per branch capable of absorbing possible supply variations. These two elements are the transistors M_2 and M_3. The drain to source voltage of M_1 and M_4 can not change freely: the former is two V_{GS} above ground the latter is one V_{GS} below V_{DD}. Any supply fluctuation is then "absorbed" by the high resistance that we have between drain and source of M_2 and M_3.

The circuit not only rejects the supply voltage variation, but also it suitably attenuates the small signal spur affecting the supply lines. We study this feature with the equivalent circuit in Fig. 4.22. By inspection of the circuit we can write

$$v_n^+ - v_n^- = [i_{n2} - g_{m3} i_{n1}/g_{m4}] r_{ds3} + [i_{n2} - g_{m1} i_{n1} R] r_{ds1} \tag{4.39}$$

$$v_n^+ - v_n^- = [1/g_{m4} + R] i_{n1} + [i_{n2} - g_{m2} v_{gs2}] r_{ds2} \tag{4.40}$$

$$v_{gs2} = [i_{n2} - g_{m1} i_{n1} R] r_{ds1} - R i_{n1} \tag{4.41}$$

which solution leads to the noise current, i_{n1}, produced by the spur $(v_n^+ - v_n^-)$. The analysis is just mathematical calculations and can be done by the reader as an exercise. The resulting expression is a bit complicated and more important it is not useful to extract design guidelines. However, it is possible to simplify the result achieved using the following approximations.

$$g_{m1} R = \frac{2 I_D R}{V_{GS1} - V_{Th1}} = \frac{2 V_{GS1}}{V_{GS1} - V_{Th1}} \gg 1 \tag{4.42}$$

$$g_m r_{ds} \gg 1 \tag{4.43}$$

observe that the first relationship holds when, as it normally happens, the overdrive voltage is smaller than the threshold. Using the above mentioned approximations we can obtain

$$i_{n1} \cong \frac{v_n^+ - v_n^-}{R g_{m1} r_{ds3}} \tag{4.44}$$

An equivalent result can be obtained from inspecting the equivalent circuit in Fig. 4.22: the voltage v_{gs1} is larger than $-v_{gs3}$; therefore the current generator $g_{m1}v_{gs1}$ will dominate with respect to $g_{m3}v_{gs3}$. Also, the current in r_{ds1} is small compared to $g_{m1}v_{gs1}$. Therefore, the drop voltage across r_{ds3} equals, in practice, $v_n^+ - v_n^-$. According to the above discussion we have

$$v_n^+ - v_n^- = i_{n1} \cdot R \cdot g_{m1} r_{ds3} \tag{4.45}$$

The approximated result given above leads to the following observations: R almost depends on the value of the desired current; g_{m1}, being given by twice the current divided by the overdrive, can be controlled within a limited range. Therefore, the designer mainly controls the spur rejection by keeping high the resitance r_{ds3}.

The achieved results hold for low frequency. As already discussed for the simple current reference, at high frequency, capacitors (designed and parasitic) control the circuit small signal behaviour.

Example 4.6

Determine the value of R in the circuit in Fig. 4.20 leading to a reference current equal to I_{ref} = 2 mA (a pretty large value). Assume $(W/L)_1 = 50$, $V_{th} = 0.6$ V and an n-channel process transconductance parameter equal to 120 μA/V².

Solution:
The use of equation (4.38) leads to

$$V_{GS1} = 0.6 + \sqrt{\frac{4 \cdot 10^{-3}}{120 \cdot 10^{-6} \cdot 50}} = 1.41\text{V}$$

that, using (4.37) results into R=708 Ω.
Observe that the achieved value of V_{GS1} is quite large compared to the threshold voltage. The reason is due to the fact that, despite the relatively large aspect ratio, M_1 needs quite a high overdrive to

carry the required current, 2 mA.
Remember that the overdrive is inversely proportional to the inverse of the aspect ratio. Increasing it to 200 would reduce V_{GS1} *to 1 V, requiring a resistance R equal to 500 Ω.*

4.2.3 Self Biased Micro-Current Generator

When a current in the micro-ampere range is required, the self-biased reference described in the previous sub-section must employ quite high resistances (around *MΩ*). This, in integrated implementation, corresponds to very large consumption of silicon area. This drawback can be overcome by using this very basic principle: the value of a resistance carrying a given current is reduced if the voltage across it is diminished. Thus, to save silicon area, a reduced voltage must be applied to the resistance defining the reference current. The circuit shown in Fig. 4.23 accomplishes this function. In fact, the voltage across the resistance *R* is not a V_{GS} but the difference between two V_{GS} (or the difference between two overdrive): the one going with M_1 minus that of M_2. Moreover, if we assume that the two transistors M_3 and M_4 are matched, they will carry the same current. If all transistors are in the saturation region we can write

$$I_1 = I_2 \tag{4.46}$$

$$\sqrt{\frac{2I_1L_1}{\mu C_{ox}W_1}} = \sqrt{\frac{2I_2L_2}{\mu C_{ox}W_2}} + RI_2 \tag{4.47}$$

Which is a system of two equations (one non linear) in the two variables, I_1 and I_2. Fig. 4.24 shows its possible graphic solution. It gives two results, one

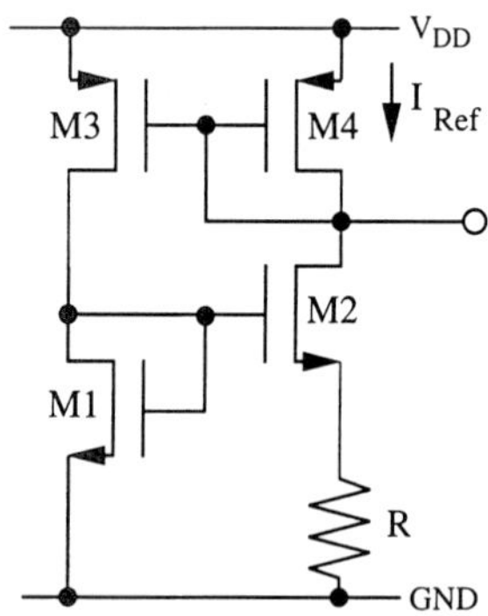

Fig. 4.23 - Self biased low current reference generator.

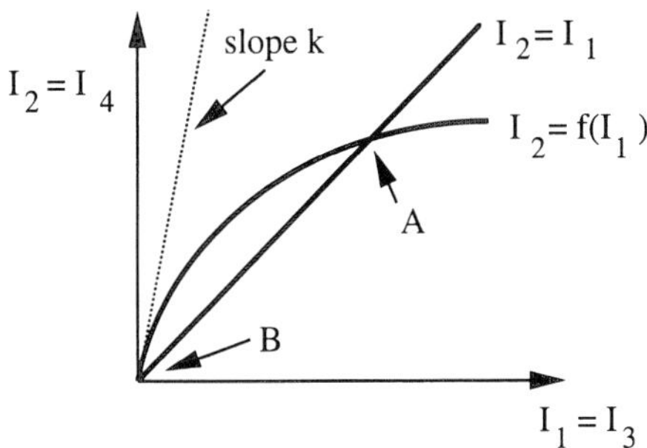

Fig. 4.24 - Graphic solution of the equations describing the circuit in Fig. 4.15.

of which, *B*, is trivial. Therefore, even for this circuit, it is necessary to use a start-up network to be sure that the current reference is driven to the proper operating point and is not stacked at zero current.

Of course the overdrive of M_1 must be suitably larger than the overdrive of M_2 to leave room for the voltage across *R*. This leads to the (obvious) condition $(W/L)_1 < (W/L)_2$. For modern CMOS technology the quantity $\mu_n C_{ox}$ is approximately $120\ \mu A/V^2$ (for *n*-channel devices); if the current to be generated is, for instance, $4\ \mu A$ and $(W/L)_1 = 3$, $(W/L)_2 = 30$, the overdrive voltage of M_1 becomes $149\ mV$ and that of M_2 becomes $47\ mV$. Therefore, the drop voltage across the resistance *R* is only $102\ mV$. The value of the resistor is consequently $25.5\ k\Omega$, approximately *6* to *10* times smaller (depending on the threshold value) than that required for the self-biased current reference studied in the subsection 4.2.2. Note that in the circuit considered here the supply independence is ensured by the high impedance elements M_2 and M_3.

The small signal equivalent circuit is similar to the one in Fig. 4.22. The only difference is that because of the diode connection M_1 is represented by a $1/g_{m1}$ resistance. (Fig. 4.25)

The current mirror $M_3 - M_4$ makes the signal currents in $1/g_{m1}$ and *R* equal ($i_{n2} = i_{n1}$). Therefore

$$v_{gs2} = \left[\frac{1}{g_{m1}} - R\right] i_{n1} \tag{4.48}$$

from which

$$[v_n^+ - v_n^-] = i_{n1}\left[\frac{1}{g_{m4}} + R + r_{ds2}\left(1 - \frac{g_{m2}}{g_{m1}}\right) + r_{ds2} g_{m2} R\right] \tag{4.49}$$

therefore

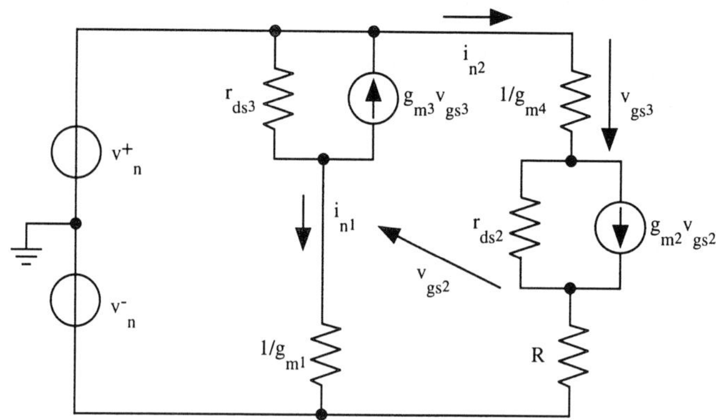

Fig. 4.25 - Small-signal equivalent circuit used to estimate the spur rejection in the self biased micro-current generator.

$$i_{n1} \cong \frac{[v_n^+ - v_n^-]}{r_{ds2} g_{m2} R} \tag{4.50}$$

The product $g_{m2}R$ (see equation (4.42)) equals the ratio between the drop voltage across R and the overdrive of M_2. Its value is not significantly lower than the one that we have in the normal self-biased current generator. Therefore, the micro-current version rejects spur with a similar extent as the scheme in Fig. 4.20 does.

Example 4.7

Determine, using Spice, the dependence on the supply voltage of the current in the self-biased micro generator shown in Fig. 4.23. Find the minimum operation value of V_{DD}. Assume the following design parameters: $(W/L)_1 = 3\mu/1\mu$; $(W/L)_1 = 30\mu/1\mu$; $(W/L)_3 = (W/L)_4 = 20\mu/2\mu$; $R = 25k\Omega$; $V_{DD} = 3\text{-}5$ V. Analyse the circuit when the p-channel transistors are changed into $(W/L)_3 = (W/L)_4 = 20\mu/8\mu$. Moreover, repeat the simulations with $R = 50\ k\Omega$. Use the model parameters given in Appendix B.

Solution:

The aspect ratios are the same used in the above section to estimate the drop voltage across the resistance R. If the process transconductance parameter of the technology used is the same used in the above section(120 $\mu A/V^2$) the voltage across R is 102 mV. Therefore, being $R = 50k\Omega$ the expected current ranges around 4 μA.

The Spice input file is the following:

```
M1 2 2 Gnd Gnd MODN L=1u W=3u
M2 1 2 3 Gnd   MODN L=1u W=30u
M3 2 1 Vdd Vdd MODP L=2u W=20u
M4 1 1 Vdd Vdd MODP L=2u W=20u
R1 4 Gnd 25K
vdd Vdd Gnd 4.0
vx 3 4 0
Rs 1 Gnd 1000000K
.dc vdd 1 5 0.1
.print dc i(vx)
```

observe that, in addition to the expected elements, the Spice net-list includes the resistance R_s (100 MΩ) connected between the node 1 and ground. Such a very high resistance forces an extremely low current (a few tens of pA) into the transistor M_4 thus avoiding the possible meta-stable condition $I_3=I_4=0$.
The results of the simulation are summarized in the figure.
The generated current is slightly greater than expected (it ranges around 5 μA). This difference comes from the approximations made in the hand calculations. Moreover the current dependence on the supply voltage is not so negligible. The reason is that the output resistance of the transistor used is far to be ideal. Therefore, increasing the supply voltage produces an increase in the current proportional to the λ factors. If the length of the p-channel transistors increases, the corresponding output resistance increases as well and the generated current is kept more independent of V_{DD}. From the simulation results we can estimate an equivalent output trans-resistance

$$r_{eq,\,out}(L = 2\mu) = \frac{\Delta V_{DD}}{\Delta I} = 3.5\text{M}\Omega \qquad r_{eq,\,out}(L = 8\mu) = 7.5\text{M}\Omega$$

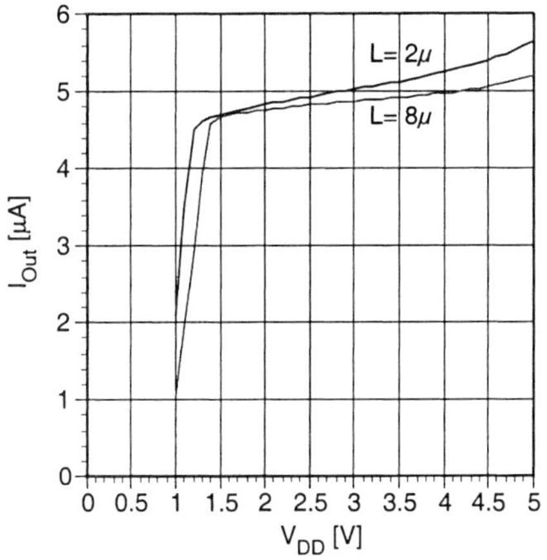

The figure also shows the lower value of V_{DD} for proper operation of the circuit: it is necessary to have a supply voltage between 1 V and 1.5 V to start-up an output current. This voltage provides the V_{GS} and the saturation voltage of the transistors in each branch of the circuit.

The version that employs p-channel transistors with length 8μ requires a slightly higher V_{DD}: increasing the length also increases the output resistance but the ensuing reduction in the aspect ratio imposes a higher overdrive voltage.

Equation (4.47), remembering that $I_1 = I_2$, leads to the relationship $R = \alpha/\sqrt{I_1}$, where α is a suitable coefficient. Therefore, increasing the resistance by a factor 2 should decrease the current by a factor of 4. Simulation leads to a different result: the current drops only by a factor of 2.5. The reason, for this specific design, is twofold: equation (4.47) is approximated and, moreover, it holds for transistors in the saturation region. The use of condition (1.61) estimates for M_2 a transition point between saturation and weak inversion at 3.3 μA. Therefore, M_2 is likely in the weak inversion region and its V_{GS} becomes lower than the threshold voltage. Therefore, the voltage across R is higher than expected.

4.2.4 Start-up Circuits

Self-biased current generators and, more in general, all the circuits with two possible *dc* operating points need a start-up circuit. Fig. 4.26 show two circuit solutions. Basically a start-up circuits checks a possible zero current condition and by injecting current in a suitable point forces the circuit to move from zero state and brings it to the correct point of operation.

The circuit in Fig. 4.26 a) operates a s follows. Transistor M_{S2} works as a current source. To avoid an extra bias the gate of M_{S2} is connected to V_{DD}. Transistor M_{S3} mirrors the current of M_4 and together with M_{S2} monitors whether the current in M_4 is lower than the level established by M_{S2} itself. In such a case node *A* drops down and M_{S1} switches on. Consequently, some current flows from M_4 starting-up the current generator. When the current in M_{S3} becomes higher than the one in M_{S2}, node *A* rises near V_{DD}; M_{S1} switches off and the start-up circuit stops and doesn't drain extra current from M_4 any more.

The saturation current of the transistor M_{S2} establishes the threshold of the start-up circuit. This current must be lower than the reference current multiplied by the M_4-M_{S3} mirror factor; if not the start-up never switches off.

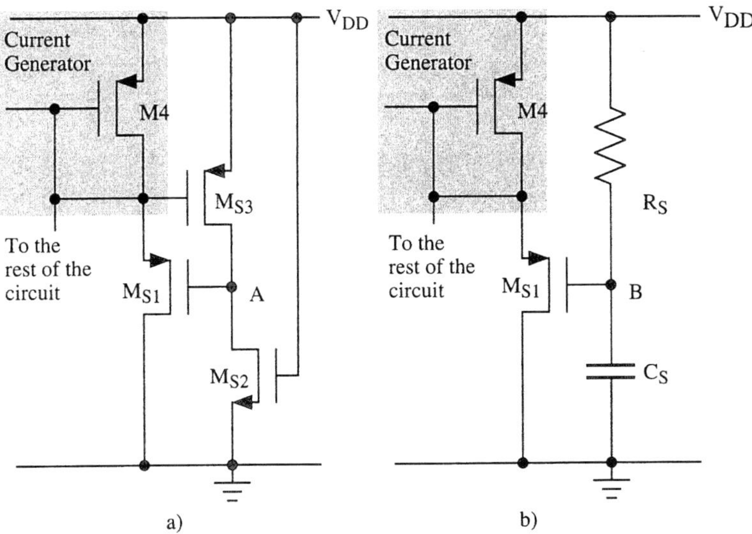

Fig. 4.26 - a) Static start-up for the current reference in Fig 4.20 or Fig 4.23 b) Dynamic start-up.

Therefore, it is necessary to check that technological changes don't bring the current *of* M_{S2} above that limit.

> **DON'T FORGET!**
>
> All the networks with two stable operative points need a start-up (even if one of the two point is meta-stable).
>
> Don't rely on computer simulation results even if they always provide the expected operative point.

A drawback of the start-up circuit in Fig. 4.26 a) is that it consumes power. The current through M_{S2} keeps node A close to V_{DD} and, therefore, prevents the switching on of the start-up circuit during the normal operation. When the consumed power is a problem the designer prefers to use a dynamic start-up circuit. Fig. 4.20 b) shows a possible circuit solution). At the power-on of the circuit the voltage on the V_{DD} line typically changes quickly from zero to the applied voltage. The discharged capacitor C_S keeps the node B close to ground for a while. Therefore, until C_S charges near V_{DD} transistor M_{S1} is "on" thus favouring the start-up process.

The dynamic start-up circuit doesn't monitor the reference current but simply produces a glitch current in M_4 that, hopefully, will bring the circuit in the desired operative point. Therefore, the circuit works properly only if the off-to-on time of the supply voltage is pretty fast, compared to the time constant R_SC_S. In addition to this, the parasitic capacitance from node B and V_{DD} should be small compared to the capacitor C_S. If not the coupling of node B with the V_{DD} line attenuates the glitch at the gate to source of M_{S1} eventually preventing it to exceed the threshold.

4.2.5 Use of Parasitic BJT for Current Reference

CMOS technology allows us to realize parasitic bipolar transistors. They can be fabricated with the substrate as collector, the well diffusion as base and source/drain diffusion as emitter. The collector, being the substrate, is always connected to the substrate biasing. Therefore, for *p-well* technology it is possible to design *npn* transistors while *n-well* technology sustains the realization of *pnp* transistors. We know that in a bipolar transistor (or more generally in a *p-n* junction) the voltage across the base-to-emitter is almost constant with a value around *0.7 V*. Moreover, the current exponentially depend on base-to-emitter voltage through the parameter $V_T = KT/q$. Therefore, we can achieve a reference current by using V_{BE} or V_T.

4.2.6 V_{BE} Based Current Reference

Here we shall study how to use the base-to-emitter voltage to generate a supply independent reference current. Fig. 4.27 shows a possible circuit schematic. It works as follows: the feedback loop established by the differential gain stage ensures that the voltages of nodes *A* and *B* are equal. The result, if the gain of the amplifier is very high, is independent of the supply voltage. Assuming $V_A = V_B$,we can write

$$I_{ref} = \frac{V_{BE}}{R} \tag{4.51}$$

Because of the high output resistance of M_1 and the exponential relationship between voltage and current, we can argue that V_{BE} weakly depends on the supply voltage, its value being about *0.7 V*. Therefore, the *dc* value of I_{ref} mostly depends on the resistor used. Moreover, the circuit rejects small signals spur affecting V_{DD} as well. In fact, by inspection of the circuit we can note that the small signal voltage at node *A* comes from the small signal spur affecting the supply line, divided by the resistive divider r_{d1} - $1/g_{m,Q1}$. Being the output resistance of the MOS transistor, r_{d1}, much higher than $1/g_{m,Q1}$ the circuit achieves a significant attenuation.

In a real circuit, we have to account for the finite gain effect and possible offset. Thus, the voltage of nodes *A* and *B* can be slightly different. A typical offset is a few *mV* while the error produced by the finite gain is given by the

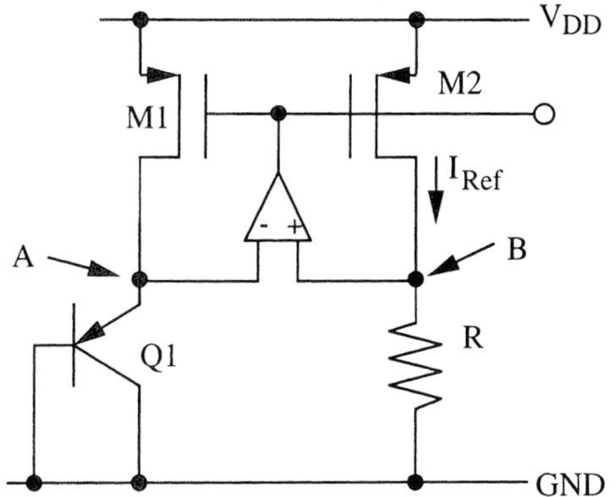

Fig. 4.27 - V_{BE}-based current reference

output voltage divided by the finite gain. Since the voltage across the diode is several hundred of *mV*, the possible error is not so large.

Observe that the circuit has two feedback paths around the gain stage. One positive, the other negative. Assuming transistors M_1 and M_2 matched, equal currents are injected into R and into the diode connected transistor Q_1. The voltage across a diode changes as the logarithm of the current; therefore, the feedback on the resistor side is typically stronger than the one on the bipolar transistor side. Transistor M_2 and resistance R form an inverting amplifier. Therefore, in order to ensure stability, node B must be connected to the positive terminal of the differential gain stage.

CALL UP

The design of good current references must ensure low sensitivity to the supply voltage even at high frequency. For this we have to ensure a minimum capacitive coupling with a careful cell layout.

The possible estimation of the temperature coefficient must include the temperature dependence of all the component used (including the resistors).

A resistor in the range of $5\text{-}10\ k\Omega$ leads to a reference current in the range of hundreds of μA, V_{BE} being around $0.7\ V$. It results that very small currents are obtained with difficulty. Moreover, the temperature coefficient of V_{BE} is negative (around - $0.3\%/°C$ at room temperature). This feature, combined with the positive temperature coefficient of typical integrated resistors, leads to a significant negative temperature coefficient in the result. Having a large temperature coefficient is not so desirable. However, in some applications a "temperature sensor" can be a positive feature.

4.2.7 V_T - Bases Current Reference

We achieve a V_T based current reference by applying a voltage proportional to V_T across a given resistance. The circuits shown in Fig. 4.28 accomplishes that function. Circuit in Fig. 4.28 a) is for *n-well* (p-substrate) technology; circuit in Fig. 4.28 b) is for a *p-well* realization. Transistors Q_1 and Q_2, whose emitter area has a ratio *1 to n*, are diode-connected. Their current-voltage characteristics are, as is known, exponential. Therefore, we can write

$$V_{BE1} = V_T ln \frac{I_1}{AI_{SS}} \tag{4.52}$$

$$V_{BE2} = V_T ln \frac{I_1}{nAI_{SS}} \tag{4.53}$$

where A is the junction area of the emitter of Q_1, and $V_T = kT/q$.

The transistors M_1 and M_2 are matched, $(W/L)_1 = (W/L)_2$. Therefore, the two branches carry the same current. Even transistors M_3 and M_4 are designed matched, thus the same voltage at nodes A and B is ensured (same current and sizes result in equal V_{GS}). From the above, the voltage across the resistance becomes

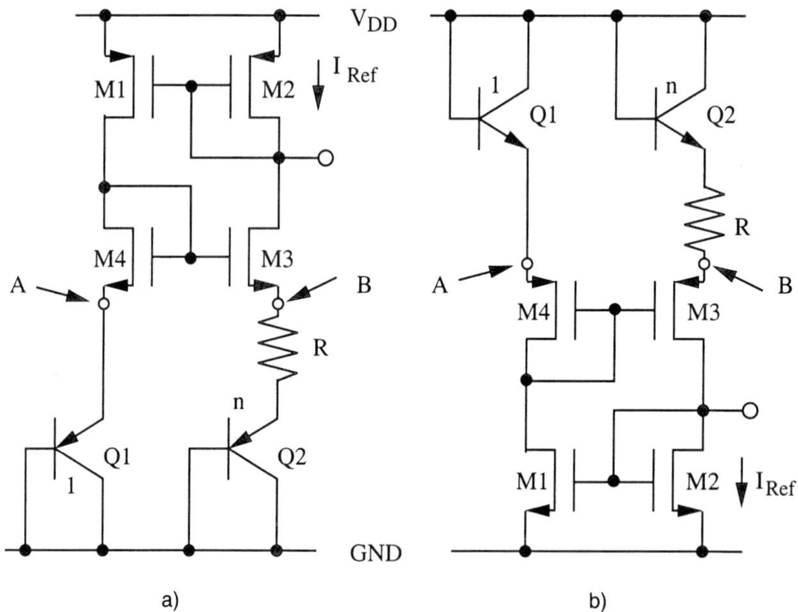

Fig. 4.28 - V_T - based current generator, using vertical parasitic bipolar transistors.

$$RI_1 = V_{BE1} - V_{BE2} = V_T ln \frac{I_1}{AI_{SS}} \frac{nAI_{SS}}{I_1} \tag{4.54}$$

which results in

$$I_1 = \frac{V_T}{R} \ln(n) \tag{4.55}$$

Note that even this circuit also admits work in the state with zero current. For this a start-up circuit is required.

Since, at room temperature, V_T is approximately *26 mV* and a suitable value for n is *8*, the drop voltage across the resistance is just *[26 ln(8)]mV= 56mV.* Therefore, the circuit is convenient for low current generations.

Above we observed that *8* is a suitable value for n. The reason for this is that, in practical realizations, transistor Q_2 is made from the parallel connection of unity elements. These, together with Q_1, are arranged in a two dimensional matrix. Using a *3 x 3* matrix, one element realizes Q_1 and the remaining *8* elements are used to realize Q_2. Fig. 4.29 shows a possible layout of the bipolar transistor matrix. Since the collector currents flow through the substrate, it is important to draw off them just after injection. This is achieved with the substrate bias ring placed around the transistor matrix. The bases of Q_1 and Q_2 are in common; therefore, the layout achieves them using a single common well. Multiple diffusions n^+*-type* (or p^+*-type*) allows us to contact the well all around the structure. The diffusion employed to realize the emitters has an octagonal shape to improve the matching between the emitter areas. Moreover the emitter of Q_1 is surrounded by the *8* emitters of Q_2, thus accomplishing a common centroid arrangement.

Finally, it is worthwhile remembering that V_T, being proportional to the absolute temperature, has a positive temperature coefficient. However, integrated resistors also have positive temperature coefficients, thus providing some compensation of the temperature dependence of the generated current.

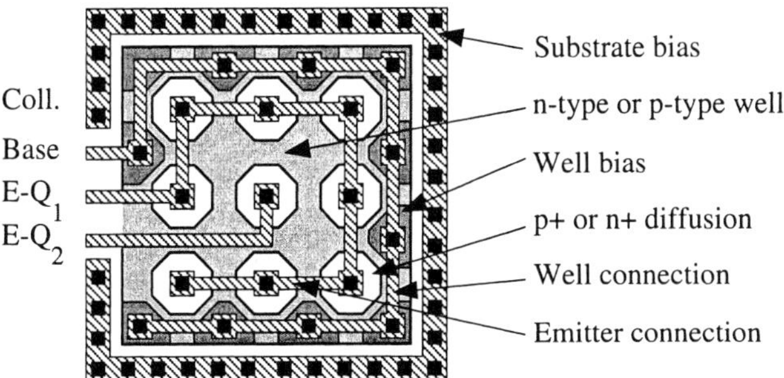

Fig. 4.29 - Typical layout of transistors Q_1 and Q_2 of the V_T based reference generator.

The circuit discussed aims at the generation of a reference current independently of the supply voltage variations. In reality the produced current changes with the rail voltage mainly because of two non-idealities: the output resistance of transistor M_1 that affects the M_1-M_2 mirroring factor, and the fact that the voltage at node B is not an exact replica of the one at node A because of possible mismatches between transistors M_3 and M_4. These two limits modify the drop voltage across R and, in turn, the value of the generated current. They account for the low frequency limitations. At high frequencies the capacitive coupling from V_{DD} and from the ground line once again conveys possible spur to the gate-to-source voltage of M_2, thus affecting the generated current. The same kind of limit affects the other current sources studied so far. A possible remedy is to use a proper layout. It is recommended to have a proper physical separation of the elements so that parasitic couplings are minimized. Moreover a capacitor across the gate and the source terminal of M_2 helps filtering possible spur that reach the drain of M_2.

Example 4.8

Design a 50 μA V_T-based current source. Use the circuit schematic shown in Fig. 4.28 a). Assume $V_{DD} = 5V \pm 10\%$, however, the circuit should be able to operate starting from $V_{DD} = 3V$. Use an emitter area ratio equal to 8. Estimate, using Spice, the sensitivity of the generated current on the supply voltage. Use the Spice models given in Appendix B and, for the bipolar transistors, use the default values excluding the parameter IS. Use instead for is 4 e-15.

Solution:
The emitter area ratio of transistors Q_2 and Q_1 is 8. Using it we obtain $\Delta V_{BE} = V_T ln(8) = 56$ mV. Therefore, the resistance R is

$$R = \frac{\Delta V_{BE}}{I_{nom}} = \frac{56 \cdot 10^{-3}}{50 \cdot 10^{-6}} = 1.08 \mathrm{k}\Omega$$

The circuit is required to operate with $V_{DD} = 3V$. By inspection of the circuit we write

$$V_{DD} = V_{BE} + V_{ds,1} + V_{ds,1} = V_{BE1} + V_{Th,n} + (V_{ov,4} + V_{ds,1})$$

$$V_{DD} = V_{BE} + V_{ds,3} + V_{ds,2} = V_{BE1} + V_{Th,p} + (V_{ov,2} + V_{ds,3}).$$

These equations determine the budget voltage available for the overdrive of the MOS transistors. Since the technology used shows a higher threshold voltage for the p-channel transistor, we will

concentrate on the second of the above conditions.
We have to keep some margin to ensure that M_3 is distant enough from the triode region (say 0.6 V). The p-channel threshold is approximately 0.8 V and V_{BE} is 0.7 V; therefore, the voltage budget for the n-channel and the p-channel overdrive is 0.9 V.
To keep the voltages of nodes 1 and 2 equal, the transconductance of the n-channel transistors must be high. Remembering that

$$g_m = \frac{2I_D}{V_{ov}}$$

we use an overdrive for the n-channel smaller than the one for the p-channel: $V_{ov,n} = 0.3$ V, $V_{ov,p} = 0.6$ V. Thus, presuming the transconductance process parameters to be equal to those given in (1.49), the following results

$$\left(\frac{W}{L}\right)_n = \frac{2I_D}{\mu_n C_{ox} V^2_{ov,n}} = \frac{100 \cdot 10^{-6}}{120 \cdot 10^{-6} \cdot 0.3^2} = 9.2$$

$$\left(\frac{W}{L}\right)_p = \frac{2I_D}{\mu_p C_{ox} V^2_{ov,p}} = \frac{100 \cdot 10^{-6}}{39 \cdot 10^{-6} \cdot 0.8^2} = 7.1$$

Note that the almost balanced aspect ratios comes from different overdrive used. Computer simulations help in defining the length. Let us use the following Spice list

```
.........
.dc Vdd 1 5 0.1
.print dc i(vx)
*.ac dec 2 1 10k
*.print ac im(vx)
M1 1 2 Vdd Vdd MODP L=4u W=28u
M2 6 2 Vdd Vdd MODP L=4u W=28u
M3 2 1 3   Gnd MODN L=2u W=20u
M4 1 1 4   Gnd MODN L=2u W=20u
R1 3 5 1.08K
Q1 Gnd Gnd 4 QMOD 1
Q2 Gnd Gnd 5 QMOD 8
Vdd Vdd Gnd 4.5 ac 1
vx 6 2 0
.model QMOD PNP IS=4e-15
```

that permits us to perform a .DC analysis and the small signal analysis around V_{DD} = 4.5 V. The DC current is 52 μA. The small signal current is 0.85 μA. This high value (compared to generated current should not be disconcerting: the result above comes from a 1 V small signal superposed to the supply voltage. A typical spur is

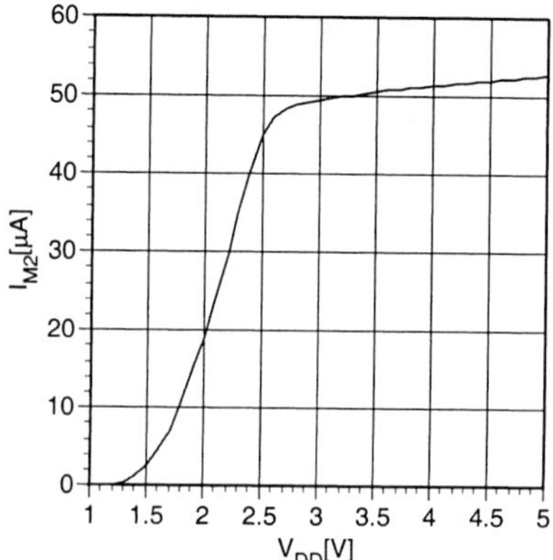

not more than few ten of mV. Therefore the actual spur component is more than hundred times smaller than the achieved current value.

The results of the .DC simulations are summarised in the figure. The circuit starts operating properly with V_{DD} higher than 2.6 V (remember we used a 0.6V margin) and generates a reference current around the required 50 mA. The supply voltage sensitivity is denoted by the slope of the curve above 2.6 V. The result depends on the large signal equivalent resistance between the supply voltage and ground. In turn, it mainly depends on the length of p-channel and n-channel transistors. The reader can verify the effect of longer devices by remembering to keep the transistor aspect ratio unchanged.

4.3 VOLTAGE BIASING

Analogue circuits typically use two connecting pins only for the voltage supply: one is V_{DD} and the other is ground (or V_{SS}). However, most applications require additional intermediate bias voltages. Since they should require additional pins the designer prefers to generate them on-chip. For example, an analog circuit often needs the analog ground. It is the intermediate voltage between and ground V_{DD} used to set the virtual ground voltage in operational amplifiers or the voltage reference used to charge or discharge capacitors in switched capacitor systems or data converters.

Even in basic blocks studied so far we need bias voltages; we have seen in the previous chapter that bias voltages control some nodes like, for instance, the gate of cascoding transistors. Often the accuracy of biasing voltages used affects the circuit performances; moreover, a possible spur leads to a worsening of the signal-to-noise ration. Therefore, the accuracy required and the immunity from spur signals can be quite high thus demanding complex solutions on chip.

In other circumstances, the accuracy required is not so critical. The voltage bias in points of the circuit with reduced sensitivity both to the absolute value and spur effects can be generated in an economical way. In this case the designer uses simple voltage dividers, diode-connected transistors, and/or level shifters.

4.3.1 Voltage Divider

A voltage divider allows us to generate a fraction of the supply voltage. A discrete components' implementation normally employs resistors and in some cases, capacitors. By contrast, in integrated technology, the resistors are not much used: they are either silicon area or power consuming. Even capacitors can be problematic: they integrate possible leakage currents that causes a change in time of the generated voltage. This limit requires a periodic reset of accumulated charge. Therefore, capacitors are normally used in circuits where the bias voltage is not continuously used: the circuit needs periodic time-slots during which a reset is performed.

For continuous-time applications the designer normally uses voltage dividers based on transistors. Fig. 4.30 shows two possible implementations of a two-transistor voltage divider. The circuit in Fig. 4.30 a) uses only n-channel devices while the one in Fig. 4.30 b) employs an n-channel and p-channel element. As the transistors are connected in the diode arrangement, they are in saturation. Since the same current is carried by both transistors, we can write the following relationships

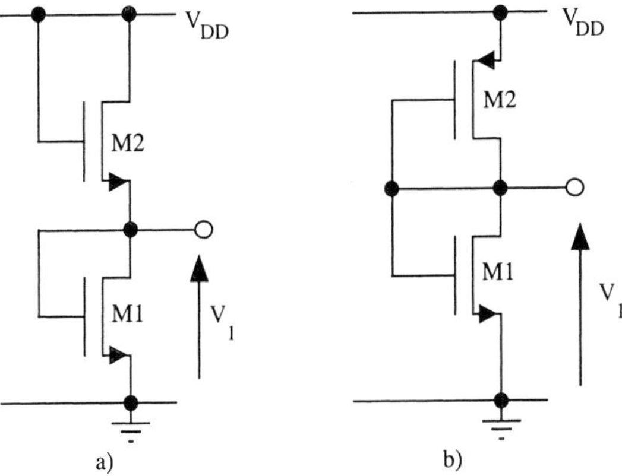

Fig. 4.30 - Voltage division of V_{DD} accomplished by transistors.

$$\frac{\mu_1 C_{ox}}{2}\left(\frac{W}{L}\right)_1 (V_{DS1} - V_{Th1})^2 = \frac{\mu_2 C_{ox}}{2}\left(\frac{W}{L}\right)_2 (V_{DS2} - V_{Th2})^2 \tag{4.56}$$

$$V_{DS1} + V_{DS2} = V_{DD} \tag{4.57}$$

from which we find

$$V_1 = V_{DS1} = \frac{\alpha_2}{\alpha_1 + \alpha_2} V_{DD} + \frac{\alpha_1 V_{Th1} - \alpha_2 V_{Th2}}{\alpha_1 + \alpha_2} \tag{4.58}$$

where

$$\alpha_1 = \sqrt{\mu_1 \left(\frac{W}{L}\right)_1}; \quad \alpha_2 = \sqrt{\mu_2 \left(\frac{W}{L}\right)_2} \tag{4.59}$$

The generated voltage V_1 is made up of a given fraction of V_{DD} plus a constant term (positive or negative). It depends on the geometrical sizes of the transistors and on their thresholds (which may be different because of the different types of transistor or because of the body effect).

The current in the circuit results from the relationship

$$V_{DD} = V_{Th1} + V_{Th2} + \sqrt{\frac{2I}{C_{ox}}}\left(\sqrt{\frac{L_1}{\mu_1 W_1}} + \sqrt{\frac{L_2}{\mu_2 W_2}}\right) \tag{4.60}$$

Thus the current increases quadratically with the bias voltage. This dependence appears to be a severe limitation. Actually, for mature technologies, in which the threshold voltage ranges close to *1 V*, a change of thresholds by *±20%* and of the bias voltage by *10%* around *5 V (4.5, 5.5 V)* can be expected. It determines a variation of the supply voltage minus the two threshold within the two limits *2.1 V, 3.9 V* (around the nominal value *3 V*). Accordingly, we have a change of the current in the divider by *-52%* up to *+69%*, which is a pretty large amount for the accuracies required in integrated technologies.

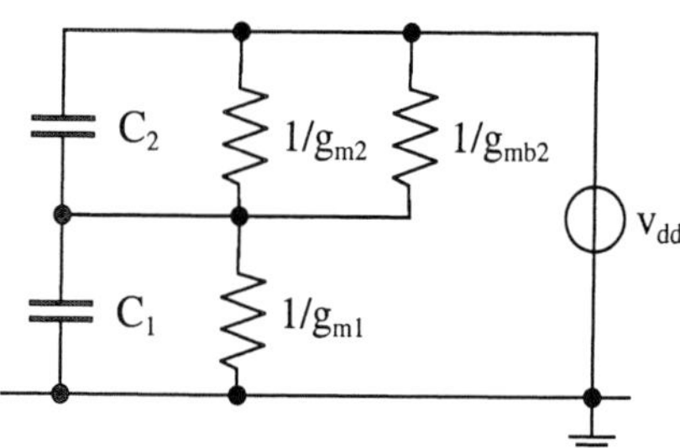

Fig. 4.31 - Small signal equivalent circuit of the voltage divider.

When we go down to *3.3 V* or even below the current spread becomes larger. It is kept to the same level only if the threshold and supply accuracies scale down as bias voltages.

We have seen that the transistor voltage divider makes available a fraction of the bias voltage. Unfortunately, the voltage divider also transfers to its output almost the same fraction of the noise and disturbances present on the supply lines. This effect can be quantified by analysing the small signal equivalent circuit shown in Fig. 4.31. It considers spur on the V_{DD} line. Readers can verify that spur signals affecting the ground line leads to the same results. For the case where the body of M_2 is not connected to its source we should account for the substrate transconductance.

KEEP NOTE

A voltage divider is used for not particularly demanding situations. Don't spend much time improving bias voltage features, but just be aware of existing limitations.

Capacitance C_1 and C_2 correspond to

$$C_1 = C_{gs1} + C_{db1} + C_{sb2} + C_L; \; C_2 = C_{gs2} \tag{4.61}$$

The transfer function of the circuit has a zero and a pole. At low frequencies the spur signal on the V_{DD} line is attenuated by

$$\frac{v_1}{v_{dd}} = \frac{g_{m2} + g_{mb2}}{g_{m2} + g_{m1} + g_{mb2}} \tag{4.62}$$

where the output conductances have been neglected. At very high frequencies the circuit becomes a capacitive divider with an attenuation of the spur present on the V_{DD} line that depends on the values of C_1 and C_2. Thus, we can have the pole before the zero or, vice-versa.

Example 4.9

Design a voltage divider like the circuit in Fig. 4.30 a). Use n-channel transistors having the same sizes. The current consumed should be lower than 30 μA. V_{DD} is 3.3 V. Estimate the current variation when the supply voltage changes by ± 10%. Find out the value of generated voltage and determine the output spur caused by a small signal affecting the V_{DD} line. Use the Spice models given in Appendix B.

Solution:

The current in the circuit depends on the overdrive voltages and

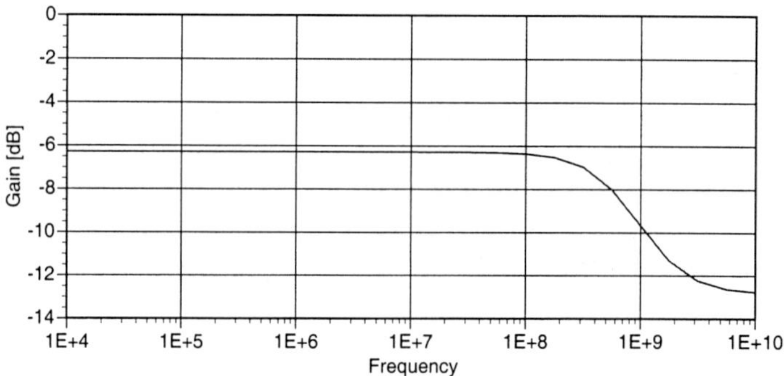

transistors' thresholds (which are different because of the body effect influencing M_2). Given the nominal threshold of transistors used (0.46 V) and the supply voltage, we can estimate the overdrive to be around 1 V. Therefore, we should use transistors with a pretty small aspect ratio. Few simulations leads to (W/L) = 1μm/ 2μm. Such a small aspect ratio leads the current 26 μA. Simulations shows also that if the supply voltage changes from 3.63 V to 2.97 V the current goes from 33.7 μA to 19.15 μA respectively. We can justify this relatively low variation (around ± 20%): simulation doesn't account for the threshold changes with technology.
With nominal V_{DD} the output voltage is 1.5 V. Moreover, the thresholds of M_1 and M_2 are 0.47 V and 0.83 V respectively. If we use equation (4.58) we obtain

$$V_1 = \frac{1}{2}V_{DD} + \frac{V_{Th1} - V_{Th2}}{2} = 1.65 - \frac{0.36}{2} = 1.47V$$

which prove the good accuracy of (4.58).
The small signal analysis leads to the Bode plot shown in the figure. The low frequency spur rejection is -6.3 dB. The use (4.62) and the parameters provided by Spice in the AC small-signal Model Table would determine -5 dB rejection. Thus, equation (4.62) is pretty precise, we achieve the same result within 1.5 dB range. The high frequency rejection is -13 dB. The use of capacitance provided by Spice leads C_1 and C_2 in Fig. 4.31 to be equal to 20 fF and 6.25 fF respectively. The resulting capacitive divider provides -11 dB of attenuation (not much different from the obtained -13 dB).
Note that a capacitance larger than few ten of fF between the output node and ground would improve the spur rejection significantly.

4.3.2 Diode-Connected Voltage Bias

In the previous section we studied a voltage divider. The result corresponds to a fraction of the supply voltage plus a possible fixed term. Often, instead of having a bias that essentially is an attenuated replica of the supply bias, it is desired to have supply voltage almost independent of the power supply. A very simple way to achieve such a feature is to use the voltage across a diode-connected transistor (Fig. 4.32). The voltage V_{out} is given by

$$V_{out} = V_{Th,n} + \sqrt{\frac{2I_x}{\mu_n C_{ox}\left(\frac{W}{L}\right)_1}} \tag{4.63}$$

Where the possible dependence on the supply voltage comes from the non-ideal behaviour of the current source. The circuit is very simple. However, it possesses the important feature that the generated voltage tracks the variations of threshold with technology drifts. This feature is important when biasing, for example, a cascode circuit. We have seen that in a cascode the biasing of the common-gate stage affects the output dynamic range. For this we should keep that bias as low as possible (for *n-channel* stages), but, at the same time, we must to avoid pushing the input transistor into saturation. The use of the diode-connected voltage bias solves the problem.

4.4 VOLTAGE REFERENCES

After considering the simple voltage references of section 4.3 this section shall study the design techniques used to generate precise voltage references. We assume that, in integrated CMOS technology, "accurate" means having

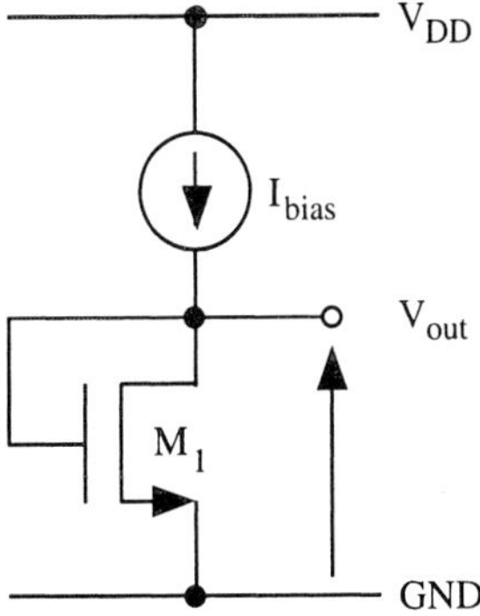

Fig. 4.32 - Diode connected voltage bias.

results around 5% of the desired value. Within this limit, it is possible to identify and use precise voltage terms with which the designer produces the required references. One important characteristics is the dependence of generated voltages on temperature: many applications require to have a temperature sensitivity as small as possible. We will see shortly how to use a proper combination of voltage terms to suitably addresses the issue. The accurate elements made available by the *CMOS* technology are:

- the base-to-emitter voltages V_{BE}, of a parasitic bipolar transistor
- the difference between the threshold voltages of an MOS transistor
- the thermal voltage, $V_T = kT/q$

all the above quantities exhibit, at a given temperature, an accuracy of more or less *5%*. For better precision on-chip trimming of elements is necessary. Trimming can be achieved by blowing up on-chip fuses or with digitally controlled arrays. Technologies with thin-film resistors permit laser trimming of the resistor value. All the above techniques are quite expensive and are used in special cases only.

4.4.1 V_{BE} Multiplier

The voltage across a directly biased *p-n* junction is, at medium-low current, close to *0.7 V*. This quantity is fairly accurate and can be used to generate adequately precise voltage references by suitable multiplication. Fig. 4.33 shows a possible circuit solution. With transistors M_2 and M_5 matched, the voltage at the output node is given by V_{BE} multiplied by the factor *k*. The circuit operates as follows: the bipolar transistor Q_1 is diode-connected. The combined action of transistors M_3 and M_4 repeats the value of V_{BE} across the resistance *R*, mak-

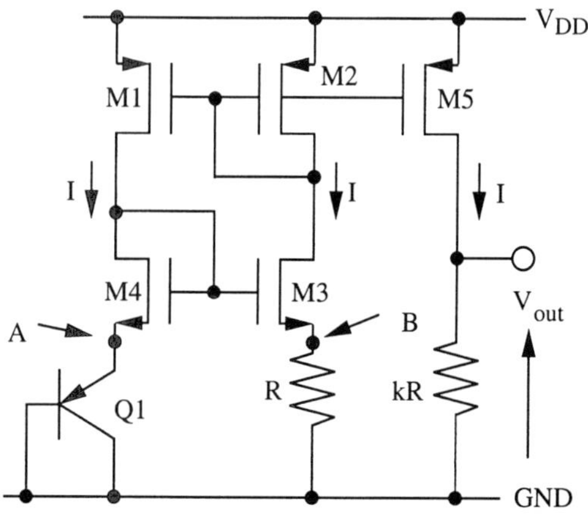

Fig. 4.33 - Voltage reference realized by multiplying V_{BE} multiplication.

ing the current on it equal to V_{BE}/R. This current is mirrored by transistors M_2-M_5 and is injected into the output resistance kR. The drop voltage across kR is therefore given by

$$V_{out} = kRI_5 = kR\frac{(W/L)_5}{(W/L)_2}I_2 = k\frac{L_2W_5}{W_2L_5}V_{BE} \tag{4.64}$$

As a further degree of freedom, we can use the mirror factor between transistors M_2 and M_5 different from one to control the V_{BE} amplification factor. Equation (4.64) shows that the V_{BE} multiplication depends on the ratio of geometric quantities and resistances. Therefore the accuracy achievable is fairly good because it depends on matching. Geometries match well and the technology provides excellent resistor matching providing that resistances are made with the same kind of material.

PROMPT

The voltage at the output of a V_{BE} multiplier has a negative temperature coefficient. At room temperature, it is equal to - 2.2 mV/°C

Again, if resistors are made with the same kind of material the multiplication factor k is almost independent of temperature. Thus, the temperature variation of the generated voltage tracks the temperature dependence of V_{BE}. Since V_{BE} has a negative temperature coefficient (*-2.2 mV/°C*), the voltage generated by a V_{BE} multiplier is affected by the same limitation. Assuming, for example, a multiplication factor equal to *2*, the generated reference is around *1.4 V*. A temperature increase by *60 °C* above the room temperature (which is quite common in integrated circuits) diminishes the reference voltage by *264 mV*, which is a significant fraction of the value achieved at room temperature.

4.4.2 V_T Multiplier

Indeed, the bias generator studied in the previous sub-section uses a V_{BE} based current to obtain a voltage. Similarly, it is possible to obtain equivalent results by using a V_T-based current generator. We already studied how to achieve a current reference based on V_T. A possible circuit configuration is given in Fig. 4.34. It is the scheme in Fig. 4.28 with an additional extra branch. Transistors Q_1 and Q_2, whose areas have ratio *1 to n*, are diode-connected. Therefore, the voltage across the resistance R is given by the V_{BE} difference expressed by equation (4.55) $I = V_T/R\ln(n)$. Assuming transistors M_2 and M_5 are matched, the same current flows in the resistance kR. Therefore, we have

$$V_{out} = kV_T ln(n) \tag{4.65}$$

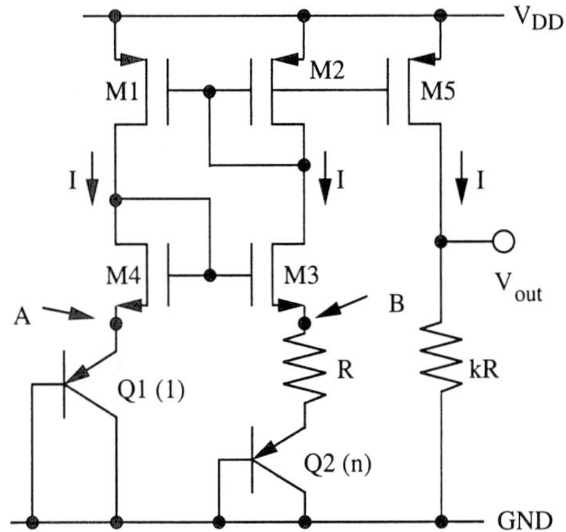

Fig. 4.34 - Voltage reference achieved with a V_T multiplication.

Since the multiplication factor is almost independent of temperature, the temperature variation of the generated voltage tracks the temperature dependence of V_T. Since V_T has a positive temperature coefficient the generated voltage will be have positive temperature coefficient as well.

The voltage generated by circuit in Fig. 4.34 as well as the one in Fig. 4.33 are sensitive to spur affecting the supply lines. We have a double limitation: one coming from the sensitivity of the current I in M_2 and the other due to the finite output resistance of M_5. Both terms go in the same direction and, therefore, they are summed up.

On the basis of previous study on current generators, by inspection of the circuit the following observation comes upon: at low frequency, spur is rejected if transistors M_1, M_3 and M_5 have an high output resistance. The reader can verify this assertion by computer simulations.

4.4.3 Voltage Reference Based on Threshold Difference

Some special technologies make available MOS transistors with different threshold voltages. The technological control of the threshold is gained by a suitable additional implant step in the technological process flow. Hence, at the expense of an additional mask it is possible to correct the threshold voltage of part of the transistors on the chip.The control over the threshold shift is fairly accurate (much better than the absolute value of the threshold). Therefore, threshold differences are good ingredients to generate reference voltages.

Fig. 4.35 shows a circuit where the transistor M_1 has a threshold voltage

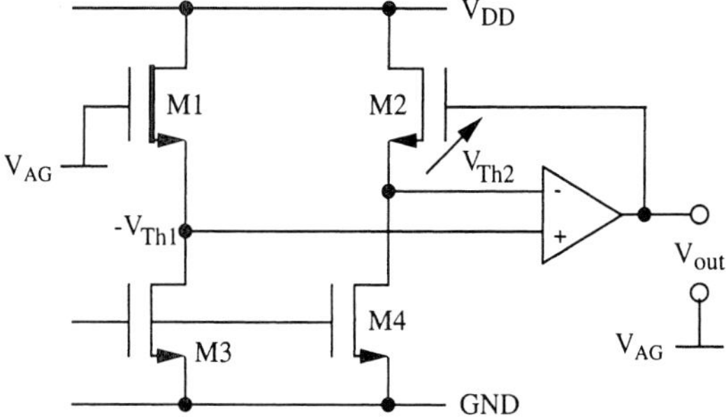

Fig. 4.35 - Voltage reference based on the difference between threshold voltages.

different from the one of other *n*-channel elements (the thicker line on the symbol denotes a special threshold voltage). The gate of M_1 is biased at V_{AG}, the analog ground, consequently its source is pulled down to -V_{Th1}. The source of M_2 (a normal enhancement transistor) is forced to the same voltage by the op-amp whose output, in feedback loop, controls the gate of M_2. The output voltage is therefore given by

$$V_{out} = -V_{Th,1} + V_{Th,2} \tag{4.66}$$

Note that V_{out} is the difference between the output voltage and the analog ground, V_{AG}. Transistors M_3 and M_4 provide the bias currents. Actually, in the above calculation it should be necessary to account for the overdrive of transistors M_1 and M_2. The currents are assumed to be such that the overdrive matches. Therefore we can neglect the overdrive contributions.

As a matter of fact, the threshold shift depends on the implant dose and is, therefore, almost independent of temperature. Thus, the quality of the reference voltage achieved is remarkable. Unfortunately, designers can use this solution only when the technology provides transistors of the same type and with different thresholds. This is a rare and expensive situation. Two different threshold voltages means that the process needs of an additional mask for the required implant. This option is used only for special applications.

4.4.4 Band-Gap Reference Voltage

In two previous subsections we studied voltage references based on the multiplication of V_{BE} and the thermal voltage V_T. The former structure has a negative temperature coefficient while the latter has a positive temperature coefficient. In this sub-section we analyse the band-gap reference generator that is able to keep the temperature coefficient to zero (or near to zero) in a given temperature range. The band-gap operates on the principle of compensating the negative temperature coefficient of V_{BE} with the positive temperature coefficient of the thermal voltage V_T. The temperature coefficient of V_{BE}, at room temperature, is *- 2.2 mV/°C*; while, the positive coefficient of the thermal voltage is *0.086 mV/°C*. Therefore, a full compensation at room temperature is obtained by combining the term with positive temperature coefficient and the one with negative temperature coefficient to give

$$V_{BG} = V_{BE} + mV_T \qquad (4.67)$$

where m must be equal to *25.6 (=2.2/0 086)*.

KEEP IN MIND

The main target of a band-gap reference is to generate a temperature compensated voltage reference. The absolute accuracy is a secondary objective.

Since the value of V_{BE} for low currents is close to *0.65 V*, and V_T at room temperature is *25.8 mV*, the value of V_{BG}, according to (4.67), is *1.31 V*. Such a value is just slightly more than the silicon energy gap (expressed in volts it is *1.21 V*). Therefore, we normally call the circuits that achieve temperature compensation *band-gap reference*.

Because many applications require a bias voltage higher or lower than *1.31 V* it is often necessary to amplify or to attenuate the band-gap voltage by a given factor. In such a cases the key feature of the band gap circuit must be preserved: the gain or attenuation factor must be temperature independent.

The implementation of relationship (4.67) implies the availability of the V_{BE} voltage and the thermal voltage V_T. Since $V_T = kT/q$ is proportional to the absolute temperature, we often refer to it using the acronym *PTAT*.

Conceptually, to achieve a band gap, it is necessary to accomplish one of the two possible functions shown in Fig. 4.36. The first scheme uses two voltages as basic elements. The arrangement directly implements the equation

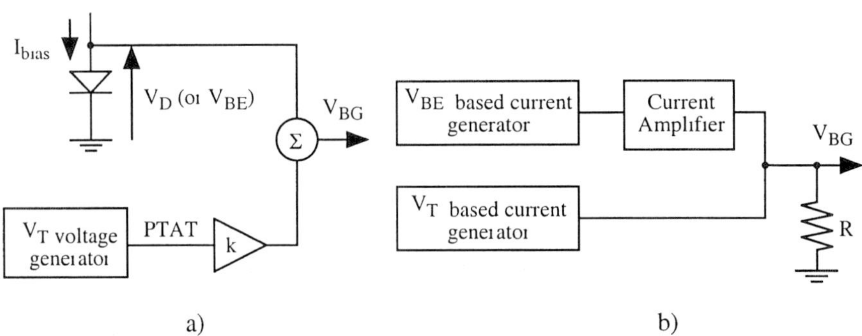

Fig. 4.36 - Two conceptual implementations of a temperature compensated reference generator

(4.67). The second scheme uses two currents proportional to V_{BE} and V_T respectively. These two currents are properly scaled and summed up. Then resistance R transforms the current into voltage. The two conceptual methods are equivalent. However, the second one permits a more easy amplification or attenuation of the generated V_D reference voltage.

The circuit shown in Fig. 4.37 belongs to the voltage processing category. This circuit is suitable for integration with *n-well* technology and operates as follows: the current mirror M_1 - M_2 defines a given ratio between the currents in Q_1 and Q_2 (normally, the mirror factor is *1*). The operational amplifier and the associated feedback loop drives the voltages of nodes *1* and *2* to an equal value. In addition, the voltage of node *1* (and *2*) stand one V_{BE} above the analog reference, V_{AG}. Designers can use any convenient value for V_{AG}. The only limit is that the voltage established at nodes *1* and *2* must permit a proper operation of the op-amp. We will se shortly that given op-amp architecture allow a common mode input voltage equal to ground (or V_{SS}). Therefore, in some cases V_{AG} is not used at all: the bases of the bipolar transistors are simply connected to *GND* to achieve a diode configuration.

The difference between the V_{BE} of transistor Q_1 and transistor Q_2 drops across resistance R_3. Therefore

$$\Delta V_{BE} = V_T ln \frac{I_1}{A_1 I_{SS}} \frac{A_2 I_{SS}}{I_2} = R_3 I_2 \tag{4.68}$$

where A_1 and A_2 are the emitter areas of transistors Q_1 and Q_2 respectively. Moreover

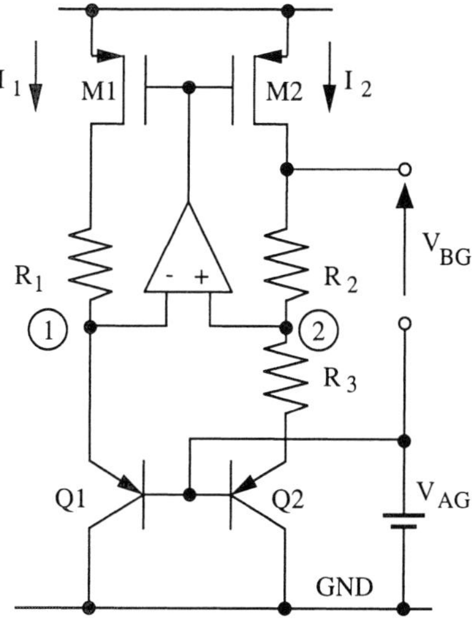

Fig. 4.37 - Band gap reference voltage for an n-well technology.

$$\frac{I_1}{I_2} = \left(\frac{W}{L}\right)_1 \Big/ \left(\frac{W}{L}\right)_2 \tag{4.69}$$

$$V_{BG} = V_2 - V_{AG} + R_2 I_2 = V_{BE1} + \frac{R_2}{R_3}\Delta V_{BE} \tag{4.70}$$

which becomes

$$V_{BG} = V_{BE1} + V_T \frac{R_2}{R_3} ln \frac{(W/L)_1 A_2}{(W/L)_2 A_1} \tag{4.71}$$

therefore, the factor multiplying the thermal voltage is given by

$$m = \frac{R_2}{R_1} ln \frac{(W/L)_1 A_2}{(W/L)_2 A_1} \tag{4.72}$$

which is thermally independent since it comes from a ratio between geometrical factors or resistances made by the same resistive layer. The resistance R_1 doesn't have a specific function. It serves for setting equal the drain voltages in transistors M_1 and M_2. In the simple case of unity mirror factor for the pair $M_1 = M_2$, $R_1 = R_2$.

Fig. 4.38 shows another implementation of the band gap reference with voltage processing. The operation of the circuit is the same as for the one in Fig. 4.37. The only difference is that the values of R_1 and R_2 control the currents in the bipolar transistors. If $R_1 = R_2$ then I_{Q1} and I_{Q2} are equal. Moreover, in the circuit in Fig. 4.38 the bipolar transistor are diode connected. Their bias current is normally provided by the output of the op-amp used.

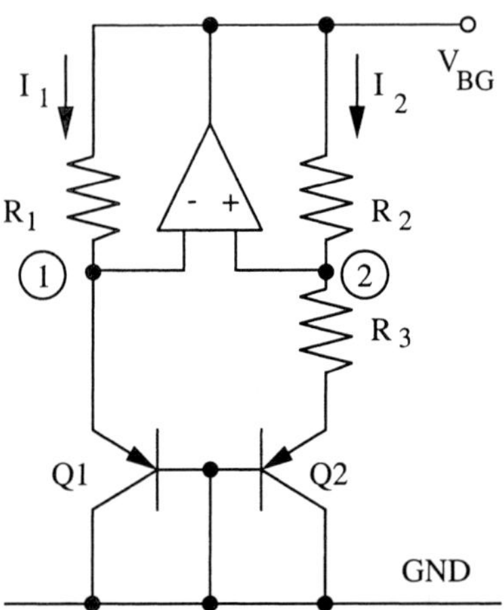

Fig. 4.38 - Another band gap reference based on voltage processing.

Both the circuits in Fig. 4.37 and Fig. 4.38 assume that the action of the operational amplifier forces the voltages of nodes *1* and *2* to an equal value. This is generally true if the gain of the op-amp is large enough. Therefore, the designer must ensure that the error, V_{out}/A_0, produced by the finite gain is negligible with respect to $R_3 I_2$. Nevertheless, a possible offset, V_{os}, affecting the op-amp leads to a systematic difference between the inverting and non inverting nodes. Thus, equation (4.68) becomes

$$\Delta V_{BE} + V_{os} = R_3 I_2 \tag{4.73}$$

If the ratio A_2/A_1 is equal to *8* and $R_1 = R_2$ then ΔV_{BE} is few tens of millivolts (*53 mV* at room temperature). An offset in the range of a few millivolts significantly affects the operation of the circuit. It is therefore recommended to dedicate special care to the offset control of the op-amp used. It is possible to attenuate the above discussed problem by increasing the ratio A_2/A_1. With an array of *5 x 5* bipolar transistors we can increase the area ratio to 24. ΔV_{BE} becomes *82 mV* at room temperature.

> **SUGGESTION**
>
> Designers achieve bipolar transistors with a given emitter area ratio by using a square array of equal elements. A N^2 number of elements (N odd) permits a common centroid layout.

Another possibility that we have to reduce the op-amp offset error is to use a double band gap with two V_{BE} in series as shown in Fig. 4.39. The emitter current in all bipolar transistors is set to equal. Therefore, the difference between

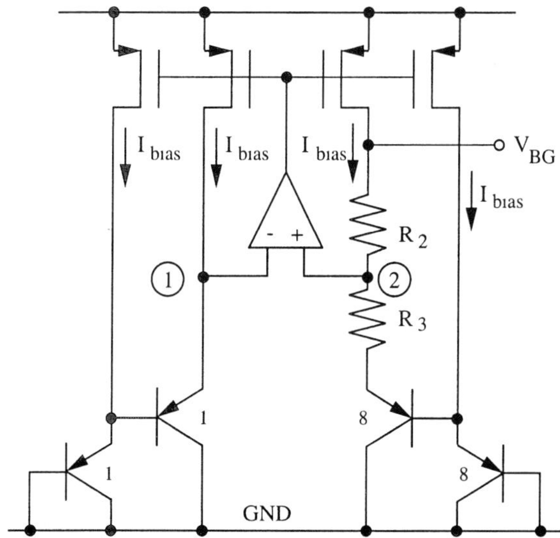

Fig. 4.39 - Double band gap voltage reference

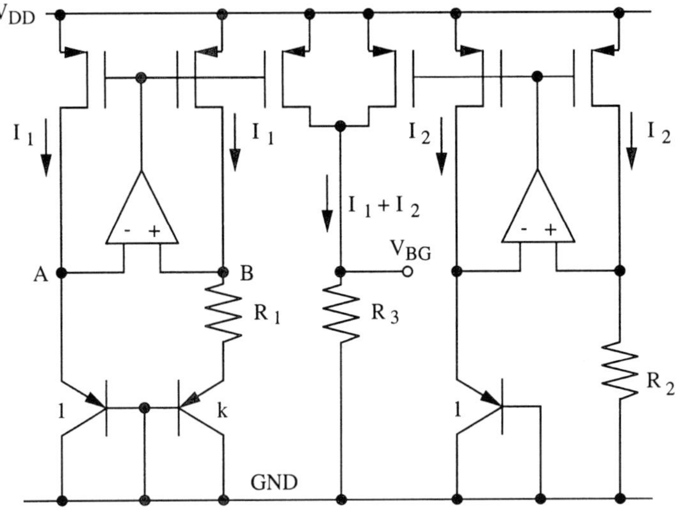

Fig. 4.40 - Band gap reference based on current processing.

the V_{BE} of transistors with area *1* and the one with area *8* applied across the resistance R_3 will be twice the one we have in the conventional structure (*106 mV* at room temperature). The result is that, in order to compensate the temperature variation of two V_{BE}, it will be necessary to multiply V_T by *51.2* instead of *26.1*. Thus the generated band gap voltage will range around *2.5 V*. Twice the normal output of a band-gap. The solution is attractive to attenuate the op-amp offset limitation. However, modern applications require reference voltages that are typically lower and not higher than *1.25 V*.

Fig. 4.40 shows the scheme of a band gap reference based on current processing. The right side of the circuit matches the scheme in Fig. 4.27. Therefore, the generated current I_2 is proportional to V_{BE}

$$I_2 = \frac{V_{BE}}{R_2} \tag{4.74}$$

The left side of the circuit, apart from the use of an op-amp to keep equal the voltages at the two nodes *A* and *B*, corresponds to the V_T based current reference shown in Fig. 4.34. The current I_1 results from

$$I_1 = \frac{V_T}{R_1} ln(k) \tag{4.75}$$

The sum of currents I_1 and I_2 flows into R_3. Therefore, the voltage across the resistance R_3,V_{BG}, becomes

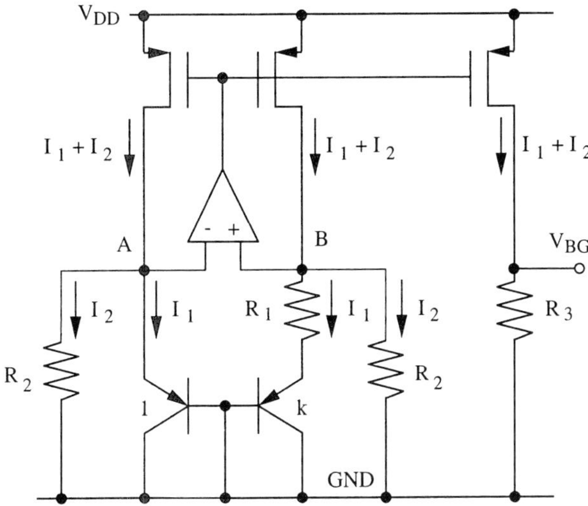

Fig. 4.41 - Single op-amp implementation of the band gap reference based on current processing

$$V_{BG} = V_{BE}\frac{R_3}{R_2} + V_T\frac{R_3}{R_1}ln(k) \tag{4.76}$$

Observe that equation (4.76) fulfils the condition (4.67) if the resistor values are such that, for instance, $R_3 = R_2$ and $R_3\, ln(k) = m\, R_2$.

In addition, the circuit can easily achieve a gain or an attenuation of the generated band gap voltage. It is enough to scale the value of resistance R_3 properly. Of course, all the resistors must be made of the same material to ensure compensation of the temperature coefficient.

As for the previously discussed band gap architecture, the offset of the op-amp in the left section is critical for the generation of the V_T-based current. Moreover, the given solution requires the use of two op-amps. Therefore, the circuit implementation demands more power and chip area. However, it is possible to use only one op-amp by incorporating the operation of the right hand section into the left hand one. Fig. 4.41 shows the resulting architecture. The voltage across the two resistors R_2 is V_{BE}. Therefore, the current that they drain to ground is expressed again by equation (4.74). It turns out that the current that

REMEMBER

The offset of the op-amp used in band-gap circuits is critical. The problem is not the imprecision of the voltage generated. What matter is the resulting inability to preserve the temperature independence.

flows in from node B is the superposition of I_2 with the V_T based term, I_1. The p-channel current mirrors have the same mirroring factor, thus current injected into node A is equal to $I_1 + I_2$. As a result, the current in the unity area bipolar transistor is, as required, I_1. The current in resistor R_3 is equal to $I_1 + I_2$, hence generating the band gap voltage as it occurs in the circuit in Fig. 4.40 given by $R_3(I_1 + I_2)$ as expected.

Since the voltage V_{BG} is temperature independent (for a given temperature range) the temperature variation of the current $I_1 + I_2 = V_{BG}/R_3$ is just determined by the temperature coefficient of resistor R_3. The use of special materials that allow R_3 to have a zero temperature coefficient permits to provide temperature independent currents as well.

4.4.5 Curvature Error

The band gap reference corrects the negative temperature coefficient of V_{BE} with a proper *PTAT* contribution. However, the compensation is limited to the linear error only. High-order coefficients of the temperature dependence of V_{BE} are not corrected by the band gap circuits studied so far. Typically, the voltage of a band gap changes with temperature as shown in Fig. 4.42. The maximum of the curve is placed at a given temperature by the designer (*25 °C* in the figure); 35 °C apart the band gap voltage drops by *1.3 mV* showing an average residual temperature coefficient of about *37μV/°C.*

The effect of non linear terms can be estimated by using the following expression of the base-emitter voltage, V_{BE}

$$V_{BE} = V_G(T) - [V_G(T_0) - V_{BE}(T_0)] \cdot \frac{T}{T_0} - (\eta - \alpha) V_T ln \frac{T}{T_0} \qquad (4.77)$$

where $V_G(T)$ is the band gap voltage of silicon at temperature T, $V_G(T_0)$ and $V_{BE}(T_0)$ are the band gap voltage and the base-to-emitter voltages, respectively, at temperature T_0. η is a process dependent parameter whose value ranges from *3* to *4*. Moreover, it is assumed that the current in the transistor varies with temperature according to the relationship

$$I_E = I_{E0}\left(\frac{T}{T_0}\right)^{\alpha} \qquad (4.78)$$

therefore, the α power coefficient describes the modification of the emitter current with temperature. Observe that the first of the two terms in (4.77) accounts for the linear variation of V_{BE} with temperature, while the latter accounts for the non-linear behaviour. It varies more than linearly with temperature because of the logarithmic factor; moreover, the multiplication factor

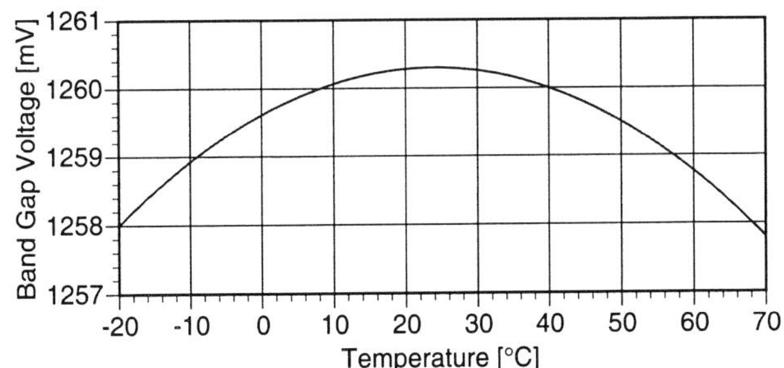

Fig. 4.42 - Typical temperature dependence of the band gap voltage.

$(\eta-\alpha)$ depends on how the emitter current varies with temperature.

An inspection of equation (4.77) shows a possible compensation strategy of the non-linear terms. It is necessary to generate a voltage contribution proportional to $V_T ln(T/T_0)$. Designers can build such a correction term by using, for example, the V_{BE} of two transistors, one biased with a temperature independent current ($\alpha = 0$) and the other biased with a *PTAT* current ($\alpha = 1$). Circuit implementations of the above strategy are not considered here. The reader can refer to the references for more details.

4.5 REFERENCES

- D. A. Johns, K. Martin, *Analog Integrated Circuit Design,* J. Wiley & Sons, New York, NY, 1997
- R. Gregorian, G. C. Temes, *Analog MOS Integrated Circuits,* J. Wiley & Sons, New York, NY, 1986
- K.R. Laker, W.M.C. Sansen, Design of Analog Integrated Circuits and Systems, McGraw-Hill Inc. New York, NY, 1994
- Z. Wang: "Analytical determination of output resistance and DC matching errors in MOS current mirrors", IEE Proceedings, Vol. 137, Part. G, 1990, pp. 397-404
- G. Palumbo: "Optimised design of Wilson and improved Wilson CMOS current mirrors", Electronic Letters, Vol. 29, 1993, pp. 818-819.
- A. K. Gupta, J.W. Haslett, F.N. Trofimenkoff: "A wide dynamic range continuously adjustable CMOS current mirror", IEEE Journal of Solid State Circuits, Vol. 31, 1986, pp.1208-1213
- P. Malcovati, F. Maloberti,C. Fiocchi, and M. Pruzzi: "Curvature-compensated BICMOS bandgap with 1-V supply voltage", IEEE Journal of Solid State Circuits, Vol. 36, 2001, pp.1076-1081.

4.6 PROBLEMS

4.1 Equation (4.3) was derived assuming $V_{DS1} = 0$. Determine the effect of the neglected term and plot I_{out}/I_{ref} as a function of I_{ref}. Assume $(W/L)_1 = (W/L)_2$, $\mu C_{ox}/2 = 120\ \mu A/V^2$, $V_{DS2} = 0.7\ V$, $V_{Th} = 0.7\ V$. Assume that I_{ref} ranges between $50\mu A$ and $300\mu A$.

4.2 Estimate the accuracy of a simple current mirror. The mirror factor is 1 and the transistors have $W = 50\ \mu m$, $L = 0.5\ \mu m$. The used layout and the technology achieve the following absolute and relative accuracies: $(\delta C_{ox})/C_{ox} = 0.025\%$; $(\delta\mu)/\mu = 0.08\ \%$; $\delta W = \delta L = 5nm$; $\delta V_{Th} = 0.6mV$. Moreover, the overdrive voltage used is $320\ mV$ and the drop voltage on metal connections is negligible.

4.3 Using Spice and the models in Appendix B, simulate the Wilson current mirror shown in Fig. 4.4. Use the same transistor sizing of Example 4.1. Plot the output resistance as a function of the resistance of the real reference current generator used.

4.4 Consider the Wilson current mirror in Fig 4.4. Add a battery shifting the voltage of the gate of M_3 away from the drain of M_1 by a variable amount ranging from 0 to $1V$. Determine the change in the output current for $I_{ref} = 100\mu A$. Use the same design parameters given in the Example 4.1.

4.5 Simulate the improved Wilson current mirror of Example 4.2 and determine the output resistance and the output current when the output voltage drops down from $2.5V$ to $0.8V$. Plot the results with $0.1V$ steps.

4.6 Using Spice simulate a cascode current mirror. Use for all the transistors $W = 50\mu m$ $L = 0.5\mu m$. Moreover, $I_{ref} = 150\mu A$. Use the Spice models given in Appendix B. Determine the output resistance as a function of the output voltage (ranging from $2.5V$ to $0.8V$ in $0.1V$ steps).

4.7 A small signal current is superposed to the reference current of the simple current mirror shown in Fig. 4.2. Determine the expression of the signal current at the output as a function of frequency. Assume that a capacitor C_p represents the parasitic between the gates of M_1, M_2 and ground.

4.8 Solve the small signal equivalent circuit in Fig. 4.5 under the following conditions: a signal current i_s is injected into node G_3; a resistance and a capacitor connects the output node to ground. Determine the transfer function i_x/i_s.

4.9 Design using Spice the Wilson current mirror shown in Fig. 4.4 the input current is $250\mu A$. The overdrive voltage of M_1 must be $200\ mV$ and the mirror factor must be 4. Use the models of Appendix C.

4.10 Determine the frequency response of the improved Wilson current mirror of Fig. 4.6. Use the models of Appendix C and the transistor sizing given in Example 4.2. Assume an output load given by $1M\Omega$ in parallel with $2pF$.

4.11 Determine using Spice the mall signal output resistance of a cascode current mirror as a function of the output voltage. Use the model of Appendix B and the following design parameters: $(W/L)_1 = (W/L)_2 = (W/L)_3 = (W/L)_4 = 100\mu/0.6\mu$. $I_{ref} = 300\ \mu A$.

4.12 Sketch the layout of a cascode current mirror. Use the transistor sizing of Problem 4.11. Split each transistor in 8 fingers and use a common centroid arrangement.

4.13 Design the modified cascode current mirror of Fig. 4.12 a). The mirror factor is 1. The minimum output voltage must be $0.5\ V$. The reference current is $I_{ref} = 250\mu A$. Divide evenly, the available output voltage budget between M_2 and M_3. Use the Spice models of Appendix B.

4.14 Design a high compliance current mirror that meet the follow design conditions: $I_{ref} = 100\mu A$; $V_{out,min}=0.4V$; $R_{out} = 20\mu\Omega$. Use the Spice models of Appendix B and select a proper V_{bias}.

4.15 Simulate using Spice and the models of Appendix B the current mirror with enhanced output impedance of Fig. 4.16 a). Use the following parameters: $I_{ref} = I_{B1} = 100\mu A$; $(W/L)_2 = (W/L)_3 = 60\mu/0.5\mu$. Determine the output resistance as a function of the output voltage. (Note that transistor M_1 is not shown in Fig. 4.16).

4.16 Simulate the circuit in Fig. 4.17 b). Use the same design parameters of Example 4.5. Design the transistors M_3 and M_4 in order to have with $V_{con3}=V_{con4}=2\ V$ $100\ mV$ across the drain-to-source of M_3 and M_4. Determine the output current control for V_{con4} varying in the range 1-$3\ V$.

4.17 Design the self-biased current reference of Fig. 4.20 capable to generates 150μA. Use the Spice models of Appendix B and the following conditions: $V_{DD}=5V$; $(W/L)_1 = (W/L)_2=50\mu/0.8\mu$; $(W/L)_3=100\mu/0.8\mu$; $(W/L)_4=50\mu/0.8\mu$. Note that the mirror factor of M_3-M_4 is not 1. Find the spur current caused by a noise signal affecting V_{DD} and ground.

4.18 Design, using the same transistor's parameters of Example 4.7. a self biased micro-current generator. The current required is $8\mu A$. Modify the circuit in order to have a current tunability in the range $4\text{-}12\mu A$. The current must be trimmed by a voltage that ranges between 2 and $3\ V$.

4.19 Use the start-up circuit of Fig. 4.26 in a self-biased current reference. Assume $V_{DD} = 3.3V$ and $I_{ref} = 120\mu A$. The supply voltage can change by ±10%. Design the transition M_{s2} in a way that it ensured the proper operation of the start-up at the two limit of variation of V_{DD}.

4.20 Simulate the V_{BE}-based current reference in Fig. 4.27 using the following design conditions: $V_{DD}=3{,}3V$; $(W/L)_1=(W/L)_2=100\mu/1\mu$, $V_{BE} = 0.75V$; $R = 10K\Omega$. Design a proper gain stage, which gain is at least 100 and that operated with the input and output quiescent levels.

4.21 Repeat Example 4.8 with the following changed conditions: $V_{DD} = 3.3V \pm 10\%$; starting of the operation at $2.7\ V$; generated current $75\mu A$. Use the Spice models of Appendix C.

4.22 Sketch the layout of two bipolar transitions which area ratio is $2/23$. Use a common centroid arrangement.

4.23 Design a V_T multiplier based on the schematic in Fig. 4.34. The output voltage is $200\ mV$. Use for the bipolar transistors a ratio 24. The supply voltage is $3.3.V$. The required total power consumption is $0.1\ mW$. Use the Spice models in Appendix C.

4.24 Design the band-gap voltage generator of Fig. 4.37. Use $V_{DD} = 3.3\ V$, $V_{AG} = 1.65\ V$. The current in the bipolar transistor is as low as $2\ mA$. The aspect ratio of the p-channel transistor is (W/L) = 30μ/0.4μ. Estimate the error caused by $2\ mV$ offset in the gain stage. Determine the variation of the generated voltage as a function of the gain of the stage used, Consider the range $20\ dB$ - $80\ dB$.

4.25 Using Spice simulations determine the temperature dependence of the band-gap voltage generated by the circuit studied in Problem 4.24. Assume an ideal op-amp with zero input offset. Determine the design conditions that sets the zero derivative of the curve at -40° and 100°.

Chapter 5

CMOS OPERATIONAL AMPLIFIERS

Operation amplifiers (usually referred to as op-amps) are key elements in analogue processing systems. Ideally they perform the function of a voltage controlled current source, with an infinite voltage gain. Operational amplifiers are extensively studied in basic electronic courses. Therefore, their function and operation should be well known to the reader. For this reason, this chapter deals with those circuit implementations that are specifically used to achieve the op-amp function in CMOS integrated VLSI systems. We shall learn that when used inside an integrated architecture, op-amps are mainly employed to drive capacitive loads, namely gates of transistors, capacitors or arrays of capacitors. This makes the request of having a low output impedance of little importance. Therefore, very often op-amps are replaced by operational transconductance amplifiers (OTAs) whose output resistance is quite high.

5.1 GENERAL ISSUES

From the basic courses of electronics we identify an op-amp with the basic schematic shown in Fig. 5.1. It is a four terminal block with two inputs and two

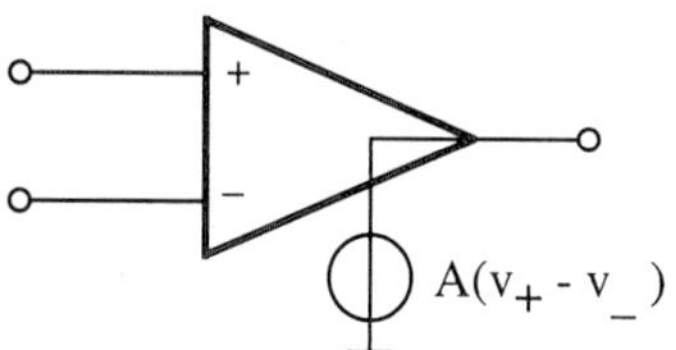

Fig. 5.1 - Symbol and equivalent circuit of the ideal op-amp

outputs. One of the outputs is the analog ground. The key function of the op-amp is to generate at the output an amplified replica of the voltage across the input terminals. Ideally, the voltage gain is infinite. Moreover, the input impedance is infinite as well and the output impedance is zero. Therefore, as we normally say, the op-amp measures the input voltage in a voltmetric fashion and generates the output in the same fashion of an ideal voltage source. From the circuit theory point of view the op-amp performs the same function as a voltage controlled voltage source (*VCVS*).

We know that an op-amp is never used as a stand alone block but has surrounding passive circuit elements to establish some feedback and achieves a given functional operation. It is because of this use that the block is called operational amplifier.

A quite general feedback connection is shown in Fig. 5.2. We have four lumped passive elements that interconnect the two inputs, the output and two input signal generators V_1 and V_2. Impedance Z_2 placed between the inverting terminal and the output ensures that the established feedback is negative. Under the assumption of ideal behaviour, the output voltage is given by

$$V_0 = V_2\frac{Z_4}{Z_3+Z_4}\cdot\frac{Z_1+Z_2}{Z_1} - V_1\frac{Z_2}{Z_1} \tag{5.1}$$

Thus, the achieved transfer function depends only on the impedances used. However if the op-amp is not ideal, the small-signal output voltage will be affected by those op-amp parameters depicting the non idealities, namely the finite gain, A_0, the finite bandwidth and the finite input and output impedance. If we consider the finite gain only, equation (5.1) is modified into

$$V_0 = \left(V_2\frac{Z_4}{Z_3+Z_4}\cdot\frac{Z_1+Z_2}{Z_1} - V_1\frac{Z_2}{Z_1}\right)\left(1-\frac{Z_1+Z_2}{A_o Z_1}\right) \tag{5.2}$$

that is the output voltage calculated previously multiplied by an error term. Observe that the error vanishes when the finite gain goes to infinite. This because the error is proportional to $1/A_0$. Therefore, if we want to limit the error we should use a very large finite gain. However, we should note that the

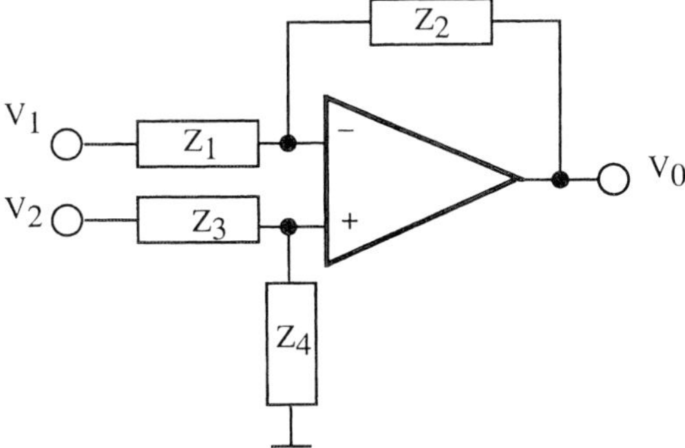

Fig. 5.2 - Typical feedback configuration used around an op-amp

input-output transfer function also depends on ratios between impedances. Possible errors affecting impedances modify too the transfer function thus contributing an additional source of error. The errors caused by the finite gain of the op-amp and that of the impedance mismatch are uncorrelated, so they must be combined quadratically. It turns out that the global error is dominated by the greater one. It follows that it is not advisable to stress the design to achieve a finite gain larger than a given amount. A proper target is to have a finite gain somewhat larger than the inverse of the impedance matching accuracy. Being the matching accuracy of integrated components around *0.1%*, a gain a bit larger than *60-70 dB* is normally enough for most applications.

> **KEEP NOTE**
>
> The finite dc gain of an op-amp matters until the error that it produces is dominant with respect to other sources of error, like the passive component mismatch.

The input impedance of a *CMOS* op-amp is not normally a problem. Since we use the gate of an *MOS* transistor as input terminals we have a very large input impedance over a wide frequency interval. Instead obtaining low output impedance can be problematic. Nevertheless, the typical use that we make of op-amps in integrated systems does not lead to a strict request for low output impedance. Even better, in some cases the output impedance may not be a problem at all. Let us see why.

In many systems the designer uses a capacitor as the feedback element while the output load and the input are capacitors as well, possibly connected by switches. Fig. 5.3 shows a typical situation. When the charged capacitor C_1 is switched onto the virtual ground, a transient in the circuit will take place. The output node supplies current during the transient but after the output node settles the current goes to zero. As a result of this operation the output impedance is no longer important.

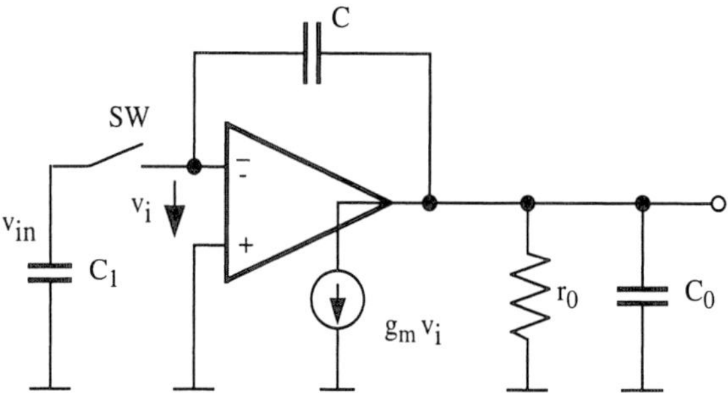

Fig. 5.3 - Typical use of an op-amp in an integrated signal processor

To verify the above statement, let us analyse in some detail the circuit shown in Fig. 5.3. It models the operational amplifier with a transconductance generator $g_m v_i$ having in parallel the output resistance r_0 and the output capacitance C_0. The capacitor C_1, charged at the input voltage V_{in} (thus, $Q_1 = C_1 V_{in}$), at time $t = 0$ is connected by the switch to the virtual ground. At time $t = 0^+$ the op-amp does not react and, assuming C and C_0 to be initially discharged (this is not really necessary since the circuit is linear and we can use the superposition principle), the charge on C_1 is shared with the series connection C - C_0. We obtain the following voltages

$$V_i(0^+) = V_{in}\frac{C_1}{C_1 + (C_0 C)/(C_0 + C)} \tag{5.3}$$

$$V_o(0^+) = V_i(0^+)\frac{C}{C_0 + C} \tag{5.4}$$

Both the virtual ground voltage and the output jump are positive (if V_{in} is positive). Then, the differential input causes a current to come from the transconductance generator that leads to the asymptotic voltages

$$v_i(\infty) = v_{in}\frac{C}{C_1 + C(1 + g_m r_0)} \tag{5.5}$$

$$v_o(\infty) = -v_i(\infty) g_m r_0 \tag{5.6}$$

Observe that the output voltage goes from positive to negative. Moreover, its final value marginally depends on the finite gain of the op-amp $A_0 = g_m r_0$. The transient of the output voltage and the one of the inverting input are described by exponential (Fig. 5.4) whose time constant is

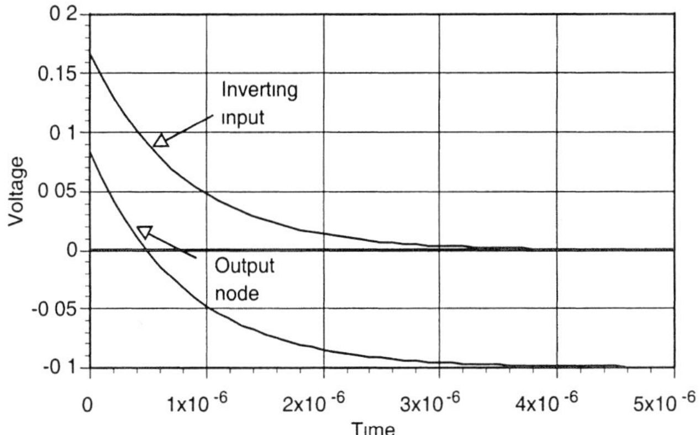

Fig. 5.4 - Transient voltage at the output node and the inverting input of the circuit in Fig 5.3. V_{in} = 1 V; C_1=0.2 pF, $C=C_0$=2pF; τ = 80 nsec

$$\tau = \left(\frac{1}{g_m}\cdot\frac{C}{C+C_1} // r_0\right)\cdot\left(C_0+\frac{CC_1}{C+C_1}\right) \tag{5.7}$$

> **KEEP IN MIND!**
>
> A low value of the op-amp output resitance is not essential when the load and the feedback network are made by capacitors and, possibly, switches. In such a situation we can use an *OTA* (operational transconductance amplifier).

Since $1/g_m \ll r_0$, the time behaviour of the output voltage is mainly controlled by the transconductance gain and is almost independent of the output resistance. The result achieved is very important for the design of op-amps used in many analog signal processors (switched capacitor circuits, track and hold, data converters). When the feedback network comprises only capacitors (and switches) the output resistance is not relevant. We can use a special class of operational amplifiers where the output resistance can be very high and, possibly, used to enhance the voltage gain. This class of operational amplifiers is called *OTA* (operational transconductance amplifier). An *OTA* achieves a high voltage gain with a given transconductance gain, g_m, by the use of a very high output resistance.

5.2 PERFORMANCE CHARACTERISTICS

This section recalls the definition of the most important features of operational amplifiers. Moreover, we shall also consider the most useful circuit con-

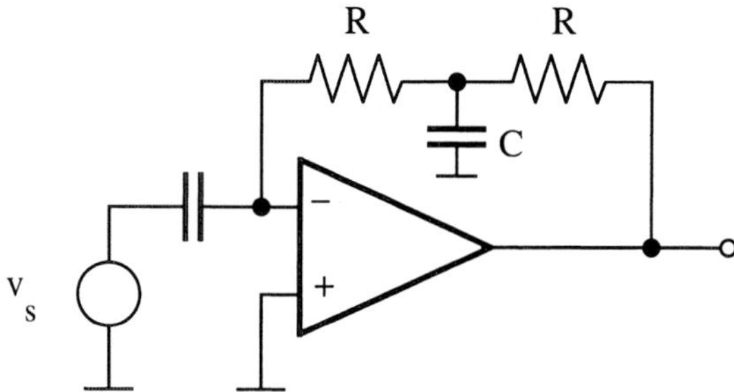

Fig. 5.5 - Schematic arrangement useful for simulating the differential gain of an op-amp.

figurations for the experimental or simulated estimation of the key parameters.

Differential gain (A_d): this is the open loop differential gain measured as a function of frequency. To estimate the differential gain we must compensate the offset (to be discussed shortly) Fig. 5.5 shows a convenient configuration for the *SPICE* estimation of the *dc* differential gain. The small signal input generator is connected between the two input terminals through a big capacitor *C*, while a *T* network made of two resistors and a capacitor establishes a feedback path around the op-amp. Observe that the *T* network acts like a unity gain configuration at very low frequencies and becomes an opened network at very high frequency. A proper choice of resistors and capacitors allows the transition between the two states to take place at a suitably frequency range. For computer simulations we can use huge elements, like *F* or many $G\Omega$ and bring the transition down to a very low frequency. Moreover, observe that the feedback network employed loads the output. Therefore, we have to use resistances *R* which value is higher than the output resistance of the op-amp.

A typical value of the differential gain, A_d, ranges from *70* to *90 dB*. For very precise functions (like high-resolution data converters), the designer needs higher gains in the *100* to *140 dB* range.

Common mode gain (A_{cm}): this is the open loop gain obtained by applying a small signal to both inputs. To measure it we can use the basic configuration shown in Fig. 5.5 with a small modification. We just have to connect the signal generator directly at the positive input and short the positive and negative terminal with a big capacitor *C*. Ideally an op-amp should amplify the differential signal only. Therefore, a low common mode gain over a wide frequency range is quite advisable. A typical value of A_{cm} at low frequency is *10 - 30 dB*.

Common mode rejection ratio (*CMRR*): this is the ratio between the differential gain and the common mode gain. A high *CMRR* is a merit factor for any op-amp.

Power supply rejection ratio (*PSRR*): if we apply a small signal in series

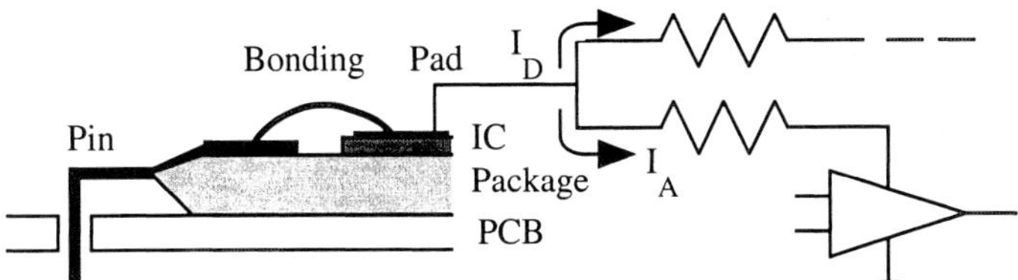

Fig. 5.6 - Typical scheme of connection between the external pin and an analog block.

with the positive or the negative power supply we obtain a corresponding signal at the output with a given amplification (A_{ps+} or A_{ps-}). The ratio between the differential gain and the power supply gain leads to two *PSRRs*. These are two merit factors showing the ability of the op-amp to reject spur signals coming from the power supply.

For mixed-signal circuits the *PSRR* is a very important issue. The power supply that we use in op-amps is not the voltage applied to external pins. As shown in Fig. 5.6, the supply voltage applied to an external pin feeds the integrated circuit through an inductive bonding from the pin to a pad. The inductance is, as a rule of the thumb, *1 μH* per *mm* of bonding length. Then, metal connections distribute the voltage to various sections of the integrated circuit. The metals display a given resistance that depends on the length of the connection. The resulting network made by a given inductance and resistances leads to a drop voltage between the external pin and the actual supply voltage. In mixed systems the current that we have to account for is not only the one in analog sections; often the dominant spur is caused by the *LdI/dt* noisy contribution related to fast current transitions in digital sections.

Having a good *PSRR* is an important merit factor. Unfortunately, especially at high frequencies, the *PSRR* achieved is typically quite poor. For computer estimation of the power supply rejection the same circuit arrangement shown in Fig. 5.5 can be used: it is required just to use a signal generator in series with a supply connections and set to zero the input signal. A typical value of *PSRR* is *60 dB* at low frequencies that decreases to *20 – 40 dB* at high frequencies.

Offset voltage (v_{os}): if the differential input voltage of an ideal op-amp is zero the output voltage is also zero. This is not true in real circuits: various reasons (that we shall study shortly) determine some umbalancement that, in turn, lead to a non-zero output. In order to bring the output to zero it is therefore required to apply a proper voltage at the input terminals. Such a voltage is the offset.

The simple unity gain configuration shown in Fig. 5.7 permits us to measure the offset. It is represented by a voltage generator in series with the non-inverting terminal. The feedback makes output and inverting input identical. Moreover, assuming that the gain is large enough, the two inputs are equal. In turn, the output measures the offset. Actually the output is not zero, as required: it is the

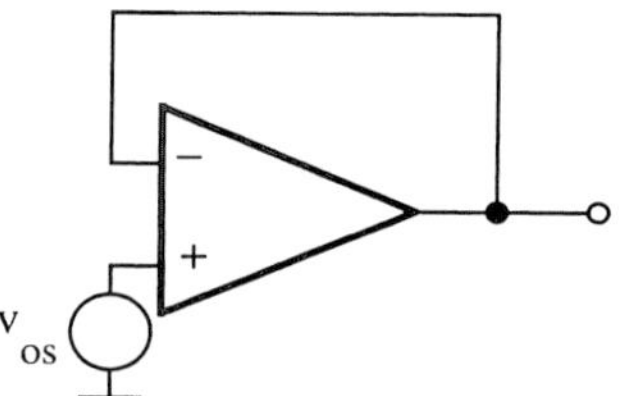

Fig. 5.7 - Schematic arrangement useful for measuring the offset.

offset. Nevertheless, assuming that the offset is relatively small the discrepancy doesn't affect the measure.

Input common mode range: this is the voltage range that we can use at the input terminal without producing a significant degradation in op-amp performance. Since the typical input stage of an op-amp is a differential pair, the voltage required for the proper operation of the current source and the input transistors limit the input swing. A large input common mode range is important when the op-amp is used in the unity gain configuration. In this case the input must follow the output. For such a conditions a so called rail-to-rail operation at the input terminals is often desired.

Output swing: this is maximum swing of the output node without producing a significant degradation of op-amp performance. Since we have to leave some room for the operation of the devices connected between the output node and the supply nodes, the output swing is only a fraction of $(V_{DD} - V_{SS})$. Typically it ranges between *60%* and *80%* of $(V_{DD} - V_{SS})$. Within the output swing range the response of the op-amp should conform to given specifications and in particular the harmonic distortion should remain below the required level.

Equivalent input noise (v_n): the noise performance of a *CMOS* operational amplifier depends on the noise of the transistors used and circuit architecture. Since the *MOS* transistor is a voltage controlled device its noise performance is well described by using an input referred voltage noise generator only. It turns out that the noise in a network caused by the interconnection of *MOS* transistors can also be represented with an input referred noise generator (see

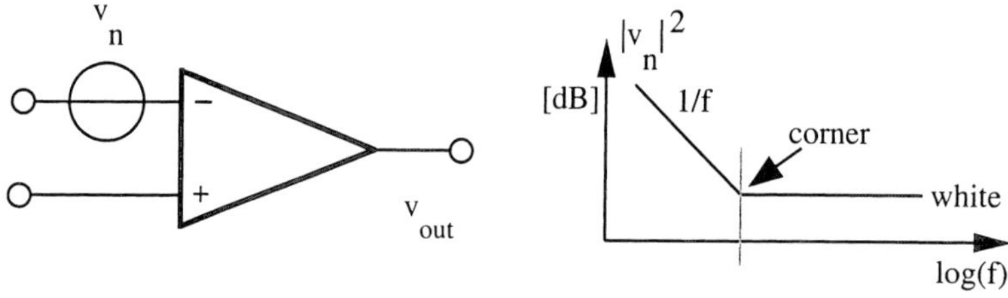

Fig. 5.8 - Equivalent input referred noise voltage generator and its typical spectrum.

Fig. 5.8) that accounts for the effect of all the *MOS* transistors of the network. The noise spectrum of an *MOS* transistor is made by a white and a *1/f* term. Thus, even the spectrum of v_n is made of the same two components: a white term and an *1/f* term. The frequency at which the *1/f* term becomes equal to the white one is called corner frequency. A typical value of the white noise is $50 nV/\sqrt{Hz}$. The *corner frequency* depends on the quality of the technology and the kind of input transistors used. It is in the range of 1 *kHz* to *10 kHz*.

Unity gain frequency (f_T) (or gain-bandwidth product, GBW): the speed performance of the op-amp are described by small signal and large signal parameters. The small signal analysis determines the frequency response sketched by a set of zeros and poles. Since we have to ensure stability, one of the poles (f_1) must be dominant. The amplitude Bode diagram will display a *20 dB/decade* roll-off until the gain reaches *0 dB*. The frequency at which the gain becomes *0 dB* is called unity gain frequency, f_T. With a constant roll-off *20 dB/ decade* of the achieved unity gain frequency equal the product of gain and bandwidth $f_1 A_o$. Therefore, f_T is also named the gain-bandwidth product, *GBW*. Other poles, f_2, f_3..., exceeding f_T, are named non-dominant poles.

We can measure f_T using the op-amp either in the open-loop or in the unity gain configuration. In the former case the circuit schematic shown in Fig. 5.5 must be used. In the latter we can use the configuration shown in Fig. 5.9. The output voltage follows the input ($A_0 = 0\ dB$) until frequency f_T at which point the gain starts rolling down by *20 dB/decade.*

Phase margin: this is the phase shift of the small-signal differential gain measured at the unity gain frequency. In order to ensure stability when using the unity gain configuration it is necessary to achieve a phase margin better than *60°*. A lower phase margin (like *45°* or less) will cause ringing in the output response. However, for integrated implementation it is not strictly necessary to ensure absolute stability. The use of an op-amp in specific configurations permits to know the value of the feedback factor, β. If β is lower than *1* (as it often happens) the *60°* phase margin should be fulfilled not at the unity gain frequency but at the frequency at which the gain is $1/\beta$.

Slew rate (*SR*): this is the maximum achievable time derivative of the output voltage. It is measured using the op-amp in the open loop or the unity gain con-

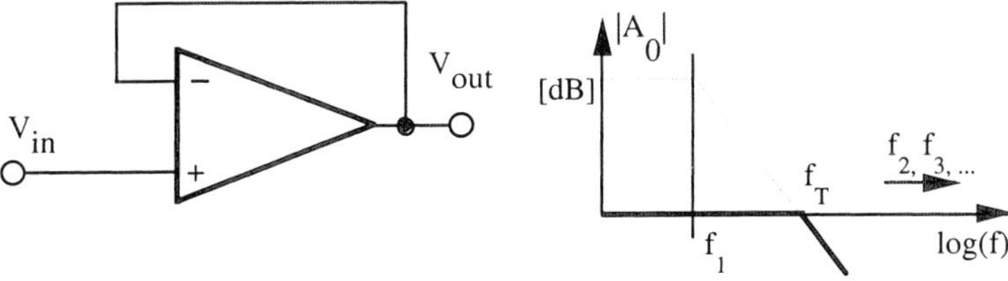

Fig. 5.9 - Schematic arrangement for estimating f_T.

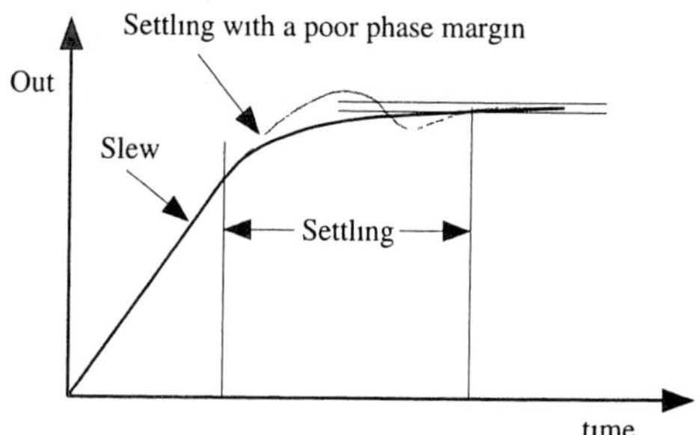

Fig. 5.10 - Typical output response of a real op-amp (slewing and settling limitations)

figuration. A large input step voltage fully imbalances the input differential stage and brings the op-amp output response into the slewing conditions. The positive slew rate can be different from the negative slew rate, depending on the specific design. Typical values ranges between *40* and *80 V/μsec*, but micro-power circuits (for whose the quiescent current available is pretty small) can show much lower figures.

Settling time: this is the time that the output voltage requires, under given operating conditions, to achieve the expected output voltage within a given accuracy (usually *0.1%* or better). The settling time is measured from the end of the slewing period. It critically depends on the phase margin: a poor phase margin leads to a ringing response that augments the settling period. Fig. 5.10 shows the combined effect of slewing and settling.

> **RECOMMENDATION**
>
> Define, analyse, and carefully "negotiate" with system engineers all the op-amp specifications before starting any design.
>
> Ambiguous parameters are always source of many design iteration. By contrast, detailed and well defined specifications permit a quick, safe and successful design.

Power consumption: this is the power consumed under stand-by conditions. Observe that the power used in the presence of a large signal can significantly exceed the one required in the quiescent conditions. Moreover, the consumed power depends on the speed specifications. Typically, higher bandwidth leads to higher power consumption. Low-power operation is a very important quality factor: more and more electronic systems are powered by batteries that should supply the system for hours or days. Therefore, a key design task is to achieve the minimum power consumption for a given required speed. In many applications, we can use a dynamic biasing of the op-amp or a power down mode to reduce power consumption.

Silicon area: specific values of performances discussed above establish a given circuit configuration that, in turn, leads to a corresponding silicon area. Typically the layout of an op-amp has a rectangular shape and includes substrate and well biases. For op-amp schemes of medium or low complexity it is possible to accommodate the entire layout within $2000\ \mu m^2$ (for a $0.25\ \mu m$ CMOS technology).

Table 1 summarises the typical performance parameters achievable with a $0.25\ \mu m$ CMOS Technology. The given figures corresponds to a relatively easy design. Specific architecture permit to improve (even significantly) one or more the given parameters.

TABLE 5.1

Typical parameters of a 0.25 μm CMOS Transconductance Op-Amp

Feature	Value	Unit
DC gain	80	dB
CMRR	40	dB
Offset	4-6	mV
Bandwidth	100	MHz
Slew-rate	3	V/μs
Settling time: 1 V, $C_L = 4$ pF	300	nsec
PSRR @ dc	90	dB
PSRR @ 1 kHz	60	dB
PSRR @ 100 kHz	30	dB
Input referred noise (white)	100	nV/√Hz
Corner frequency	1	kHz
Supply voltage	3.3	V
Input common mode voltage	1.5	V
Output dynamic range	2.2	V_{pp}
Power consumption	1	mW
Silicon area	2000	μm^2

5.3 BASIC ARCHITECTURE

The typical gain that an op-amp (or an *OTA*) should achieve is around *80 dB*. A simple gain stage has, according to the results of a previous chapter, a gain in the order of *40 dB*. Therefore, the cascade of two stages is normally enough. Alternatively, we can use a cascode with a cascode load configuration that alone obtains around *80 dB*. The choice between a single stage and a two stage architecture depends on a number of design issues that will be considered in some detail later in this chapter. These include dynamic range, bandwidth and power consumption.

As we shall see shortly, the generic circuit schematic of an op-amp can be represented by the functional diagram shown in Fig. 5.11.

> ***REMEMBER***
>
> Operational amplifiers that, when used in an integrated circuit, drive capacitive load don't strictly need an output stage. However, circuits with an output stage improve the output current capability, thus reducing the time that the output node takes for slewing.

The first block is a differential amplifier. It provides at the output a differential voltage or a differential current that, essentially, depends on the differential input only. The next block is a differential to single-ended converter. It is used to transform the differential signal generated by the first block into a single ended version. Some architecture don't require the differential to single ended function; therefore, the block can be excluded. Possibly, a second gain stage enhances the differential gain. Finally, we have the output stage. It typically provides a low output impedance or improves the slew rate of the op-amp. Even the output stage can be dropped: many integrated applications do not need a low output impedance; moreover, the slew rate permitted by the gain stage can be sufficient for the application. When the output stage is not used the circuit is not, strictly speaking an op-amp. It is, instead, an operational transconductance amplifier, *OTA*. The name remember us that the circuit achieves the voltage gain using an input transconductor and a relatively large output resistance. The product of transconductance and output resistance fixes upon the voltage gain.

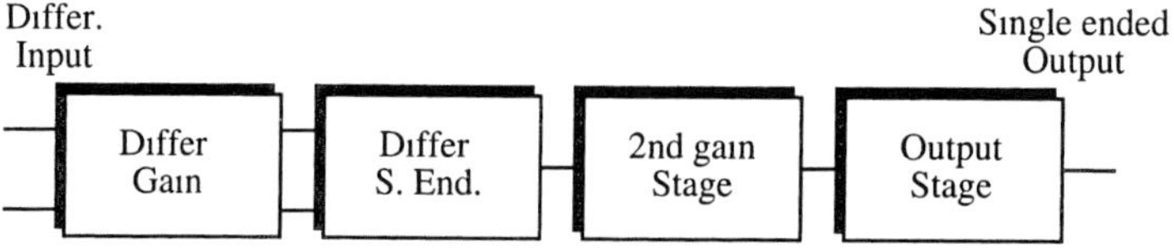

Fig. 5.11 - Functional diagram of a typical op-amp.

5.4 TWO STAGES AMPLIFIER

Fig. 5.12 shows the circuit configuration of a two stage transconductance amplifier. The scheme uses *p*-channel transistors at the input. Of course, it is possible to use a complementary scheme with *n*-channel input transistors. As specified by the name, the circuit is the cascade of two stages: the first is a differential amplifier with differential to single ended transformation, the second is a conventional inverter with active load. The circuit in Fig. 5.12 uses the same reference current for the differential amplifier and the second stage. Therefore, the bias currents in the two stages will be controlled together. Observe that the conversion from differential to single ended is achieved in the first stage with a current mirror (M_3-M_4). As a matter of fact, the signal at the output of the differential pair is current. The current from M_1 is mirrored by M_3-M_4 and subtracted from the current from M_2. The signal contributions of the two currents multiplied by the output resistance of the first stage give the single-ended first stage output voltage. The resulting signal constitutes the input of the second gain stage. Capacitor C_c (or possibly a more complex network) takes care of compensation requirements.

In the next subsection we shall study the features of the two stage op-amp in some detail. Our present target is not to perform complicated analysis but to derive equations appropriate for guiding the computer simulation design. Once again, remember that the design activity is based on computer simulations: the models used by the computer program are much more precise than the ones we use for hand calculations. Nevertheless, it is important to acquire a number of "rules of thumb" that direct the design activity and avoid using computer with a dangerous "try and see what happens" approach.

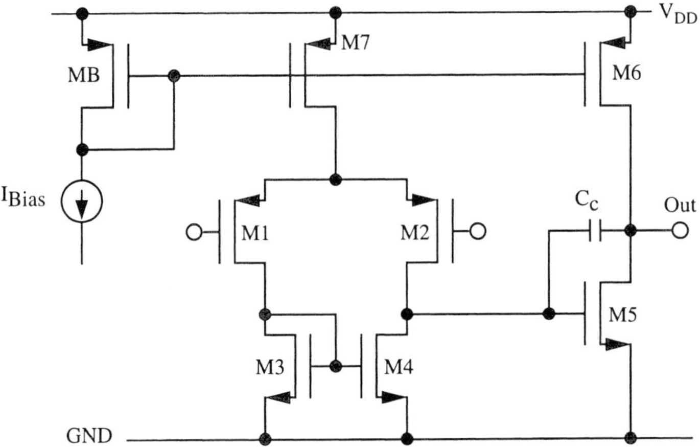

Fig. 5.12 - Basic configuration for a two stages OTA (p-channel input devices).

5.4.1 Differential Gain

Small signal differential gain results directly from the basic block analysis performed in Chapter 3. We can observe that, at low frequency, the second stage does not load the output of the first. Therefore, the low frequency gain is given by the product of the two gains. By inspection of the circuit we obtain

$$A_v = A_1 A_2 = \frac{g_{m1} g_{m5}}{(g_{ds2} + g_{ds4})(g_{ds5} + g_{ds6})} \tag{5.8}$$

$$A_v = \alpha \cdot \frac{\sqrt{\left(\frac{W}{L}\right)_1}\sqrt{\left(\frac{W}{L}\right)_5}}{\sqrt{I_7}\sqrt{I_6}} = \frac{\alpha}{I_{Bias}} \cdot \frac{\sqrt{\left(\frac{W}{L}\right)_1}\sqrt{\left(\frac{W}{L}\right)_5}\left(\frac{W}{L}\right)_B}{\sqrt{\left(\frac{W}{L}\right)_7}\sqrt{\left(\frac{W}{L}\right)_6}} \tag{5.9}$$

Where α is a proper constant that depends on mobility, specific oxide capacitance and the λ factor of the technology used. Moreover, we assume all transistors to be in saturation. Equation (5.9) states that the low frequency gain is inversely proportional to the bias current. The result is obvious: we have already seen that, for transistors in saturation, the gain of a single stage is inversely proportional to the square root of the bias current. Being the circuit the cascade of two stages we achieve immediately the result.

Looking again at equation (5.9) we observe that gain depends on the aspect ratio of four transistors and, through α, on the length of the active load devices of the first and the second stage. This is apparently a fairly large number of degrees of freedom. We shall see shortly that aspect ratios and lengths cannot be defined freely. In order to achieve given basic features (like a zero systematic offset or a symmetrical slew rate) the designer must respect certain constraints.

5.4.2 Common Mode dc Gain

We determine the common mode *dc* gain by applying the same signal to both inputs. Under those conditions the electrical behaviour of the first stage becomes symmetrical. We therefore can derive the common mode *dc* gain using, for the first stage, half circuit. As shown in Fig. 5.13, transistor M_7 is divided into two parts. Moreover, because of the symmetry the drain voltage of M_4 equals the one of M_3. Therefore, we use that voltage to control the gate of M_5. The output of the first half stage is then amplified by the second stage.

The gain of the first half stage should be calculated using the small signal

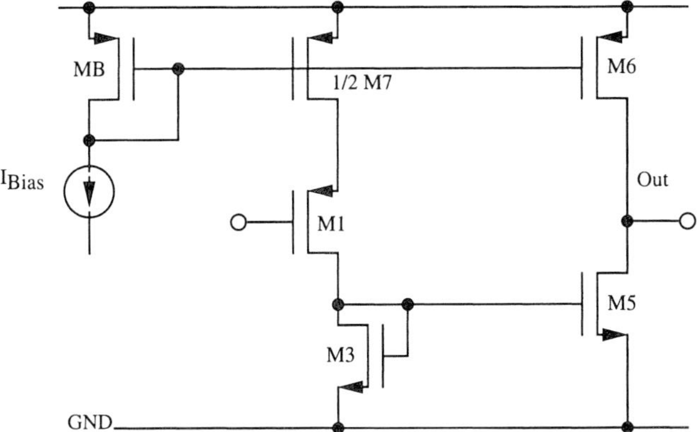

Fig. 5.13 - Circuit schematic useful for the common mode gain estimation.

equivalent circuit. However, by inspection of the circuit an approximated relationship directly results. The drain of $1/2M_7$ follows the input. Therefore, the signal current is $v_{in}/(2r_{ds7})$. Since this current flows in the diode connected transistor M_3 it results that the common-mode gain of the first stage is

$$A_{CM,1} = -\frac{1}{2g_{m3}r_{ds7}} \tag{5.10}$$

therefore, the total common mode gain is given by

$$A_{CM} = \frac{g_{m5}}{2g_{m3}r_{ds7}(g_{ds5} + g_{ds6})} \tag{5.11}$$

that leads the low frequency value of the *CMRR* to be

$$CMRR = \frac{A_d}{A_{CM}} = \frac{2g_{m3}r_{ds7}g_{m1}}{(g_{ds2} + g_{ds4})} \tag{5.12}$$

showing that the CMMR comes from the first stage design only. Such a result is obvious: the second stage amplifies in the same way any signal that it receives from the first stage; no matter if the signal results from a common mode or a differential input.

5.4.3 Offset

The supply voltages in the circuit in Fig. 5.12 are V_{DD} and ground. The analog ground, V_{AG}, is somewhere in between these two levels. Depending on the application the designer defines the value of the analog ground used in the project. Typically it is $V_{DD}/2$. (if V_{SS} is used instead than ground, the analog ground lies midway between V_{DD} and V_{SS}). For an ideal circuit it is expected that a zero input signal (input terminals shorted) determines an analog ground voltage at the output. Instead, an improper design or possible technological mismatches in the circuit deviates the output from the expected value. Often, in open loop conditions, the output voltage saturates very close to one of the supply rails. In order to bring the output to V_{AG} it is necessary to apply a proper input signal that counterbalances the existing mismatches. This inputs signal is the offset.

We have two additive contributions to offset. They are:

- systematic offset
- random offset

The former component depends on the circuit's design. We can minimize it with suitable cell design. The second contribution comes from random fluctuation of physical and technological parameters along the chip. We can minimize it with a careful layout that minimizes the mismatch between critical components.

We estimate the systematic contribution with the following approach: the differential input is set to zero. A small signal analysis determines its effect at the output. Then, a differential input capable to counterbalance that output voltage is determined.

Connecting the inputs at the same voltage makes, as already discussed in the common mode gain evaluation, the input stage electrically symmetrical. This permits us to split the input stage into two equal parts to achieve again the circuit in Fig. 5.13. Assuming that the common mode input voltage is within the normal range of operation, $1/2\ M_7$ works in the saturation region and, at first approximation, its current and, consequently, the one in M_3 can be assumed a replica of the bias current, I_{Bias}.

We can regard the output stage as being made by two current generators: one (M_5) sinking the current injected by the other (M_6). Both transistors M_5 and M_6 are assumed to be in the saturation region. Since the two current sources have a high output impedance the output voltage approaches the analog ground only if the two currents I_{M5} and I_{M6} are almost equal. As M_5 mirrors the current in M_3 that in turn is a replica of the current in M_7, the following relationships must be verified

$$I_{Bias}\frac{(W/L)_6}{(W/L)_B} = \frac{I_{Bias}}{2} \cdot \frac{(W/L)_7}{(W/L)_B} \cdot \frac{(W/L)_5}{(W/L)_3} \tag{5.13}$$

that leads to the following constraint between aspect ratios of transistors

$$(W/L)_3 \cdot (W/L)_6 = \frac{1}{2} \cdot (W/L)_7 \cdot (W/L)_5 \tag{5.14}$$

The above analysis assumes the use of ideal current mirrors. Because of the transistor finite output resistance the current in one branch and its mirrored replica don't match perfectly. Thus, even if condition (5.14) is respected, the slight mismatch between the currents in M_5 and M_6 flows into the output resistance of M_5 and M_6 and causes a residual offset. The designer could trim the transistor sizing in order so as to bring the residual offset to zero with simulations. However, the practice is not advisable for two reasons: the accuracy and the process variation affecting the output resistance are both wide: zeroing results achieved with a given simulation condition are not exactly verified in the experimental verifications. A second, and more important reason, is that for optimizing the layout we need transistors with properly ratioed widths. Unrelated transistor sizing as could result from the trimming process makes the layout problematic and irregular.

If we use the condition (5.14) in (5.9) we have

$$A_v = \frac{\alpha}{I_{Bias}} \cdot \frac{2\sqrt{(W/L)_1\sqrt{(W/L)_3}}}{(W/L)_7}(W/L)_B \tag{5.15}$$

Therefore, when we take care of the systematic offset the DC gain can be no longer adjusted by operating over the second stage of the op-amp.

> **LESS DESIGN FREEDOM**
>
> In order to achieve a nominally zero systematic offset the relationship (5.14) between transistor sizing must be fulfilled. The condition costs the designer one of the degrees of freedom available.

The second contribution to offset is random. Since it comes from unavoidable mismatches, we have to understand where the critical points of the design are and, possibly, concentrate on the sensitive points to improve the matching performance.

The circuit has two stages, therefore, any possible mismatch will determine offset at the input of the first and the second stage (Fig. 5.14). The two offsets are uncorrelated and are combined quadratically. However, because of the gain of the first stage, when we refer the

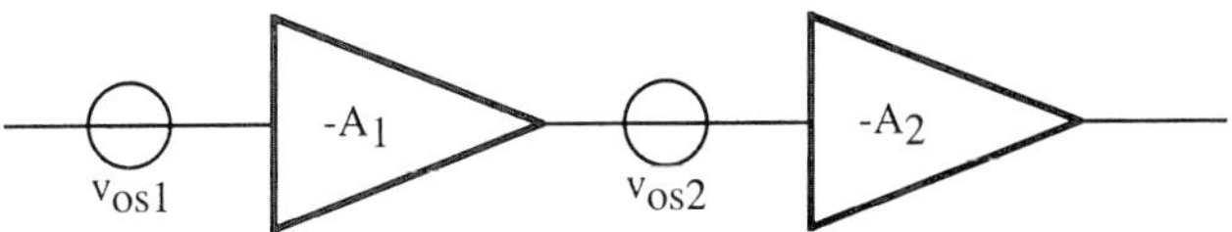

Fig. 5.14 - Input referred offset generators in a two stages amplifier

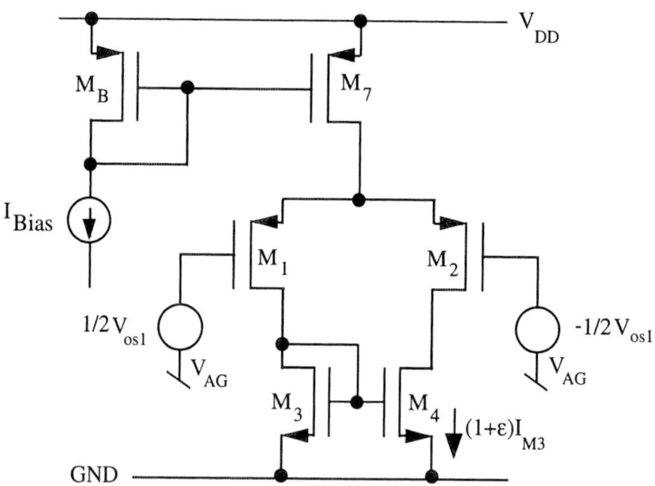

Fig. 5.15 - A mismatch in the mirror factor M3-M4 causes random offset.

offset of the second stage at the input terminal we have to divide it by the gain of the first stage to obtain

$$V_{os} = \sqrt{V_{os1}^2 + (V_{os2}/A_1)^2} \tag{5.16}$$

Assuming that the two offset contributions have similar value, since A_1 is pretty large, the input stage practically determines the random offset.

Nominally, the transistors M_1 and M_2, as well as M_3 and M_4, in Fig. 5.15 are equal. However, because of fabrication errors they show differencies. To simplify the analysis we assume that a possible mismatch between M_1 and M_2 can be transferred in a mismatch between M_3 and M_4. Therefore, we assume that M_1 and M_2 match perfectly and we study the effect on a mirror factor different to *1*, say $(1+\varepsilon)$, in the pair M_3-M_4. The current in M_4 is a mismatched replica of the current in M_1. Therefore, to make equal the current in M_4 with the one in M_2 it is necessary to suitably imbalance (by the offset voltage) the differential input. Assuming that the offset is small and that the total current in the input pair is I_{Bias} we have

$$\left(\frac{I_{Bias}}{2} - g_{m1}\frac{V_{os1}}{2}\right)(1+\varepsilon) = \left(\frac{I_{Bias}}{2} + g_{m2}\frac{V_{os1}}{2}\right) \tag{5.17}$$

that becomes, being $g_{m1} = b_{m2}$

$$V_{os1} \cong \frac{I_1}{g_{m1}} \cdot \varepsilon \tag{5.18}$$

Therefore, the voltage offset is proportional to the mismatch ε through the multiplying coefficient I_1/g_{m1}. For the input pair in saturation and in the sub-threshold conditions we have, respectively

$$\frac{I_1}{g_m} = \frac{V_{gs1} - V_{Th}}{2} \quad \text{in saturation}$$
$$\frac{I_1}{g_m} = nV_T = \frac{nkT}{q} \quad \text{in sub-threshold} \tag{5.19}$$

Therefore, for a given mismatch the random offset is lower for input stages operated in sub threshold. A possible global mismatch is $\varepsilon = 0.02$ and a typical overdrive voltage in saturation is $(V_{gs} - V_{Th}) = 300\ mV$. Thus, the offset in saturation can be as large as $3mV$.

From (5.19) we note that the offset in *CMOS* circuits is always worse than for their bipolar counterpart. Since in many design the transconductance of CMOS transistors in saturation is from *5* to *10* times smaller than for equivalent bipolar counterpart, we have to expect in bipolar circuits a random offset which is *5* to *10* times smaller than *CMOS* circuits.

DESIGNER, TAKE NOTE

A large systematic offset discloses an inadequate circuit design.

A large random offset marks an improper layout.

5.4.4 Power Supply Rejection

Fig. 5.16 shows possible spur generators affecting a *p-channel* input two-stages amplifier. We assume that the source of M_3-M_4-M_5 and M_7-M_8 are physically one next to the other and a low impedance achieves their common connection. Therefore the sources of M_3-M_4-M_5 and those of M_7-M_8 are at the same voltage. By contrast, the transistor M_B can be located somewhere far from the amplifier (transistor M_B can be, for example, used to provide the bias voltage of many op-amps on the same chip). The wire connecting the source of M_B and the source of M_6-M_7 can collect some noise as represented by Δv_n^+. In addition, the spur signal on the supply lines possibly alters the current reference, I_{Ref}. The spur current source, $i_{n,Ref}$, models this effect.

We estimate the output noise voltage due to the spur generators by a small signal analysis. The various contributions are considered separately and are properly combined afterwards. Fig. 5.16 considers two possible outputs: the first, $v_{o,loc}$, applies when the output is used locally: it is referred to the source of M_5. The second one, $v_{o,far}$, is when the output signal is used far away from the op-amp location. In this case, the voltage spur affecting the wire connection

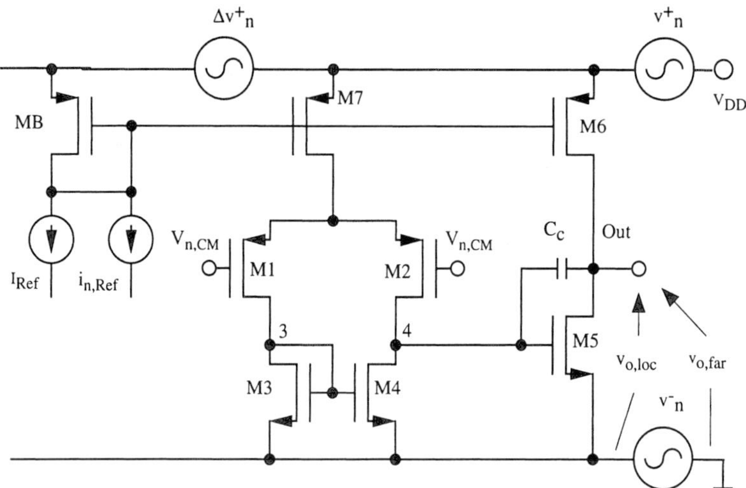

Fig. 5.16 - Two stages amplifier with possible voltage and current spur generators

from the output to the far away location should supplement the result achieved. Here we consider the local output only: the far away output differs from it by just v^{-}_{n}.

We can observe that the noise component Δv^{+}_{n} modifies the gate-to-source voltage originated by the diode connected transistor M_B

$$V_{GS6} = V_{GS7} = V_{GS,M_B} + \Delta v^{+}_{n} \tag{5.20}$$

As a result, some noise will affect the currents of M_6 and M_7. Assuming the spur Δv^{+}_{n} small enough, the noise currents are given by

$$\frac{i_{n,6}}{(W/L)_6} = \frac{i_{n,7}}{(W/L)_7} = \mu C_{ox}(V_{GS,M_B} - V_{Th})\Delta v^{+}_{n} \tag{5.21}$$

Moreover, since transistors M_6 and M_7 mirror the current in M_B, the noise terms in (5.21) are added to the ones caused by $i_{n,Ref}$. Therefore, we can combine the effect of Δv^{+}_{n} with the one of $i_{n,Ref}$. Assuming those two contributions to be uncorrelated the combination should be quadratic. However, the resulting currents affecting M_6 and M_7 are fully correlated.

The current of M_7 is equally split by M_1 and M_2. Moreover, because of the symmetry of the circuit, at low frequency, the voltage signal at node *4* equals the one at node *3*. Therefore M_5 mirror half of the spur current in M_7. The spur currents from M_5 and M_6 flow in the output node. Their superposition multiplied by the small signal output resistance lead to the low frequency output

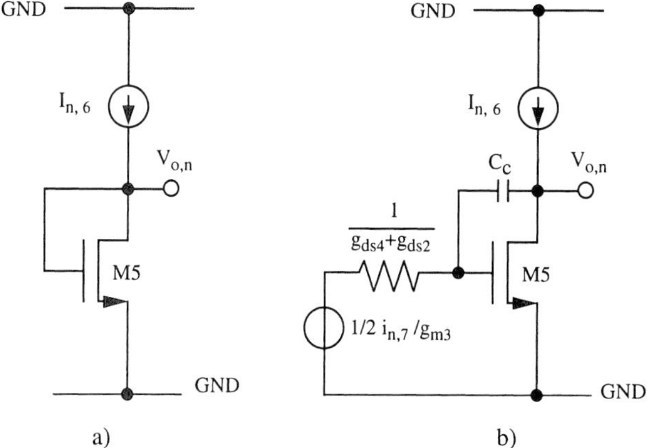

Fig. 5.17 - Small signal circuits for calculating the output spur produced by noise current

noise caused by noise on the current reference

$$v_{o,n,1} = i_{n,tot}\left[\frac{(W/L)_6}{(W/L)_B} - \frac{1}{2}\frac{(W/L)_5(W/L)_7}{(W/L)_4(W/L)_B}\right]\frac{1}{g_{ds6} + g_{ds7}} \tag{5.22}$$

Observe that if we use in (5.22) the same condition derived to minimize the systematic offset (equation (5.14)) the output noise becomes zero. Therefore, the rule for nulling the systematic offset also permits us to reject the low frequency components of $i_{n,Ref}$ and $\Delta v^+{}_n$. Unfortunately the compensation almost vanishes at high frequencies. Capacitor C_c, whose function will be discussed shortly, at high frequencies connects the gate and drain of M_5 making the second stage of the op-amp like the schematic shown in Fig. 5.17 a). The contribution of the first stage fades out and the second stage behaviour fully controls the output voltage. Therefore, we have

$$v_{o,n,1} = i_{n,Ref}\frac{(W/L)_6}{(W/L)_B} \cdot \frac{1}{g_{m,5}} \tag{5.23}$$

The transition between the results given by equation (5.22) and that in (5.23) occurs at frequencies for which the impedance of C_c, amplified by the Miller effect, becomes comparable to the output resistance of the first stage. Fig. 5.17 b) shows the circuit that we can use to study the issue. A voltage source and its output resistance depicts the small signal operation of the first stage. Remember that the spur currents $i_{n,6}$ and $i_{n,7}$ are fully correlated being produced by the same noise generator, $i_{n,Ref}$ (in Fig. 5.16). Analysis of the circuit leads to the transition frequency mentioned above. We don't provide here the details of calculations. They can be a useful exercise for the reader.

BE AWARE

The power supply rejection is a very difficult issue. Suitable circuit solutions provide acceptable protections at low frequency. Unfortunately, at very high frequency the circuit defences are almost vanquished because of unpredictable capacitive couplings.

In order to study the low frequency effect of v^+_n and v^-_n we furthermore exploit the symmetry properties. Assuming that any spur affects the differential inputs to the same extent $(v_{n,CM})$, for the first stage we can use the half schematic shown in Fig. 5.18 a). Fig. 5.18 b) displays its small signal equivalent circuit and also includes the small signal equivalent circuit of the second stage. The voltage at the output of the first stage, $v_{out,n1}$ is amplified by the gain of the second stage, $A_2 = g_{m5}/(g_{ds5}+g_{ds6})$. This term is superposed to the direct spur affecting the second stage.

The common-mode noise $v_{n,CM}$ comes from a possible coupling between the noisy supply lines and the input terminals. It can be represented by

$$v_{n,CM} = k_+ v^+_n + k_- v^-_n \tag{5.24}$$

where k_+ and k_- denote attenuation factors.

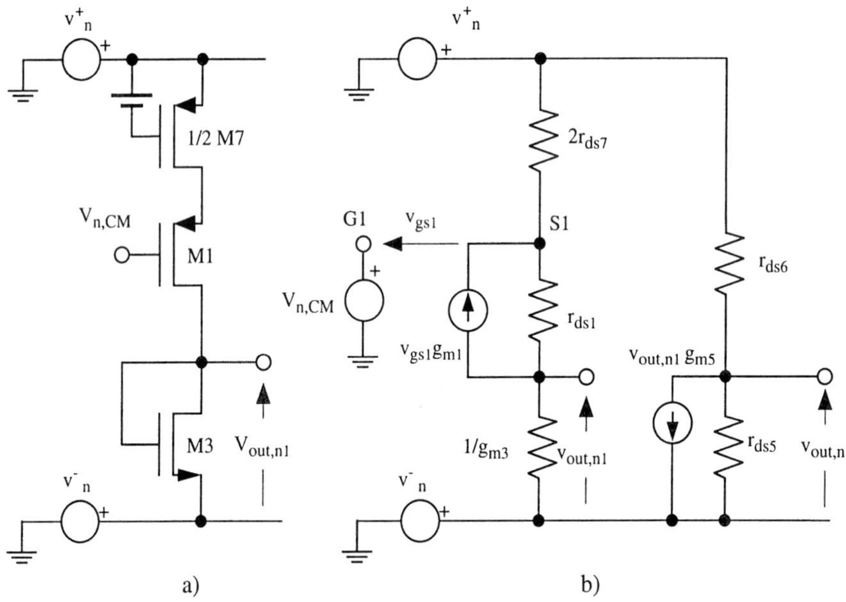

Fig. 5.18 - Circuits for calculating the effect of spur signals on the supply lines.

Therefore, we have to account for four terms: the two components that make $v_{n,CM}$, v^+_n and v^-_n. With the system being linear we can handle the problem by considering separately the contributions of the four spur generators. Then, the achieved outputs must be properly superposed. The terms produced by the same noise generator are correlated; therefore, their effects must be superposed linearly. By contrast, v^+_n and v^-_n are uncorrelated and their contributions must be superposed quadratically. The reader can do, as an exercise, the analysis using the equations that describe the small signal equivalent circuit. Here we study the problem by inspecting the circuit. Before, we shall estimate the noise currents; then, we shall calculate the effected the noise voltage.

Let us consider the first stage before. Observe that the source of M_1 follows its gate. Therefore, the spur current produced by $v_{n,CM}$ is approximated by

$$i_{n,CM} = \frac{v_{n,CM}}{2r_{ds,7}} = -\frac{k_+ v_n^+ + k_- v_n^-}{2r_{ds,7}} \tag{5.25}$$

The equivalent resistance seen from the source of M_3 is the series connection of diode-connected transistor M_3 and the cascode structure $M_1-(1/2)M_7$. Since $1/g_{m3}$ is negligible with respect to the cascode resistance, we assess the noise current developed by v^-_n as

$$i_n^- = -\frac{v_n^-}{r_{ds,7} g_{m1} r_{ds,1}} \tag{5.26}$$

The equivalent resistance seen from the source of $(1/2)M_7$ is given by the drain resistance of $(1/2)M_7$ in series with the resistance seen from the source of M_1. Here, $1/g_{m1}$ is negligible with respect to $2r_{ds,7}$; therefore, the effect of v^+_n is

$$i_n^+ = \frac{v_n^+}{2r_{ds,7}} \tag{5.27}$$

Note that the noise currents due to v^+_n in (5.25) and (5.27) have opposite sign.

The superposition of the four noise currents flows into $1/g_{m3}$ and develops a noise voltage given by

$$v_{o,n1}^2 \cong \frac{(1-k_+)^2 (v_n^+)^2 + \left(k_- + \frac{1}{g_{m1} r_{ds1}}\right)(v_n^-)^2}{(2g_{m3} r_{ds,7})^2} \tag{5.28}$$

it, amplified by $-g_{m5}/(g_{ds5}+g_{ds6})$, comes out across the drain-to-source of M_5.

The noise due to the first stage is superposed to the one given by the second stage. By inspection of the equivalent circuit in Fig. 5.18 b), one obtains

$$(v_{o,n2})^2 = \frac{g_{ds6}^2((v_n^+)^2 + (v_n^-)^2)}{(g_{ds5} + g_{ds6})^2} \tag{5.29}$$

and, superposing the effects

$$(v_{o,tot})^2 \approx \left(\frac{g_{ds6} - \dfrac{g_{m5}(1-k_+)}{2g_{m3}r_{ds3}}}{g_{ds5} + g_{ds6}}\right)^2 (v_n^+)^2 + \left(\frac{g_{ds6} - \dfrac{g_{m5}k_-}{2g_{m3}r_{ds3}}}{g_{ds5} + g_{ds6}}\right)^2 (v_n^-)^2 \tag{5.30}$$

An examination of (5.30) leads to the following observations:

- the noise from the first stage dominates the total result;
- the coupling coefficient k_+ reduces the outcome of $v^+{}_n$. It is therefore advisable to tight the common-mode input with the positive supply voltage (for p-channel input pair);
- the coupling coefficient k_- materializes a noise term. It is therefore advisable to well disjoint ground from the common-node input (for p-channel input pair);
- for n-channel input pair the two above recommendations must be switched.

High frequencies require a more simple study: the compensation capacitor establishes a connection from the output of the second stage to its input. Therefore, the contribution of the first stage vanishes. The spur from the supply lines is transferred to the output through the network made by $1/g_{m5}$ and r_{ds6}. Therefore, at high frequencies, it yields

$$(v_{o,loc})^2 = [(v_n^+)^2 + (v_n^-)^2]\left(\frac{1}{1 + g_{m5}r_{ds6}}\right)^2 \tag{5.31}$$

At high frequencies the local output rejects of the supply noise significantly because of the connection established by the compensation capacitor. However, this happens at frequencies around the gain-bandwidth product.

5.4.5 Effect of External Components on the *PSRR*

Power supply rejection additionally depends on the circuit that uses the op-amp. To discuss the issue, let us consider the scheme in Fig. 5.19. It illustrates a simple switched capacitor integrator. The input signal is injected by a capac-

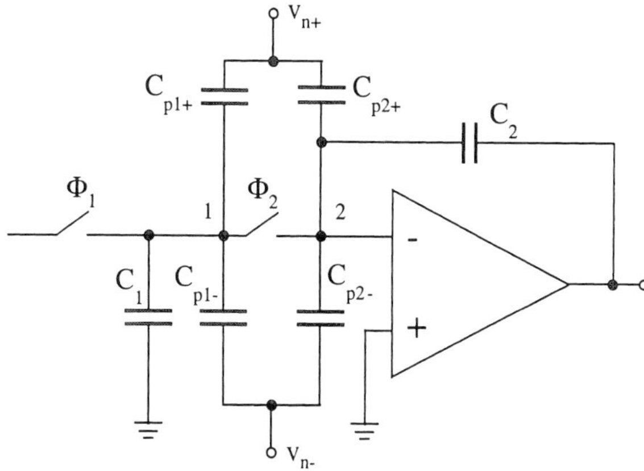

Fig. 5.19 - Spur signal injection caused by external parasitic elements.

itor C_1 and two switched driven by complementary non-overlapped phases. During phase *1* an input voltage charges capacitor C_1. During phase *2* the action of the op-amp transfer the charge on C_2. Four parasitic capacitances represent possible parasitic couplings between node *1* and *2* (the virtual ground) and the power supply lines. The parasitic capacitances are, in general, non linear elements often associated to *p-n* junctions. They include the parasitic given by the switches and the parasitic coupling created by the input transistors of the op-amp.

The spur signal at the output of Fig. 5.19 can be easily estimated. The result is different when the injecting switch is *on* (phase 2) or when it is *off* (phase 1). In the two cases we obtain, respectively

$$(v_{o,n})^2 = (v_{n+})^2 \cdot \left(\frac{C_{p1+} + C_{p2+}}{C_2}\right)^2 + (v_{n-})^2 \cdot \left(\frac{C_{p1-} + C_{p2-}}{C_2}\right)^2 \tag{5.32}$$

$$(v_{o,n})^2 = (v_{n+})^2 \cdot \left(\frac{C_{p2+}}{C_2}\right)^2 + (v_{n-})^2 \cdot \left(\frac{C_{p2-}}{C_2}\right)^2 \tag{5.33}$$

Therefore, depending of the phase in which we sample the output, we have a larger or smaller coupling with the spur supply lines.

All the components' parasites must be reduced to a minimum. In particular, it is essential to minimize the coupling between the connection to the virtual ground and the substrate. Long interconnections collect significant spur signals because of their substrate coupling. When long wiring to the virtual ground is necessary it is recommended to use electrically shield, achieved by a proper use of the metal layers available.

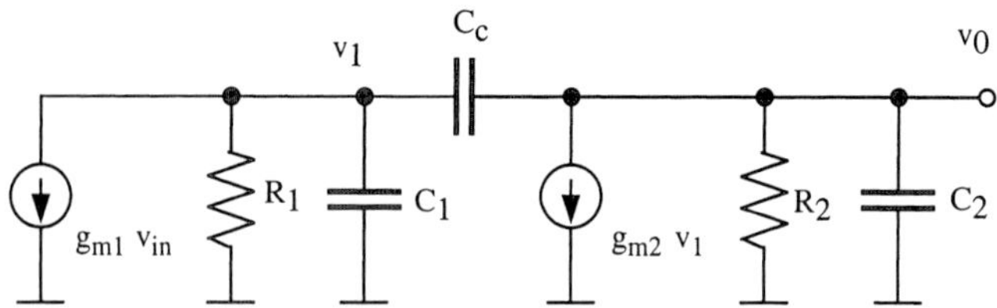

Fig. 5.20 - Small signal equivalent circuit of a two stages OTA.

5.5 FREQUENCY RESPONSE AND COMPENSATION

The small signal equivalent circuit of the two stage amplifier in Fig. 5.12 can be represented with the simplified diagram in Fig. 5.20. Each stage is represented by a transconductance generator and the parallel connection of an output resistance and a load capacitance. The *dc* gain is given by the product of the two gains $g_{m1}R_1$, $g_{m2}R_2$. The two *RC* networks contribute with two poles whose angular frequencies are

$$p'_1 = \frac{1}{\tau_1} = \frac{1}{R_1 C_1} \tag{5.34}$$

$$p'_2 = \frac{1}{\tau_2} = \frac{1}{R_2 C_2} \tag{5.35}$$

We know that the output resistances of the two stages are given by the parallel connection of two r_{ds}. The load capacitances result from the parasitic elements of the transistors used and, possibly, the capacitance loading the output. It turns out that the two time constants of the first and the second stage are not significantly different.

However, since the transfer function has two poles, we can ensure stability only if the poles are sufficiently far apart that when the module of the gain becomes *1* the second pole has not been achieved yet. The first pole is called the *dominant pole* since it dominates the frequency response behaviour in the region where the gain is larger than one.

Unfortunately, in the two stage amplifier, the poles p'_1 and p'_2 are relatively close one to the other. The circuit does not have a dominant pole and *compensation* becomes necessary.

One possible (and widely used) compensation network is achieved by the capacitor C_c connected between input and output of the second stage (see Fig. 5.20). We account for its effect in the small signal transfer function by

$$v_1(g_1 + sC_1) + (v_1 - v_0)sC_c + g_{m1}v_{in} = 0 \qquad (5.36)$$

$$v_0(g_2 + sC_2) + (v_0 - v_1)sC_c + g_{m2}v_1 = 0 \qquad (5.37)$$

they are the node equations at the output of the first and the second stage. Moreover, $g_1 = 1/R_1$ and $g_2 = 1/R_2$. Solving and leads to the frequency response

$$\frac{V_0}{V_{in}} = g_{m1}R_1R_2 \frac{g_{m2} - sC_c}{1 + sR_1R_2g_{m2}C_c + s^2R_1R_2[C_1C_2 + (C_1 + C_2)C_c} \qquad (5.38)$$

that shows one zero and two poles. The position of the poles is approximately given by

$$p_1 \cong \frac{-1}{g_{m2}R_2R_1C_c} \qquad (5.39)$$

$$p_2 \cong \frac{-g_{m2}C_c}{C_1C_2 + (C_1 + C_2)C_c} \qquad (5.40)$$

and the zero is located at

$$z = +\frac{g_{m2}}{C_c} \qquad (5.41)$$

therefore, the action of the compensation capacitor C_c on the poles is twofold. Pole p_1 is pushed at low frequency. In fact, it is $g_{m2}R_2C_c/C_1$ lower than p'_1 in (5.34). Pole p_2 is brought to a high frequency: it is approximately $g_{m2}R_2\,(C_2/C_1+C_2)$ times p'_2. The double action on the poles is called *pole splitting*.

To better memorize the achieved result we observe that the action of the compensation capacitor can be analysed by using the Miller theorem. Since the gain of the second stage is $g_{m2}R_2$, capacitor C_c is amplified by $1+g_{m2}R_2$ when transferred across the output of the first stage and ground and remains almost unchanged when transferred to the output of the second stage. Therefore the capacitive load of the first stage becomes $(C_1+g_{m2}R_2C_c)$ and the one of the second stages slightly increased: (C_2+C_c). Moreover the compensation capacitor establishes a negative feedback around the second stage of amplification. Therefore, as we know from basic

Two Stages OTA Compensation

A capacitor in feedback around the second stage splits the poles. The pole of the first stage goes at low frequency and the one of the second stage is pushed at high frequency. The action is named pole splitting.

electronic courses, the bandwidth of the second stage is enlarged by the loop gain.

The two poles of the transfer function are in the left s-plane, as result from the minus sign in (5.39) and (5.40). By contrast the zero is in the right s-plane. Therefore, the phase shift produced by the zero will be negative like a pole in the left s-plane. As a result the phase shift of the zero does not improve the phase margin but instead, worsens it. Consequently, the zero can be a problem if it is located close to the unity gain frequency. Assuming that the first pole is dominant, multiplying it by the *dc* gain obtains the expected unity gain angular frequency

$$\omega_T = 2\pi f_T = \frac{g_{m1}}{C_c} \tag{5.42}$$

REMEMBER THESE EQUATIONS

Dominant pole:

$$f_1 = \frac{1}{2\pi} \frac{1}{g_{m2} R_2 R_1 C_c}$$

Unity gain frequency:

$$f_T = \frac{1}{2\pi} \frac{g_{m1}}{C_c}$$

Zero in the right s-plane:

$$f_1 = \frac{1}{2\pi} \frac{g_{m2}}{C_c}$$

by comparing (5.41) with (5.42) we see that the ratio between the zero and ω_T is equal to the ratio of the transconductance gain of the second and the first stage. Therefore, the zero is far away from the expected unity gain frequency if the transconductance of the second stage is much higher than the one of the first stage. With a *CMOS* circuit it is difficult to achieve large differences between g_{m2} and g_{m1}: we have an increase in g_m with the square root of the current and the aspect ratio. By contrast, with a bipolar technology the transconductance is proportional to the bias current. Therefore, in bipolar implementations g_{m2} can be designed suitably larger than g_{m1}, while with *CMOS* we cannot achieve the same result easily. If the zero is near the expected ω_T, its effect modifies significantly the frequency response near the *zero dB* crossing degrading the stability conditions. Thus, in practical cases, we cannot leave the circuit as it is, but we need to find some remedy to the unsatisfactory phase margin.

The problem of the zero in the right s-plane can be solved by three techniques:

- use of a unity gain buffer
- use of the zero nulling resistor
- use of a unity gain current amplifier

If we examine equations (5.36) and (5.37) we identify the term $-sC_cv_1$ in

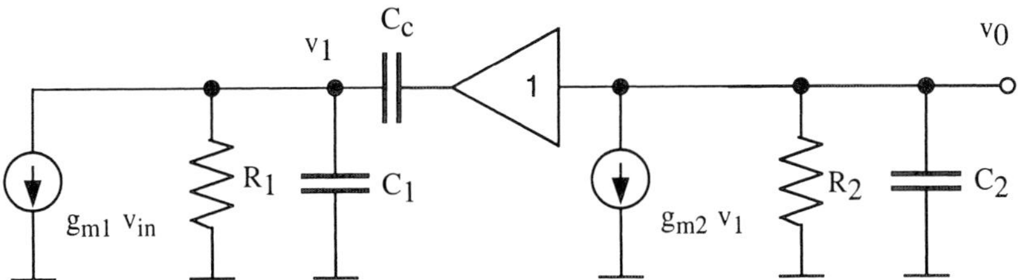

Fig. 5.21 - Use of a unity gain buffer to cancel the right-plane zero

(5.37) as the responsible that lead to the zero. If we are able to eliminate such a term the zero disappears from the solution. We achieve the result with a unity gain buffer connected to the output node and feed the compensation capacitor (Fig. 5.21). The second term in (5.37) disappears, as the current from the output node to the compensation capacitor is provided by the buffer. Actually, in addition to the term $-sC_c v_1$, we also eliminate $sC_c v_0$; therefore, we will have some changes in the pole positions. If we use instead of (5.37) the relationship

$$v_0(g_2 + sC_2) + g_{m2}v_1 = 0 \tag{5.43}$$

the solution of the new system made by (5.36) and (5.43) becomes

$$\frac{V_0}{V_{in}} \cong \frac{-g_{m1}g_{m2}R_1R_2}{1 + sR_1R_2g_{m2}C_c + s^2R_1R_2(C_1 + C_2)C_c} \tag{5.44}$$

whose denominator is only slightly different from that in (5.38). Therefore, the poles will approximately remain unchanged but the zero disappears (5.37).

Fig. 5.22 shows the possible circuit implementation of the technique discussed above. It is applied for an *n-channel* input transistors architecture. In the right side circuit a source follower achieves the required buffer. Note that, to operate properly, the buffer needs at least $V_{th}+2V_{sat}$ at its input. If the output voltage is lower than this limit the current source is pushed into the triode region and the buffer does not operate properly any more. Therefore, the use of a unity gain buffer would limit the achievable output swing of the two stages op-amps.

Another important point to observe is that proper operation of the buffer matters around the unity gain frequency of the op-amp: the compensation network must correct an undesired behaviour at high frequency. At very high frequency the buffer can suffer some frequency limitations with a consequent phase shift from input to output. The resulting effect is that the zero is not eliminated but a doublet zero-pole, each close to the other, appears in the transfer function. The doublet (in the left *s*-plane) is not particularly problematic. How-

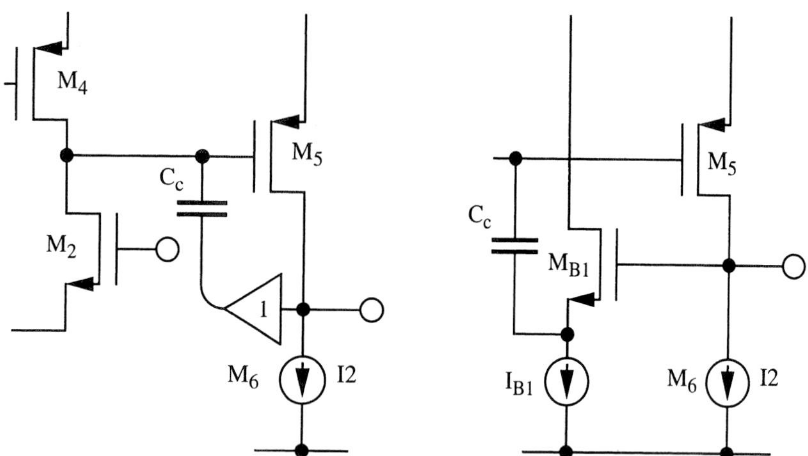

Fig. 5.22 - Eliminating the right-plane zero with a unity gain buffer Left use of a source follower.

ever, under given feedback factors it can be responsible for a worsening of the phase margin. In addition to the above limitation, we have to account for an obvious increase in power consumption and chip area.

> **CALL UP**
>
> The unity gain buffer possibly used in the compensation network must be very fast. It should work properly at frequencies higher than the unity gain frequency. Otherwise the right-half plane zero is not cancelled but it is replaced by a doublet pole-zero.

The second technique that handles the problem of the zero in the right s-plane is the zero nulling. The compensation network is made, instead of a capacitor, by an impedance: it is the series connection of a resistance and the compensation capacitance. Fig. 5.23 show the small signal schematic. If we use this kind of solution we have to replace sC_c in equations (5.36) and (5.37) with

$$sC_c \Rightarrow \frac{sC_c}{1 + sR_2C_c} \tag{5.45}$$

The solution of the modified equations (5.36) and (5.37) does not show any significant change in the denominator of the transfer function. Thus, the poles are substantially unchanged. By contrast the zero position is modified into

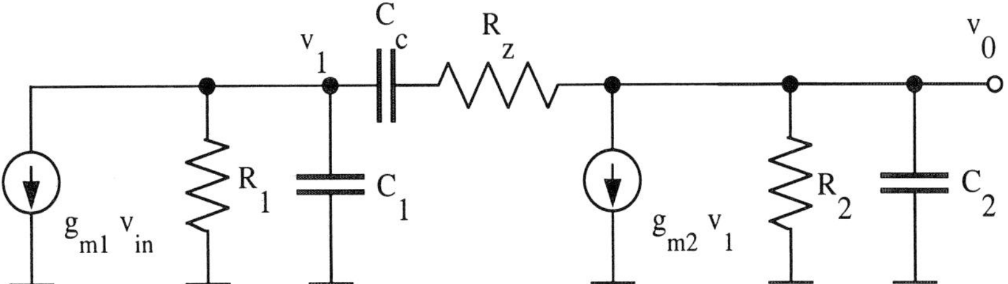

Fig. 5.23 - Use of the zero nulling resistor to move the right-plane zero.

$$z = \frac{1}{C_c\left(\frac{1}{g_{m2}} - R_z\right)} \tag{5.46}$$

We can observe that depending on the sign of $(1/g_{m2} - R_z)$ the zero is in the right or left side of the s-plane. Moreover for $(1/g_{m2} - R_z) = 0$ the zero goes to frequency ∞, or as we normally say, it is nulled. Having a zero in the left s-plane can be attractive and possibly, achieving a pole-zero cancellation we can enlarge the op-amp bandwidth. Nevertheless, it is not advisable to try this design avenue. The accuracy of integrated resistors is not particularly high; thus, a pole zero cancellation is never achieved in practical cases. Instead, it is advisable to use a value of R_z that nominally accomplishes the nulling operation. Possible variations due to technological changes will move the zero to around infinite in the positive or negative half plane but, always sufficiently far away from the critical unity gain frequency.

The zero nulling resistance should match the inverse of the transconductance gain of the second stage that, in turn, depends on the electrical and technological parameters. A possible way to track some of the parameter variations is to

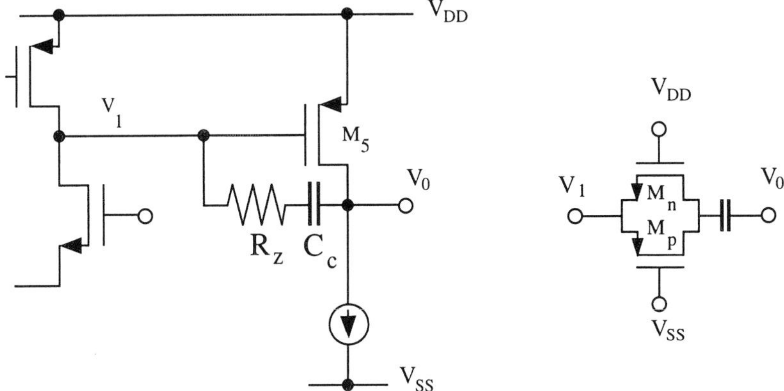

Fig. 5.24 - Zero nulling resistor achieved with complementary transistors.

use one or two transistors to realize the zero nulling resistor. Since they are in series with the compensation capacitor, the *dc* current is zero. Therefore, the transistor(s) is (are) in the triode region. Assuming we use the compensation network in right side of Fig. 5.24 the resistance R_z will be expressed by

$$R_z = \frac{R_n R_p}{R_n + R_p} \tag{5.47}$$

where

$$\frac{1}{R_n} = k'_n \left(\frac{W}{L}\right)_n [V_{DD} - V_1 - V_{Th,n}] \tag{5.48}$$

$$\frac{1}{R_p} = k'_p \left(\frac{W}{L}\right)_p [V_1 - V_{SS} - V_{Th,p}] \tag{5.49}$$

ZERO-NULLING TECHNIQUE

The resistance that cancels the zero in the right-half plane is the inverse of the transconductance gain of the second stage.

The best strategy is just to send the zero at infinite and not trying to enlarge the bandwidth by some pole-zero cancellation.

observe that the two transistors implementing the zero nulling resistor are connected toward the node V_1. From the electrical point of view it is irrelevant if we connect the zero nulling resistor on the V_1 side or on the V_0 side. However, we have a significant practical difference. Since we have amplification between V_1 and output, the voltage of V_1 will show little variation with respect to the output. Therefore, if we put the equivalent resistor on the V_1 side, the variation in R_n and R_p will be kept to a minimum. We should note that the *dc* level of V_1 is below V_{dd} by the gate to source voltage of M_5. Therefore, when we use the equivalent resistance on the V_1 side we have

$$V_{gs,n} = V_{gs,5} \tag{5.50}$$

Transistor M_5 is a *p-channel* device while M_n is a *n-channel* one. Remembering that the thresholds of the n-channel and *p-channel* can be different, and their technological variation uncorrelated, we observe that in nominal or worst case situations the voltage $V_{gs,5}$ cannot be enough to drive M_n properly. Therefore, when using the equivalent R_z on the V_1 side, many designers prefer to employ only one *p-channel* transistor (they use an *n-channel* for a complementary architecture).

If we put the equivalent resistor on the V_o side, when the voltage of the transistor channels changes, the resistance R_n moves in one direction while R_p moves in the opposite direction. Consequently, if

$$k'_n\left(\frac{W}{L}\right)_n = k'_p\left(\frac{W}{L}\right)_p \tag{5.51}$$

we can keep constant R_z by compensating the two changes.

$$\frac{1}{R_z} = k_n\left(\frac{W}{L}\right)_n [V_{DD} - V_{SS} - V_{Th,n} - V_{Th,p}] \tag{5.52}$$

However, since the condition (5.51) relies on the accuracy of technological parameters, the above benefit can be difficult to achieve in practice.

Example 5.1

Design a two stages OTA having more than 70 dB dc gain and 70 MHz GBW. The phase margin must be higher than 55° with a 2 pF output stage load. Use a zero nulling compensation network. The supply voltage is 1.8 V and the target power consumption is 0.6 μW. The desired output swing is 0.3 1.5 V. Use the transistor models given in Appendix C.

Solution: *The total bias current permitted by the specifications is around 300 μA. We assume that the power budget doesn't include the power required by the bias network. Moreover, we suppose that the current in the second stage is twice the one in the input differential pair. Accordingly, the current in M_7 (see below diagram) becomes 100 μA and the one in M_6 200 μA.*
The load of the OTA is 2 pF capacitor. This value is pretty large compared to all the parasitics affecting the output of the first and the second stage. Therefore, in the small signal equivalent circuit in

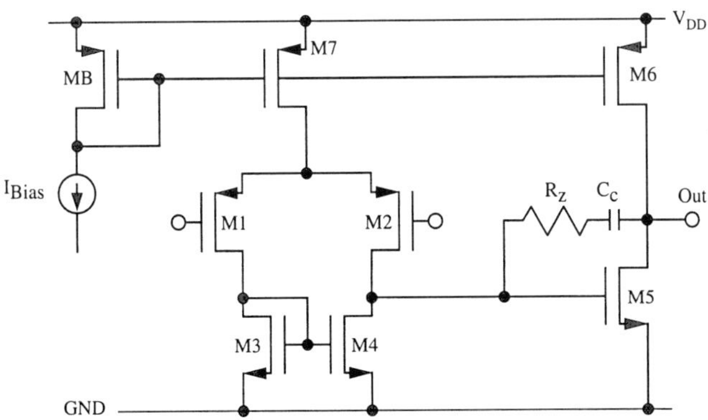

Fig. 5.23 it is allowable to use the condition $C_1 \ll C_2$. Equations (5.40) and (5.42) state

$$\omega_T = \frac{g_{m1}}{C_C} \ll \omega_2 \approx \frac{g_{m2}}{C_2}$$

where the inequality ensures a proper phase margin. Since the transconductance of the input pair is typically comparable to the one of the second stage, the compensation capacitance must be equal or, when necessary, larger than the load capacitance. Using C_C =2 pF we have

$$g_{m1} = 2\pi f_T C_C = 6.28 \cdot 70 \cdot 2 \cdot 10^{-6} = 0.879 \cdot 10^{-3} \Omega^{-1}$$

The minimum feature of the technology is 0.25 μm. However, since the speed specification doesn't impose to go to the minimum allowed, a 0.4 μm length in input pair provides some safety margin. The Spice model parameters gives μ_p = 120 cm^2/V-sec and t_{ox}=7.5 nm ($C_{ox} \approx 4fF/\mu^2$). The use of the relationship

$$g_m = \sqrt{2\mu C_{ox} \frac{W}{L} I_D}$$

leads to $(W/L)_1$ = 154. However, a Spice simulation reveals that the above aspect ratio leads to a transconductance lower than expected ($0.62 \cdot 10^{-3}$ Ω^{-1}). This is likely due to the limits of the approximate expression of g_m used. Spice simulations show that the transistor width capable to achieve the requested g_m using 50 μA bias current is W=160 μm, (W/L)= 480. The g_m becomes equal to $0.86 \cdot 10^{-3}$ Ω^{-1}. The saturation voltage is 94 mV. Observe that the use of the relationship $g_m = 2I_D/V_{sat}$ produces $g_m = 1.07 \cdot 10^{-3}$ not much different from the simulated result. Thus, the second expression of g_m is more accurate than the previous one.
The length of the active loads, M_3 and M_4 should be longer than the input pair (we derived such condition for the optimization of the noise performance of the inverter with active load. It hold more in general and we use it here). We choose L_3=L_4=0.8 μm. The width of M_3 and M_4 are designed according to the following observations: the width should lead to a reasonably low overdrive voltage; moreover, the sizing and the current in M_5 are related to the ones of M_3 and M_4. Since g_{m5} should be large, g_{m3} and g_{m4} should be reasonably large as well. In addition the overdrive of M_5 must be low enough to accommodate the required output swing (saturation 0.2 V). The above thoughts and Spice simula-

tions lead to W_3=W_4 = 40 μm.
Next step of the design process involves M_6 *and* M_7. *They provides the bias current in the two stages and, in addition,* M_6 *works as the active load of the second stage. Possible design guide-lines are the following:*

- *the output resistance of* M_6 *must be high*
- *the overdrive voltages must be low (around 0.2 V)*
- *the length of the two transistor must be the same (to ensure matching)*
- *the current in* M_6 *is twice the one in* M_7.

Using the above propositions and Spice simulations we can end on the following transistor sizing: $L_{6,7}$ = 0.6 μm; W_6= 200 μm W_7 = 100 μm.
The aspect ratio of M_5 *is defined by the condition of zero systematic offset. Therefore, using the same length of* M_4 *we have* L_5= 0.8 μm *and* W_5= 160 μm.
Observe that all the designed widths are round numbers. Attempts to optimize some feature can lead to widths with fraction of microns. That is not advisable because the layout would become problematic.
A .op *and a* . ac *simulation of the circuit using the designed transistor sizing lead to the following results:*

- *the transconductance of the input pair is* $0.85 \cdot 10^{-3} \Omega^{-1}$ *(as expected)*
- *the transconductance of* M_5 *is as high as* $3.2 \cdot 10^{-3} \Omega^{-1}$
- *the saturation voltages of all the transistors is below 0.2 V*
- *the low frequency gain is 73 dB (satisfactory)*
- *the residual systematic offset is only 0.14 mV*

Therefore, simulations confirm the validity of the design as far as the operating point and low frequency gain is concerned. The simulated output conductances are $g_{ds1}=15.31 \cdot 10^{-6}$; $g_{ds3}=2.28 \cdot 10^{-6}$; $g_{ds5}=9.81 \cdot 10^{-6}$; $g_{ds6}=121.55 \cdot 10^{-6}$. *The use of these figures in (5.8) leads to a dc gain of 74 dB, just a bit more than the simulated value.*
Finally we have to study the frequency compensation. The value of g_{m5} *determine the zero nulling resistor:* R_Z =320 Ω; *with a Spice simulation we verify that a compensation capacitance of 2 pF leads to* f_T = 100 MHz *but the phase margin is only 45°. This result comes from the high value of* g_{m5}. *The low phase margin requires to increase the compensation capacitance. Increasing it to 3 pF leads to* f_T = 75 MHz *and phase margin 57°. We can observe that the sec-*

ond pole is at 107 MHz (where the phase is margin (45°), it is about 1.5 times larger than the unity gain frequency.

The third technique used to control the right s-plane zero is again based on the observation that the zero comes from the ($-sC_cv_1$) component of the current flowing from the output node to the node *1*. We cancel that component by using instead of a unity gain buffer a unity gain current amplifier as shown in Fig. 5.25.

The matched current generator I_{comp}, compensates the current injected into node *A* thus ensuring that the biasing condition of the first stage is not disturbed. Moreover, transistor M_c establishes a low impedance at node *A*. Therefore, the small signal current flowing in the compensation capacitance approximately results

$$i_c = v_0 sC_c \tag{5.53}$$

and, in particular is not affected by the voltage at node *1*. Because of the low impedance at its source transistor M_c delivers the signal current i_c to the node *1*, acting like a current buffer. Equations (5.36) and (5.37) are slightly modified in

$$v_1(g_1 + sC_1) + g_{m1}v_{in} - v_0 sC_c = 0 \tag{5.54}$$

$$v_0(g_2 + sC_2) + g_{m2}v_1 + v_0 sC_c = 0 \tag{5.55}$$

where we neglect the resistance seen from the source of M_c.

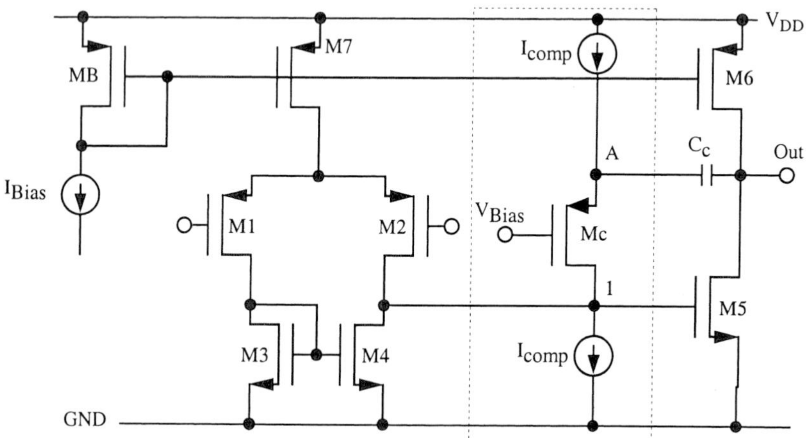

Fig. 5.25 - Cancellation of the right-plane zero by using a unity gain current amplifier.

The solution of (5.54) and (5.55) (that can be done by the reader as a useful exercise) reveals that the transfer function is no longer affected by a zero and that the two poles are approximately located as in the previously discussed solutions. In designing the bias voltage V_{bias} we have only to account for the biasing needs of transistor M_c and the current generators used. In order to avoid pushing the current generator on the upper part of the schematic in the triode region, we should use $V_{bias} \leq V_{DD} - V_{Th,p} - 2V_{sat}$.

Example 5.2

Design the unity gain current amplifier used to compensate a two stages amplifier, as shown in Fig. 5.25. Use the same transistor sizing derived in Example 5.1. Assume ideal the current generators required.

Solution: *The effectiveness of the technique rely on the use of a good unity gain current amplifier: the input conductance should be pretty high with respect to the admittance of the compensation capacitance at frequencies around the GBW. If not, the current in the compensation capacitor will exhibit a phase shift that, in turn, will lead to a zero-pole doublet. Assuming that the compensation capacitance is 2 pF, at 70 MHz the capacitive admittance is*

$$\omega C_C = 2 \cdot \pi \cdot 70 \cdot 10^6 \cdot 2 \cdot 10^{-12} = 8.79 \cdot 10^{-4} \Omega^{-1}$$

The simulation done in the Example 5.1 showed that the p-channel transistor M_6 (W/L = $200\mu m/0.6\mu m$, I_D=200 μA) showed a transconductance $g_m = 21 \cdot 10^{-4} \Omega^{-1}$. therefore, we expect that in the current buffer it will be necessary to use such a pretty large aspect ratios and current level.

In order to investigate the effect of a limited input conductance it is helpful doing a simulation with, for instance, a p-channel transistor with aspect ratio $W/L = 200\mu m/0.4\mu m$ and a bias current of only 50 μA. The figure in the next page shows the Bode diagrams of the small signal response. Observe that the relatively low input conductance at the source of M_C produces the expected zero-pole doublet; the zero is in the left s*-plane (as shown by the phase behaviour) and it is located around 65 MHz; the pole is near 170 MHz. The effect of the doublet is not beneficial because the zero lies before the unity-gain frequency. It sustains the gain above 0 dB and when the non-dominant poles occur the frequency response displays a fast phase drop.*

In order to push the doublet at a frequency higher than the unity gain bandwidth it is necessary to increase the g_m of M_C. Bringing

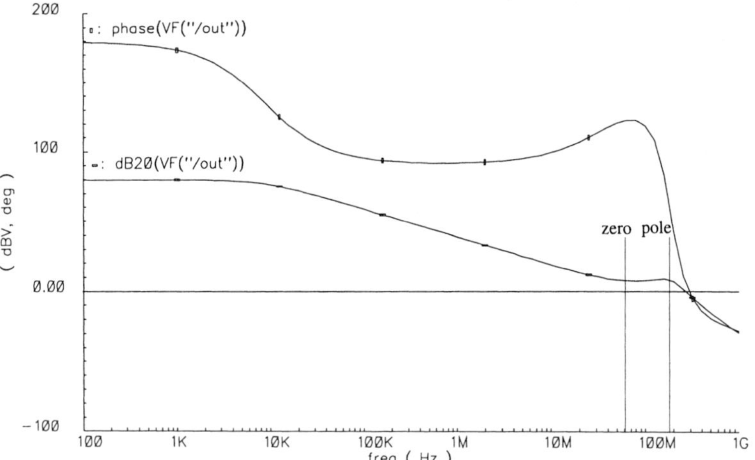

I_{comp} to 250 μA, we obtain the Bode diagram shown in the below Figure. The zero-pole doublet is moved at high frequency and it doesn't affect the Bode diagram any more. The reader can acquire familiarity with the circuit and the technique with some Spice simulations. The result shown in the Figure achieves a 72° phase margin. The compensation capacitance is 2 pF and the unity gain frequency is as good as 159 MHz. The superior performances of the circuit compared to the zero nulling resistor technique is paid with a significant increase of power consumption.

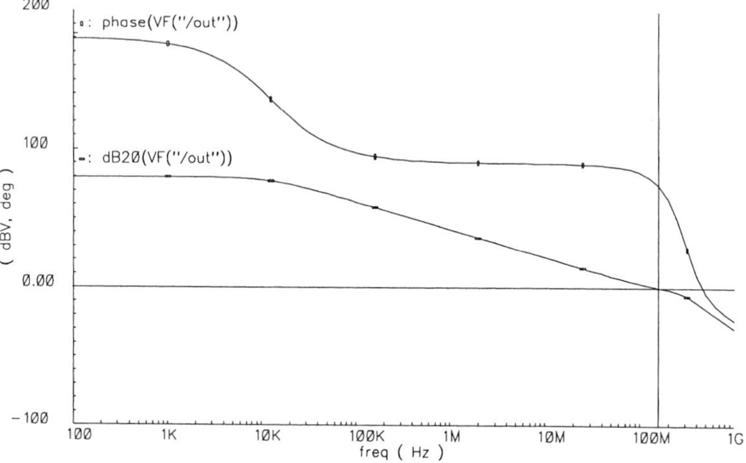

5.6 SLEW RATE

A large signal applied to the input pair causes a complete umbalancement of the stage: the bias current completely flows through M_1 or M_2. Under such conditions the feedback network loose the control of the output node; the swing of the output voltage depends on the load capacitance and the compensation capacitance, that must be both charged or discharged, and the current available for this operation in the first and the second stage of the amplifier.

Let us assume the input pair unbalanced so that the current I_{M7} totally flows through M_2 (Fig. 5.26); M_1 is off. M_3 is consequently off and M_4 is off as well since it mirrors the current in M_3. As a result, the entire current I_{M7} flows through the compensation network. Therefore, current I_{M7} discharges capacitor C_c with a constant pace. The voltage across C_c decreases and, if the voltage of the drain of M_2 remains constant, the output voltage goes down with a slew rate given by

$$SR- = \left.\frac{\Delta V_-}{\Delta t}\right|_{max} = -\frac{I_{M7}}{C_c} \tag{5.56}$$

The discharge current at the other plate of C_c flows through M_5. The voltage of the drain of M_2 drops-up by a given extent so that the current furnished by M_5 increases and likely provides the current discharge of both C_c and C_L. If the increase of I_{M5} is not large enough the operation slows down. Therefore, the slew rate expressed by (5.56) is the maximum achievable.

In the other symmetrical case, when the current I_{M7} flows throw M_1, M_4 mirrors the I_{M7} current while M_2 is off. Therefore the I_{M7} current from M_4 flows comes from the compensation network. Under the assumption that the drain of M_2 jumps down by the overdrive of M_5, transistor M_5 turns off and the current

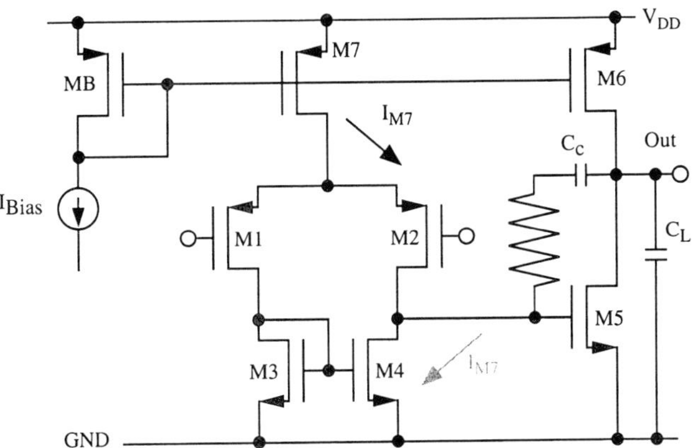

Fig. 5.26 - OTA in the positive and negative slew rate conditions.

I_{M6} is fully available to charge C_c and C_L. Therefore

$$SR_+ = \left.\frac{\Delta V_+}{\Delta t}\right|_{max} = \frac{I_{M6}}{C_c + C_L} \tag{5.57}$$

In addition, we have the current limit through C_c. It gives $SR_+ = I_{M7}/C_c$. So, the smaller one will hold. Note that the above equations are an approximation: M_5 does not switch on or off immediately and the transient of this process moderates the output node slewing up and down.Therefore, the use of equations (5.56) and (5.57) provide just guidelines in the preliminary design phase. Computer simulations achieve a more accurate estimation.

Often it is necessary to have symmetrical positive and negative slewing. This design requirement leads to the condition

$$\frac{I_{M7}}{C_c} \leq \frac{I_{M6}}{C_c + C_L} \tag{5.58}$$

that establishes a design constraint over the choice of the bias current of the first and second stage.

Since $\omega_T = \frac{1}{\tau_T} = \frac{g_{m1}}{C_c}$, the slew-rate can be expressed as

$$SR = \frac{I_{M1}}{g_{m1}}\omega_T = (V_{GS1} - V_{Th})\omega_T = \frac{V_{GS1} - V_{Th}}{\tau_T} \tag{5.59}$$

REMEMBER THAT

A symmetrical slew-rate requires a current in the second stage larger than the tail current of the input pair by the factor $(C_C+C_L)/C_C$.

therefore, the slew-rate is proportional to the overdrive voltage of the input pair and the unity gain frequency. If f_T is *40 MHz* and the overdrive is *300 mV* the slew rate becomes *75.4 V/μs*, Moreover, using the same figures, the time required for a *1 V* slewing turns out to be *3.3* times τ_T.

Example 5.3

Given the op-amp designed in Example 5.1, determine with a Spice simulation the positive and negative slew-rate. Use the OTA in the open-loop conditions and apply a large step signal at the inputs. Compare the achieved result with what expected using the approximate expressions (5.56), (5.57) and (5.59).

Solution: *Example 5.1 uses a capacitive load* $C_L = 2\ pF$ *and a compensation capacitance* $C_c = 3\ pF$*. Moreover, the bias current* *in the first stage is I_{M7} = 100 μA and the current in the second stage is I_{M6} = 200 μA. The use of equation (5.58) predict that the slewing in the positive and negative directions are both limited by the $SR = I_{M7}/C_c$ condition. In fact, using (5.56), (5.57) we obtain*

$$SR_- = -\frac{I_{M7}}{C_c} = 33V/\mu sec \qquad SR_+ = \frac{I_{M6}}{C_c + C_L} = 40V/\mu sec$$

The current of the input transistors is 50 μA while their g_m is $0.88 \cdot 10^{-3}\ \Omega^{-1}$. Since the f_T achieved in Example 5.1 is 75 MHz, the use of (5.59) predicts

$$SR = \frac{I_{M1}}{g_{m1}} 2\pi f_T = 23.56V/\mu sec$$

which is lower than the value predicted by the use of equations (5.56) and (5.57).
A time-transient simulation with a big step (± 100 mV) applied to the inputs lead to the responses shown in the below figure. Since the OTA is in the open loop conditions we observe only the slewing transients: they bring the output from the positive to the negative (ground) rails and vice-versa. The two slopes are, as expected, equal and their value is 22 V/μsec. This value is not much different from the results of the hand calculations. However, the use of equation (5.59) leads to a more accurate forecast.
Once again the reader should be aware that the equations derived in this book are the result of many approximations. It should not be surprisingly if the difference with simulation results are as large as

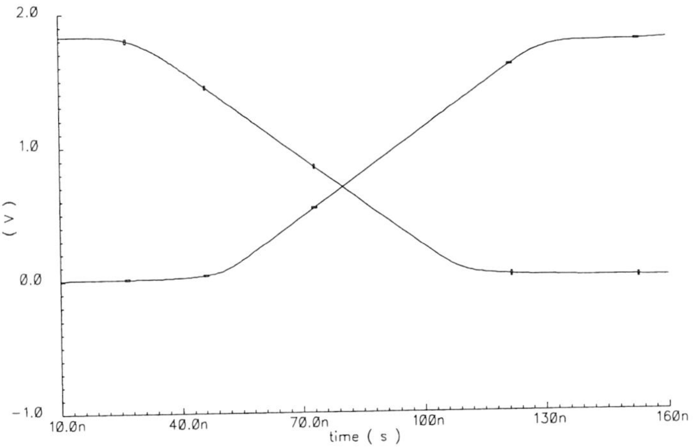

50% or more.
The transient responses in the figure mark some delay from the time at which the impulse is applied and the time the output starts slewing. Since the impulse start from 100 mV (or -100 mV) the initial condition of the output is a strong saturation at 0 or 1.8 V. The transition between saturation and normal condition of operation leads to some delay. For the specific circuit studied the delay required to recover from zero is higher than the one required to exit from the V_{DD} *saturation.*

5.7 DESIGN OF A TWO STAGE *OTA*: GUIDELINES

The study of the features of a two-stage *OTA* determined a number of design recommendations. Some of them are conflicting one each other. So, the designer should find the best trade-off to match the design specifications. Some recommendation are statement other are expressed by an equation. The most important of them are the following:

- a reduction of the bias current increases the *dc* differential gain;
- in order to optimize the systematic offset the transistor aspect ratios should comply equation (5.14);
- the random offset depends on matching; however, for a given mismatch offset is proportional to $I_{bias}/g_{m,in}=(V_{GS1}-V_{Th})$; thus, in saturation the random offset is reduced by reducing the overdrive voltage of the input pair;
- the unity gain frequency ω_T is equal to $g_{m,in}/C_c$; additionally, in order to ensure a proper phase margin, ω_T must be smaller that $p_2 \approx g_{m2}/(C_1+C_2)$. Therefore, the compensation capacitance must be larger (by at least a factor 2) than $(C_1+C_2)g_{m,in}/g_{m2}$ [C_1 and C_2 are the capacitive load of the first and the second stage];
- since ω_T must be smaller than the non-dominant pole, p_2, the achievable gain bandwidth depends on the angular frequency of p_2. The designer achieves high speed positioning the non-dominant poles at pretty high frequencies;
- the zero nulling resistor should be equal to $1/g_{m2}$;
- symmetrical slew rates require $I_{B1}/C_c \leq I_{B2}/(C_c+C_L)$, [$I_{B1}$ and I_{B2} are the bias currents in the first and the second stage];
- with a given bandwidth we maximize the slew-rate with an high input pair overdrive [remember the relationship $SR=(V_{GS,1}-V_{Th})\omega_T$];
- optimum noise performances (topic that we will study shortly) require an

input pair transconductance higher than the one of the input stage active load. Moreover, the length of the active load should be higher than the one of the input pair.

5.8 SINGLE STAGE SCHEMES

The previous sections discussed in a detail the features of a two stages op-amp (or better an *OTA*). The gain that we can achieve with a simple stage of amplification is around *40 dB*. Thus, in order to achieve *80 dB* or so it was necessary to use the cascade of two stages. However, we have seen that two stages bring about two poles one close to the other and this requires compensation. In turn, a compensation network, beside increasing the global complexity, reduces the design flexibility.

A cascode with cascode load permits us to achieve a high gain (around *80 dB*) without the disadvantage of having two poles one close to the other. Therefore, the use of cascode based *OTA* is an interesting solution alternative to the two stages *OTA*. This section discusses the features of a "single stage" *OTA* so that the outcomes of the study will permit us to properly decide which solution is convenient for given design constraints.

5.8.1 Telescopic Cascode

The simplest version of a single stage *OTA* is the telescopic architecture, shown in Fig. 5.27: the input differential pair injects the signal currents into common gate stages. Then, the circuit achieves the differential to single ended conversion with a cascode current mirror. Note that the transistors are placed one on the top of the other to create a sort of telescopic composition (from this the name of the circuit). The small signal resistance at the output node is quite high: it is the parallel connection of two cascode configurations. Such a high resistance benefits the small signal gain without limiting the circuit functionality when we require an *OTA* function.

By inspection of the circuit one finds the low-frequency small signal differential gain

$$A_0 \cong g_{m1} \frac{r_{ds8} g_{m6} r_{ds6} \cdot r_{ds2} g_{m4} r_{ds4}}{r_{ds8} g_{m6} r_{ds6} + r_{ds2} g_{m4} r_{ds4}} \tag{5.60}$$

it is proportional to the square of the product of a transistor transconductance and an output resistance. Therefore, as expected, the telescopic cascode achieves a gain similar to the one of the two stages architecture. Moreover, by

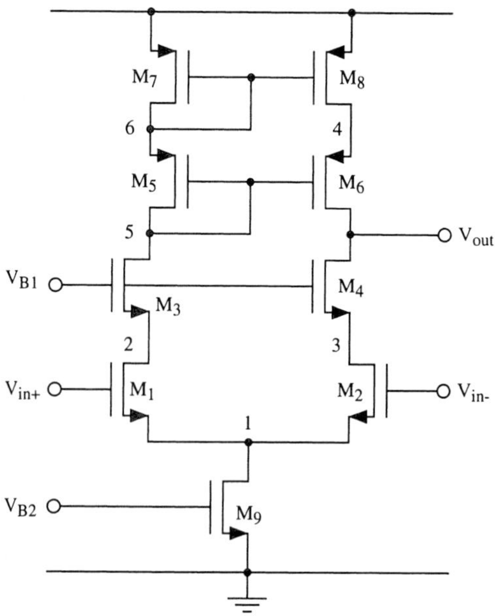

Fig. 5.27 - Telescopic cascode single stage OTA.

inspection of the circuit, one observes that all the nodes, excluded the output, shows a pretty low small signal resistance. Node *1* is an equivalent ground for differential signals; node *2, 3* and *4* are sources of a transistor, node *5* is coupled with node *6* by a diode connected transistor. Assuming the capacitance affecting the given nodes all due to parasitic contributions, the resulting time constants are all much smaller than the one associated to the output node. Therefore, the output node can easily become the dominant pole of the circuit. The other nodes should be at least A_0 apart being the non dominant ones.

Since the circuit shows one high impedance node only, it is not possible (and not necessary) to exploit the Miller effect to procure pole splitting. A possible capacitance loading the output node permits to make dominant the related pole and ensure stability.

Fig. 5.28 shows the simplified small signal equivalent circuit for a single pole approximation. The current generated by the transconductance generator $g_{m1}v_{in}$ is injected into the output resistance with in parallel the capacitance at the output node (C_L). The time constant $C_L\, r_{out}$ gives rise to a pole that causes a roll-off of the Bode plot with a slope of *20 dB/decade*. Therefore, the *0 dB* axis is crossed at the angular frequency $\omega_T = g_{m1}/C_L$.

An accurate study of the frequency response is not done here: it should involve the analysis of a pretty complex equivalent circuit. Since we have many nodes, we expect that the transfer function has a high order polynomial

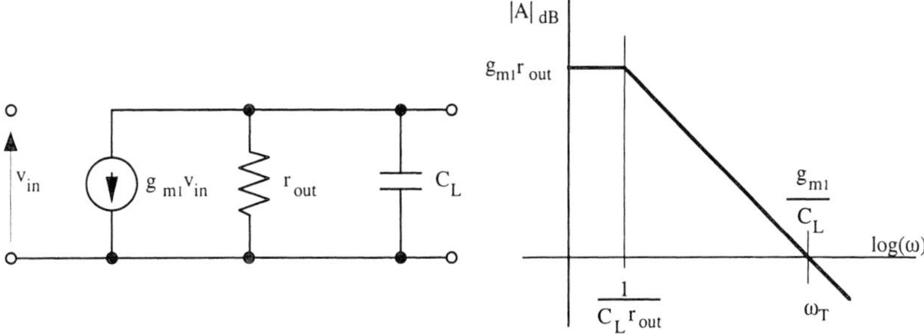

Fig. 5.28 - Simplified equivalent circuit of a single stage OTA and its Bode diagram.

denominator which requires approximations to be solved. Typically, the results achieved after approximations are not much different from the ones resulting from the intuitive estimation given above.

The telescopic configuration uses only one bias current. It flows through the differential input stage, the common base stage and the differential to single ended converter. Therefore, for a given bias voltage, the power is used at the best. By contrast, we have disadvantages: they concern the limited output dynamic range and the request to have an input common mode voltage pretty close to ground (or V_{SS}).

The triode limit of M_6 establishes the maximum allowed output voltage. By inspection of the circuit it is given by

$$V_{out,max} = V_{DD} - V_{GS7} - V_{GS5} + V_{GS6} - V_{sat,6} \cong V_{DD} - V_{Th,p} - 2V_{sat,p} \tag{5.61}$$

For typical situations it is *1 V* or more below the positive supply voltage.

The lower boundary of the output voltage depends the triode limit of M_4 that, in turn depends on V_{B1}

$$V_{out,min} = V_{B1} - V_{GS4} + V_{sat,4} = V_{B1} - V_{Th,n} \tag{5.62}$$

Normally the designer broadens the output swing by keeping low V_{B1}. However, the value of V_{B1} affects the minimum level of the input common mode voltage

$$V_{in,CM} \le V_{B1} - V_{GS4} - V_{sat,2} + V_{GS2} \cong V_{B1} - 2V_{sat,n} \tag{5.63}$$

In turn, the input common mode voltage should allow M_9 to be in the saturation region.

$$V_{in,CM} \ge V_{sat,9} + V_{GS2} \tag{5.64}$$

Therefore, we can achieve an optimum negative swing ($3V_{sat,n}$ above ground) keeping the input common mode voltage as low as $V_{sat,9} + V_{GS2}$ (approximately equal to $V_{Th,n} + 2V_{sat}$). Assuming a symmetrical output swing around $V_{out,max}$ and $V_{out,min}$, the output common mode voltage becomes

$$V_{out,CM} \cong V_{DD} + V_{B1} - V_{Th,n} - V_{Th,p} - 2V_{sat,p} \tag{5.65}$$

KEEP IN MIND!

The use of five transistors one on the top of the other burdens the output dynamic range. A proper choice of the bias voltage and a minimum input common mode (for n-channel input pair) enlarge the output swing.The cost is that the common mode output differs from the input.

that for a typical design is quite a bit higher than V_{B1}. Therefore, the output common mode voltage is different (higher) than the input common mode voltage. This, in some applications is a limit: for instance, it is not possible to connect the telescopic cascode in the unity gain buffer configuration.

An interesting feature of the telescopic cascode is that it needs only two wires for the biasing: one for V_{B1} and the other for V_{B2}. This is more than the two stages amplifier that needs only one wire, but is less than other single stage architecture that will be studied shortly. Having a reduced wiring is important since it limits the chip area and, more important, prevents possible spur injection.

Example 5.4

Design a telescopic cascode single-stage OTA (Fig. 5.27) having about 70 dB dc gain and GBW 60 MHz. The required phase margin is 55° with a capacitive load of 3pF. The circuit must operate with a 1.8 V single supply voltage. Since it is necessary to procure the greatest the output swing, suitably modify the scheme in Fig. 5.27. Use the transistor models in Appendix C.

Solution:

It is required to procure the largest output swing with only 1.8 V supply voltage. The scheme in Fig. 5.27 has from the output to ground three drain-to-source jumps. Therefore, a proper choice of V_{B1} and an appropriate input common-mode voltage would permit a minimum output swing just at three saturations above ground. Assuming that the saturation equals 0.1 V, it results

$$V_{out,min} = 3V_{sat,n} \approx 0.3V .$$

By contrast it is not possible to have the output swing close to the positive rail. Equation (5.61) bounds the upward swing to

$$V_{out,\,max} = V_{DD} - V_{Th,\,p} - 2V_{sat,\,p} \approx (1.8 - 0.7)V = 1.1V$$

Where $V_{Th,p} \approx 0.5$ V and, again, the saturation voltage is expected to be 0.1 V.
Since the major limit to the upward swing comes from the cascode current mirror, it is advisable to replace it with an equivalent counterpart that can reach a wider swing. The high compliance current mirror studied in Section 4.1.7 is a good candidate: a proper choice of the bias voltage used permits the output swing to approach the positive supply voltage by two drain-to-source drop voltages: 1.6 V, (the assumed saturation is 0.1 V). The diagram below shows the modified version of the telescopic cascode. The figure details the transistor sizes and the dc voltages.
The first step of the design consists in the estimation of the bias current. Assuming the GBW established by the dominant node, we have

$$2\pi f_T = \frac{2I_D}{(V_{GS} - V_{Th})C_L},$$

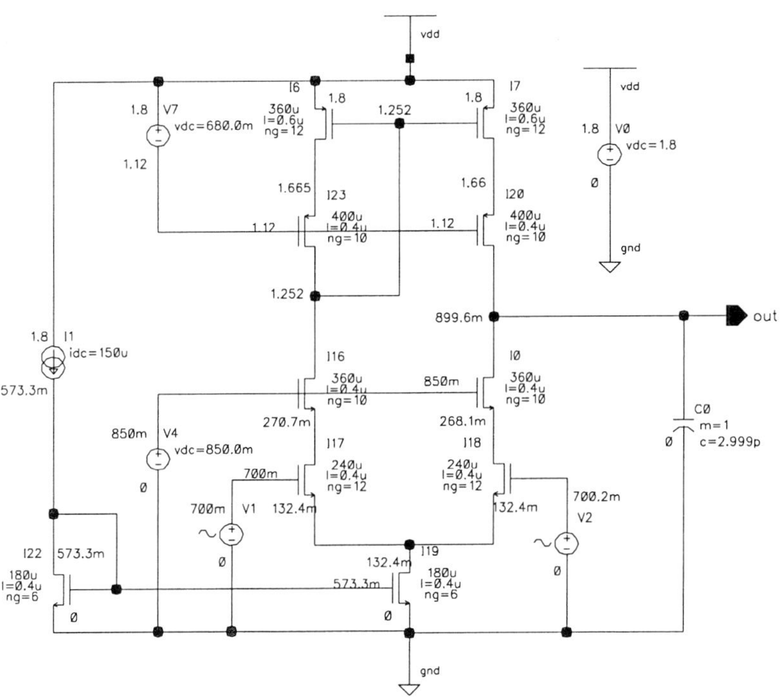

and assuming for the input pair V_{GS} - V_{Th} = 0.1 V, we calculate

$$I_D = \pi f_T(V_{GS} - V_{Th})C_L = 56.5\mu A.$$

The current used is a bit higher (150 μA tail current); it accounts for some margin and for different V_{DS} in the reference generator and the tail current source. The circuit employs large aspect ratios for all the transistors. This because the overdrive voltage must be pretty low (the expect value is 0.1 V).

The drain-to-source voltages can not be pushed at the limit of saturation (0.1 V). The output resistance of transistors drops at low value affecting, in turn, the overall gain. Simulations show that drain-to-source voltages around 0.13 V permit to achieve the target of 70 dB. Moreover, in order to increase the output resistance of the p-channel cascodes, we use the length of the transistors connected to V_{DD} equal to 0.6 μm.

The voltage at the input of high compliance cascode is 1.25 V while the (quiescent) output is 0.9 V. This difference justify the small systematic offset (0.2 mV) used at the input.

The schematic shows that all the transistors are split into a given number of parts (ng, number of gates). The numbers used are suitable for a proper layout: the input pair and the p-channel elements connected to V_{DD} are divided into 12 elements to be possibly composed in common centroid or/and an inter digitized arrangement. The cascoding transistors can not form inter digitized pairs: they are divided into 10 elements to allow some gap in the layout.

The below figure shows the Bode diagrams of the output voltage. The achieved gain is 71 dB. The unity gain frequency is 65 MHz and the phase margin is 58°.

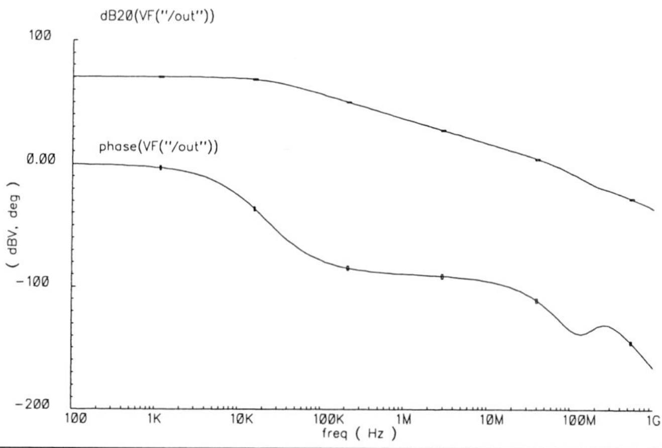

5.8.2 Mirrored Cascode

The mirrored cascode circuit, shown in Fig. 5.29, allows adaptability in the input common mode range. The signal currents generated by the input pair is mirrored by M_{11}-M_{10} and M_{12}-M_{13} and delivered to the source of M_3 and M_4 respectively. The operation of the rest of the circuit is identical to the telescopic version shown in Fig. 5.27.

To explain why the circuit offers a flexible input common mode range note that above the input pair drains we have two diode connected elements. Therefore, the common emitter of the input pair, node *1*, can approach the drains of M_{11} and M_{12} by just the saturation of the input pairs. Therefore, we have

$$V_{in,CM} \leq V_{DD} - V_{GS11} - V_{sat,1} + V_{GS1} \tag{5.66}$$

If the threshold of the *n-channel* transistors and the one of the *p-channel* matches V_{GS1} and V_{GS11} are comparable: the maximum value of $V_{in,CM}$ becomes close to V_{DD}. The minimum value of $V_{in,CM}$ is the one that brings M_9 in the triode region

$$V_{in,CM} \geq V_{GS,1} + V_{sat,9} \tag{5.67}$$

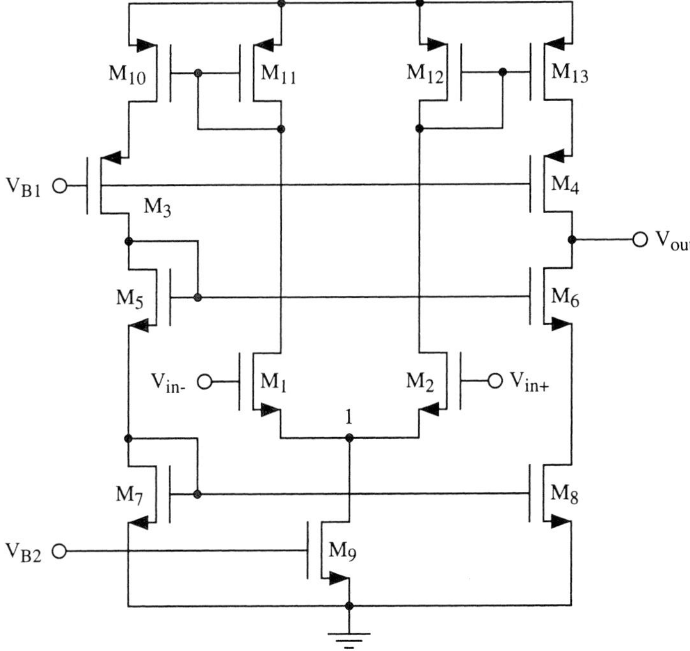

Fig. 5.29 - Mirrored cascode single stage OTA.

Therefore, the permitted range of the input common mode voltage, $V_{in,CM}$, becomes pretty large.

The output voltage range is limited by the conditions that bring M_4 and M_6 in triode. In turn, the saturation limit for M_4 depends on the V_{B1} used. By inspection of the circuit, we have

$$V_{B1,max} = V_{DD} - V_{sat} - V_{GS4} \tag{5.68}$$

this upper limit for V_{B1} leads the drain of M_{13} just one V_{sat} below V_{DD}. Therefore

$$V_{out,max} = V_{DD} - 2V_{sat} \tag{5.69}$$

that is only one saturation less than what we can achieve with a simple inverter with active load. By contrast, the cascode arrangement M_6-M_8 limits the negative swing. The voltage at the drain of M_8 is a replica of the one of M_7. Thus, we have

$$V_{out,min} = V_{GS7} + V_{sat} \tag{5.70}$$

Therefore, the output swing range is not symmetrical with respect to the positive and negative rail.

> **OBSERVATION**
>
> The use of current mirrors to lead the current to the output node improves the output swing. However, the additional non-dominant nodes affect the speed. As a general rule, the simpler is the circuit the faster is the response,

The mirror factors of M_6-M_7, M_{10}-M_{11} and M_{12}-M_{13} are additional design degrees of freedom. Of course, it is required to match the currents of M_{13} with the one of M_7; this determine a constraint between the mirror factors. However, the designer can use the remaining degrees of freedom to establish the currents in the branches with M_6 and M_7.

The current mirrors used to alleviate the constraints on the input common voltage limit the speed of the circuit. Compared to the telescopic cascode the mirrored version has two additional non dominant poles. They, possibly, reduce the phase margin. Therefore, in order to ensure stability, the designer has to likely increase the capacitance that loads the output, thus the speed (bandwidth and slew-rate) diminishes.

Observe that transistors M_3 and M_5 are used to improve symmetry in the circuit. They are not strictly necessary; by removing them the circuit is less symmetrical, but two non-dominant poles are removed as well.

Example 5.5

Design the single-stage mirrored cascode OTA shown in Fig. 5.29. The required dc gain and bandwidth are 80 dB and 50 MHz respectively. The phase margin must be better than 55°. Determine the load capacitance that complies with specifications. Study the behaviour of the small signal dc gain as a function of the output voltage. Use 3.3 V as single supply voltage and keep the power consumption below 1 mW. Use the transistor models of Appendix C.

Solution:
The available bias current is 1mW/3.3V = 300 μA. Assuming a 1:1 mirror factor for the pairs M_{10}-M_{11} and M_{12}-M_{13} the bias current is evenly distributed between the four transistors. Therefore, the current in the output branches is 75 μA and the tail current generator of the differential pair, M_9, drains 150 μA.
Example 5.4 used an input pair with W/L 240μ/0.4μ. With a tail current equal to150 μA the transconductance is 1.81 mA/V. This value is reasonably high: for achieving the 80 dB of gain the output resistance must be just 5 MΩ.
The output resistance of a cascode configuration is given by an r_{ds} multiplied by the gain of a transistor. A typical value of r_{ds} ranges

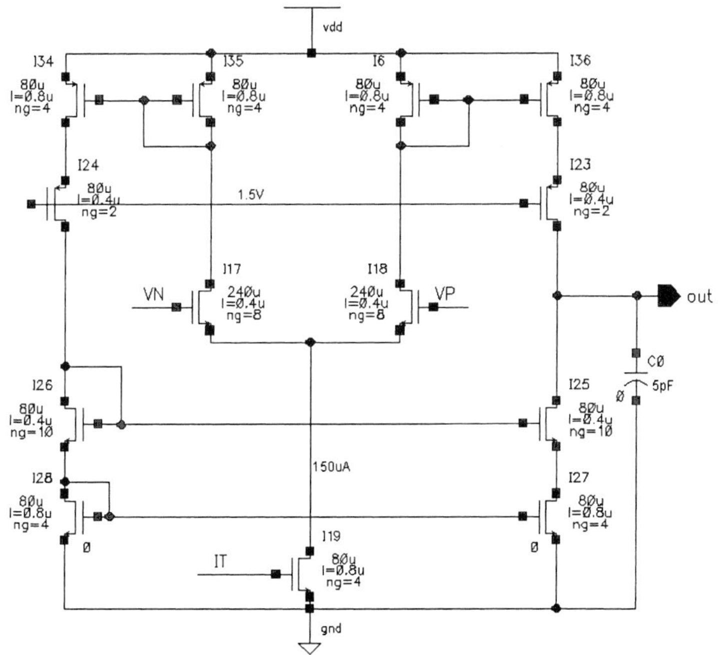

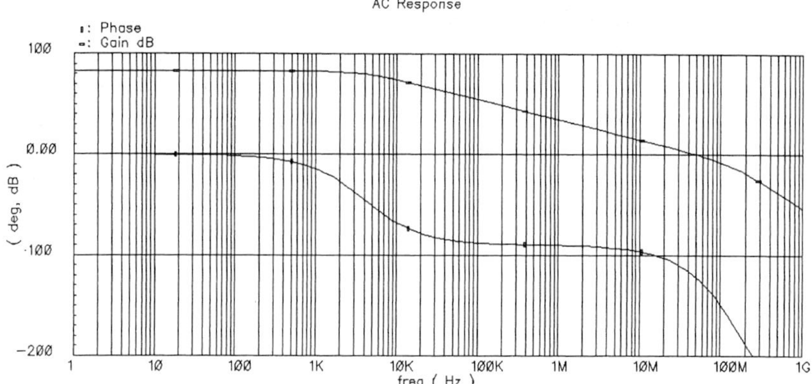

from 100 KΩ to 300 KΩ while the gain of a transistor is around 50-100. Therefore the output resistance of each cascode branch is easily larger than 5 MΩ. The transistor sizing of the schematic used leads to an equivalent resistance of 8.5 MΩ for the upper cascode and 27 MΩ for the lower cascode. The two values are unbalanced; a proper resizing of transistors would lead to more even values. Nevertheless, the used sizing make the geometry of the n-channel and the p-channel output cascodes equal. This feature can be attractive because it facilitates the layout.

The small signal analysis of the circuit leads to the results shown in the figure of the previous page. The dc gain is 83 dB, the phase margin is 59° and the gain-bandwidth is 54 MHz.

A sweep of the input voltage carried out by a .dc analysis permits us to determine the input output transfer characteristic. A simple

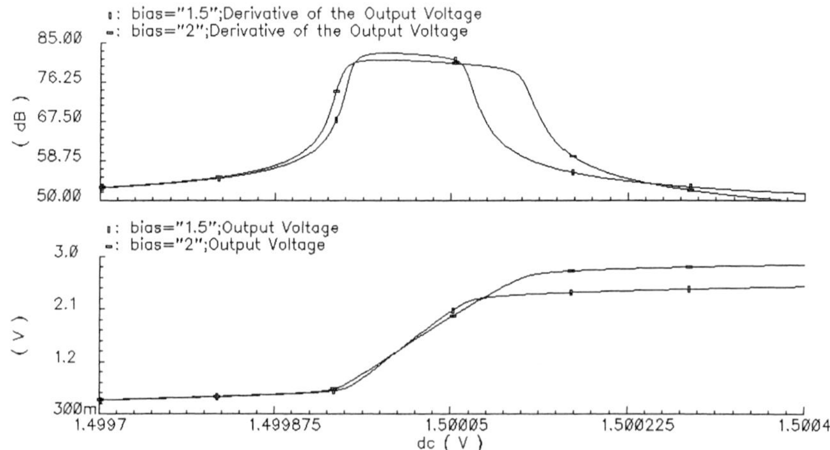

post-processing obtains the dc gain (its derivative). The results (see the bottom figure of the previous page) show that two different bias voltages used in the upper cascode: 1.5 V and 2 V lead to quite different performances. In the case of 1.5 V bias the gain is a bit larger but the output dynamic range is lower by approximately 0.5 V than the condition with 2 V bias.

5.8.3 Folded Cascode

The basic difference between the mirrored cascode and the telescopic one resides in the disconnection between the biasing of the input and the output branch. This results from the current mirrors that provide replica of the signal currents produced by the input stage and furnishes them to the output branch. The folded cascode, shown in Fig. 5.30 achieves the same conceptual function by a direct transfer of the small signal currents from the input to the output branch. The low impedance shown by the sources of M_3 and M_4 ensure the signal currents transfer. Moreover, the biasing of the input and output stages don't

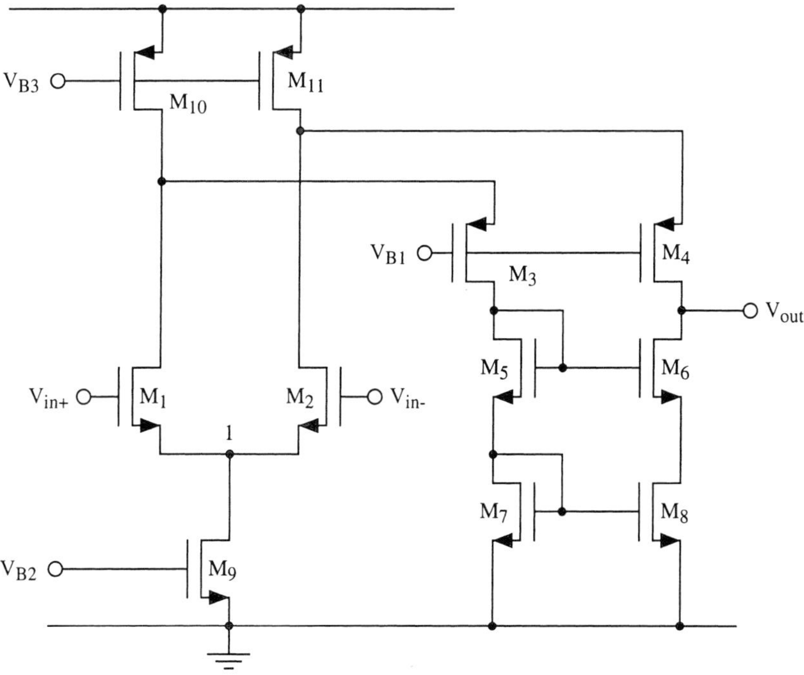

Fig. 5.30 - Folded cascode transconductance amplifier.

affect one each other. Therefore, the circuit achieves the same benefits for the common mode input and output range as for the mirrored cascode.

The folded version requires two additional current sources (M_{10} and M_{11}) to provide the current necessary for the input and the output branch. Therefore, the folded cascode demands for an additional wire connection to properly bias the two current sources.

The current source M_{11} affects the resistance of the upper cascode of the output branch. In fact, the resistance seen from the source of M_4 to ground (and node *1*, ground for differential signals) is the parallel connection of the drain resistance of M_2 and M_{11}

$$r_{eq} \cong \frac{g_{ds2} g_{ds11}}{g_{ds2} + g_{ds11}} \tag{5.71}$$

therefore, to calculate the small signal *dc* gain we should replace the pertinent term with above expression in equation (5.60).

Current sources M_9 and M_{10} (or M_{11}) set independently the currents of the input pair and the output branch. The degree of freedom available permits the designer to establish the value of the two currents freely. Assuming that the input stage is balanced we have

$$I_{M1} = \frac{I_{M9}}{2}; \quad I_{M7} = I_{M10} - \frac{I_{M9}}{2} \tag{5.72}$$

Thus, it is possible to maximise the gain by properly setting the input transconductance and the output resistance.

> **NOTICE**
>
> The mirrored cascode and the folded cascode need supplementary biases that, in turn, demand for extra bias generators and new bus lines.
>
> Recall that a capacitive coupling with noisy lines can corrupt the bias voltages.

The constraints on the output dynamic range of the folded cascode are similar to the ones studied for the mirrored version. The bias voltage V_{B1} limits the upward swing. The cascode structure M_6-M_8 bounds the downward swing, and, as already discussed, the value of V_{B1} should be low enough to consent the saturation of M_{11}. Therefore, if we use for V_{B1} the maximum value expressed by equation (5.68) we obtain the same output swing constraints given by (5.69) and (5.70). However, it is possible to improve the downward swing. We should replace the plain cascode current mirror with an high swing version, like the modified cascode (see § 4.1.6) or the high compliance cascode (see § 4.1.7).

Example 5.6

Assume to come across the data-base of a folded cascode OTA. The circuit has been designed using a 0.8 μ CMOS technology which Spice models are given in Appendix A. The folded cascode operates with a 5 V supply voltage, drives 10 pF, and consumes 3 mW. The following input list describes the circuit.

```
Folded cascode
M1  2 10 1 0 MODN W=400U  L=1.5U
M2  3 11 1 0 MODN W=400U  L=1.5U

M3  5 9 2 4  MODP W=200U  L=1U
M4  8 9 3 4  MODP W=200U  L=1U
M5  5 5 6 0  MODN W=200U  L=1U
M6  8 5 7 0  MODN W=200U  L=1U
M7  6 6 0 0  MODN W=100U  L=2U
M8  7 6 0 0  MODN W=100U  L=2U
M9  1 12 0 0 MODN W=100U  L=2U
M10 2 13 4 4 MODP W=100U  L=2U
M11 3 13 4 4 MODP W=100U  L=2U

MBN 12 12 0 0  MODN W=100U  L=2U
MBP 13 13 4 4  MODP W=100U  L=2U

IBN 0 12 0.5M
IBP 13 0 0.3M
VDD 4 0 5
V10 10 0 2.5
VB1 9 0 3

CL 8 0 10P

RF1 8 99  1G
RF2 99 11 1G
C1 99 0 1
C2 98 11 1
VIN 98 0 0 AC 1
```

Identify the key points of the design and try to understand the strategy followed by the designer.

Solution:

The technology is a 0.8μ CMOS. The lentos used for the devices M_3, M_4, M_5, and M_6 are close to the minimum allowed. The possible explanation of the choice is that the transistors must provide gain to enhance the cascode output resistance. The expected value of g_m/g_{ds} must be in the range of 50 to 100. A relatively low length sustains the transconductance and likely permits us to achieve the expected gain.

The length of the input pair transistors is 1.5 μ. This figure is almost twice the minimum allowed. This choice probably answer to the need of a low output conductance. The resistance of the upper output cascode depends on $g_{sd2}+g_{ds11}$. The length of the input transistor should lead to comparable values for output conductances of M_2 and M_{11}.

The length of all the transistors connected to the supply lines is 2 μ: for all these transistors it is essential to have pretty low g_{ds}. The Spice simulation of the circuit furnishes the results given below. Observe that the saturation voltage of most transistors is lower than 150 mV. The saturation voltage of M_9 is not relevant. What surprises us is the large saturation voltage of M_3 and M_4. Moreover, the used valuer of V_{B1} is much lower than the maximum possible. Presumably, the design aims at a symmetrical output swing around 2.5 V. Since the downward swing is bounded by the lower cascode it doesn't make sense to expand the upward dynamic range.

	m1	m2	m9	m5	m6	m7	m8
model	modn	modn	modn	modn	modn	modn	modn
id	2.49E-04	2.49E-04	4.99E-04	4.81E-05	4.81E-05	4.81E-05	4.82E-05
vgs	1.259	1.259	1.341	1.095	1.093	0.973	0.973
vds	2.632	2.636	1.241	1.095	1.525	0.973	0.975
vbs	-1.241	-1.241	0.000	-0.973	-0.975	0.000	0.000
vth	1.152	1.152	0.866	1.066	1.064	0.862	0.862
vdsat	0.119	0.119	0.362	0.060	0.060	0.117	0.117
gm	3.38E-03	3.38E-03	1.95E-03	1.28E-03	1.29E-03	6.04E-04	6.04E-04
gds	1.44E-05	1.44E-05	1.23E-05	1.01E-05	9.10E-06	2.64E-06	2.64E-06
gmb	6.59E-04	6.59E-04	5.91E-04	2.52E-04	2.53E-04	1.98E-04	1.98E-04
cbd	0.00E+00	0.00E+00	0.00E+00	0.00E+00	0.00E+00	0.00E+00	0.00E+00
cbs	0.00E+00	0.00E+00	0.00E+00	0.00E+00	0.00E+00	0.00E+00	0.00E+00
cgsovl	1.48E-13	1.48E-13	3.70E-14	7.40E-14	7.40E-14	3.70E-14	3.70E-14
cgdovl	1.48E-13	1.48E-13	3.70E-14	7.40E-14	7.40E-14	3.70E-14	3.70E-14
cgbovi	2.33E-16	2.33E-16	3.10E-16	1.55E-16	1.55E-16	3.10E-16	3.10E-16
cgs	8.63E-13	8.63E-13	2.88E-13	2.88E-13	2.88E-13	2.88E-13	2.88E-13
cgd	0.00E+00	0.00E+00	0.00E+00	0.00E+00	0.00E+00	0.00E+00	0.00E+00
cgb	0.00E+00	0.00E+00	0.00E+00	0.00E+00	0.00E+00	0.00E+00	0.00E+00

m3	m4	m10	m11	mbn	mbp	
model	modp	modp	modp	modp	modn	modp
id	-4.81E-05	-4.81E-05	-2.98E-04	-2.98E-04	5.00E-04	-3.00E-04
vgs	-0.873	-0.876	-1.252	-1.252	1.341	-1.252
vds	-1.805	-1.376	-1.127	-1.124	1.341	-1.252
vbs	1.127	1.124	0.000	0.000	0.000	0.000
vth	-0.809	-0.811	-0.711	-0.711	0.866	-0.711
vdsat	-0.090	-0.092	-0.469	-0.469	0.363	-0.469
gm	9.53E-04	9.32E-04	1.03E-03	1.03E-03	1.96E-03	1.04E-03
gds	8.68E-06	9.44E-06	2.07E-05	2.07E-05	1.18E-05	1.94E-05
gmb	8.26E-05	8.09E-05	1.57E-04	1.57E-04	5.92E-04	1.58E-04
cbd	0.00E+00	0.00E+00	0.00E+00	0.00E+00	0.00E+00	0.00E+00
cbs	0.00E+00	0.00E+00	0.00E+00	0.00E+00	0.00E+00	0.00E+00
cgsovl	7.40E-14	7.40E-14	3.70E-14	3.70E-14	3.70E-14	3.70E-14
cgdovl	7.40E-14	7.40E-14	3.70E-14	3.70E-14	3.70E-14	3.70E-14
cgbovl	1.05E-16	1.05E-16	2.15E-16	2.15E-16	3.10E-16	2.15E-16
cgs	2.74E-13	2.74E-13	2.81E-13	2.81E-13	2.88E-13	2.81E-13
cgd	0.00E+00	0.00E+00	0.00E+00	0.00E+00	0.00E+00	0.00E+00
cgb	0.00E+00	0.00E+00	0.00E+00	0.00E+00	0.00E+00	0.00E+00

The width of all transistors is multiple of 100μ. This choice permits the designer to split all the transistors into fingers 25μ wide and to achieve a modular and compact layout. The .ac simulation leads to 80 dB dc gain, unity gain frequency equal to 42 MHz and phase margin equal to 63°. The procured results are quite conventional for a general purpose OTA found in a 0.8 μ CMOS cell library.

5.8.4 Single Stages with Enhanced dc Gain

We studied in Chapter 4 that it is possible to increase the output resistance of a cascode by using a multiple cascode architecture or a local feedback. The use of the same approaches permits us to enhance the gain of a single-stage *OTA*: the gain comes from the product of the input pair transconductance and the output resistance. Therefore, for a given transconductance we increase the gain if we enhance the output resistance. Nevertheless, we have to remember that the supplementary circuits used bring about additional nodes that, in turn, lead to new poles that worsen the phase margin.

Fig. 5.31 shows a mirrored *OTA* configuration with a double-cascode load. The new circuit has four additional transistors (M_{3a}, M_{4a}, M_{5a} and M_{6a}) for achieving four common-gate stages. The output resistance increases by approximately a $g_m r_{ds}$ factor and the *dc* gain increases by the same factor.

The circuit has four additional nodes (the sources of the four added transistors). Therefore, the compensation of the stage can result critical. The small signal resistance of the added nodes is in the order of $1/g_m$; the associated capacitance comes from parasitics; therefore, the corresponding poles will be at

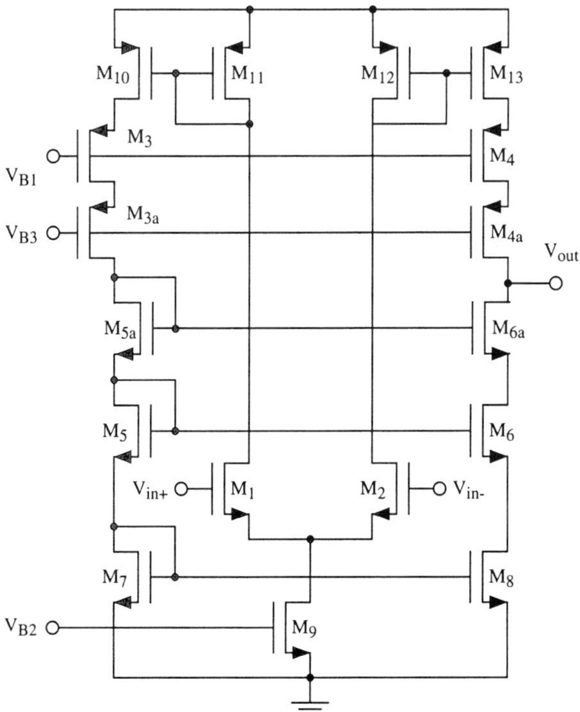

Fig. 5.31 - Mirrored double-cascode single stage OTA

a relatively high frequency. Nevertheless, the additional poles are one close the other and the phase margin is worsened because of the combined action of the cluster of non-dominant poles. Consequently, we have to bring the dominant pole at a lower frequency to ensure a more solid non-dominant position of the additional non-dominant poles.

The circuit necessitates an additional bias voltage, V_{B3}. The value of value V_{B3} determines the upward output swing constraint. Moreover, V_{B3} must be designed in combination with V_{B1}. Assuming that the gate-source voltages of M_3 and M_{3a} and the one of M_4 and M_{4a} matches, the drain-to-source voltage of M_3 and M_4 are equal to V_{B1} - V_{B3}. Therefore, if transistors should be kept at the limit of the triode condition, the two bias voltages must differ by just a saturation voltage and, more important, their difference should not change with temperature and process variation.

Fig. 5.32 shows the use of local feedback to increase the *dc* gain in a cascode architecture. The circuit includes two gain stages. One is used to enhance the output resistance of the upper cascode, the other to boost the output resist-

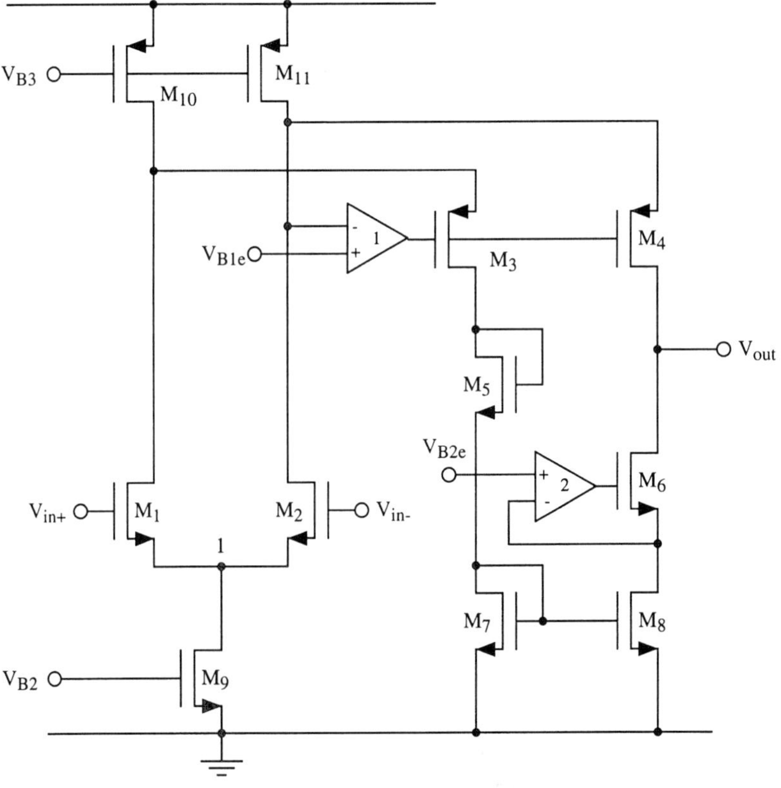

Fig. 5.32 - Use of local feedback to enhance the *dc* gain of a folded cascode.

ance of the lower one. Both gain stages control the source of the common gate elements (M_4 and M_6) to be constant. The bias voltages V_{B1e} and V_{B2e} correspond to the desired source voltages. In order to maximize the output swing these bias voltages should just be one saturation apart from the positive and negative rails.

Transistor M_5 in the normal cascode scheme biases the gate of M_6; in Fig. 5.32 it provide a drop voltage to reduce the V_{DS} of M_3 and a possibly attenuate a systematic difference between the currents in M_5 and M_4.

The frequency behaviour of the circuit depends on the used gain stages and on their frequency performance. We observe that the feedback factor in the local loops is equal to one. Therefore, the bandwidth of the feedback loop equals the unity gain frequency of the gain stage used. In turn, to ensure stability, the unity gain frequencies of the gain stage must be higher than the unity gain frequency of the cascode *OTA*.

The common-mode input of the upper gain stage is close to V_{DD} while the one of the lower gain stage is close to ground. The design of a gain stage which common mode input is close to the supply rails is not an easy task. Typically, the upper gain stage uses an *n-channel* input pair and the lower gain stage uses a *p-channel* input pair.

Example 5.7

Design an OTA able to drive a capacitive load of 2 pF with 100 dB dc gain and f_T = 70 MHz (phase margin › 50º). The supply voltage is 1.8 V. The power consumption must be less than 0.8 mW. In order to achieve high gain and low power consumption use a two stages configuration with a telescopic cascode as the first stage and an inverter with active load as second stage. Use the Spice models of Appendix C.

Solution:
The low value of supply voltage doesn't provide room for a conventional telescopic cascode. It is necessary to use a high compliance current mirror that allows room for the input pair and its current tail generator. Moreover, all the overdrive voltages must be kept at a minimum level to procure an acceptable dynamic swing. This will lead to transistors with a pretty large aspect ratio.
The above reflections lead to the schematic shown in the figure. The tail current in the first and the bias of the second stage is 150 μA. The circuit uses a p-channel input pair which large W/L (800μ/ 0.6μ) achieves a transconductance as large as 2.42 mA/V. The value of input transconductance leads, with a compensation capacitance equal to the capacitive load equal to 2 pF, to a gain band-

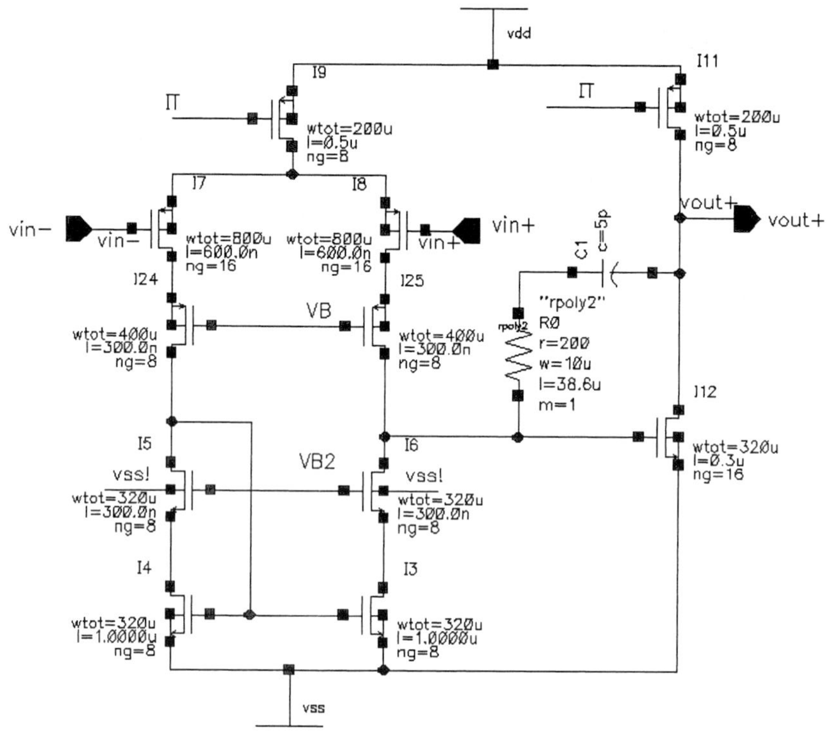

width equal to

$$f_T = \frac{1}{2\pi}\frac{g_m}{C_C} = \frac{1}{6.28}\frac{2.42 \cdot 10^{-3}}{2 \cdot 10^{-12}} = 192 \text{ MHz}$$

that is about three times the specification. This seems a good margin. However, as shown by simulations, the non-dominant poles require a pretty large compensation capacitance. To have an acceptable phase margin it is required to use $C_C = 5$ pF. The expected f_T drops by a factor 2.5 and becomes 76.8 MHz.
The large input transconductance permits to achieve a good first stage gain. The high compliance cascode current mirror keeps the transistors close to the saturation region. This prevents to obtain a very large output resistance. However, the used design leads to a first stage output resistance a bit smaller than 1 MΩ providing a first stage gain equal to 68 dB.
The second stage should provide at least a dc gain of 32 dB. The target is not particularly difficult. The sizing of input transistor equal to (W/L) = 320μ/0.3μ and an active load with (W/L) = 200μ/0.5μ fulfil the request.

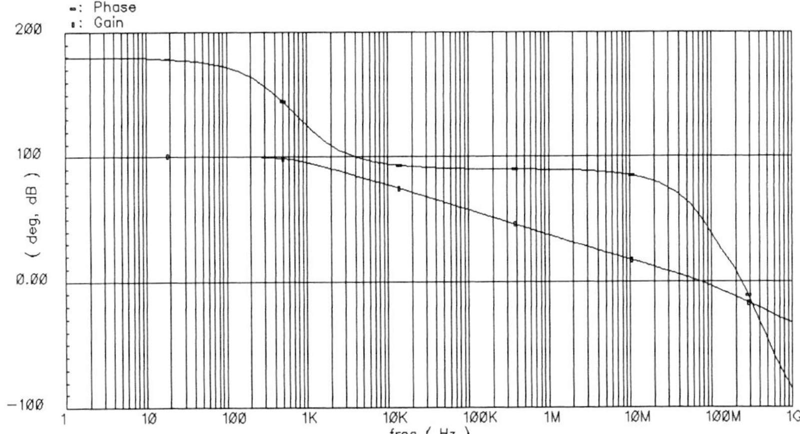

As already mentioned five low-impedance nodes in the input stage burden the phase margin. Computer simulations reveal that in order to procure a phase margin better than 50° it is necessary to use $C_C = 5\ pF$.
The above figure shows the Bode diagram of the output voltage. Other relevant results are summarized in the Table.

Parameter	Value	Unit
dc gain of the first stage	67	dB
dc gain of the second stage	34	dB
f_T	75.3	MHz
phase margin	52°	-
input referred offset	7.33	μV

5.9 CLASS AB AMPLIFIERS

Usually *AB* class amplifiers drive small resistive loads. Since in integrated *CMOS* applications only capacitors constitute the load, the objective of the *IC* designer is different than when addressing discrete component solutions. For an off-the-shelf op-amp it is necessary to ensure a low output resistance for any possible feedback and load. The aim here is to be able to drive large capacitance's quickly and with a minimum harmonic distortion. For high speed or large capacitor, the target is difficult to achieve, especially when using circuits like the *A* class *OTA's* studies so far. The *AB* class amplifier better addresses the problem since it permits to deliver currents larger than the quiescent value.

5.9.1 Two Stages Scheme

The two-stages amplifier of Fig. 5.34 achieves the class *AB* operation in the second stage.The figure shows how this function results from a modification of the already studied two stages amplifier. The output of the first stage drives both the *n-channel* and the *p-channel* transistor of the second stage. Driving of transistor M_5 equals the one in the *p-channel* input counterpart of Fig. 5.12. Transistor M_6 is no more a current source but the control of its gate is a shifted down replica of the input-stage output. Transistor M_8 and the bias current M_9, possibly controlled by the same voltage used for M_7 achieve the level shift. Both transistors M_5 and M_6 contribute to the transconductance gain of the second stage. Therefore, being the gain of the second stage

$$A_2 = \frac{g_{m5} + g_{m6}}{g_{ds5} + g_{ds6}} \tag{5.73}$$

as a side benefit, the low frequency gain will result somewhat increased.

The network C_{c1}-R_z ensures the compensation of the two stages amplifier. As already discussed, the zero nulling resistor acts to remove the zero in the right s-plane. However, at high frequency, the *0°* phase shift provided the

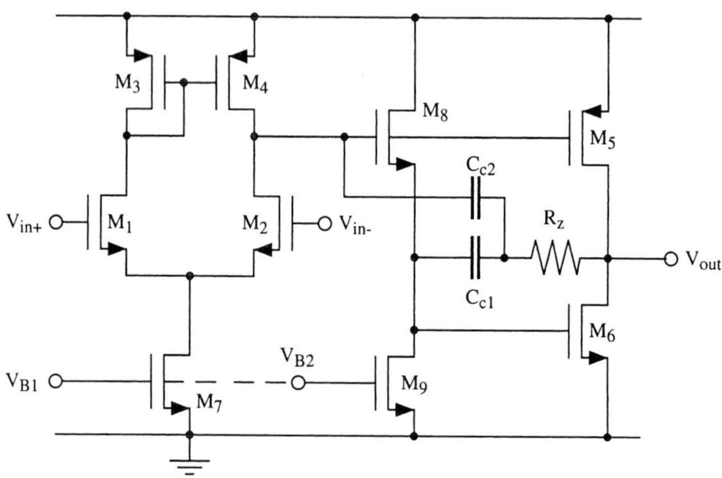

Fig. 5.33 - The two stages push-pull OTA.

buffer M_8-M_9 can deteriorate the phase response. The use of capacitor C_{c2} attenuates this possible problem.

A critical point of the design involves the control of the quiescent current, I_6 of the second stage. By inspection of the circuit we can find the following relationship between the gate-to-source voltages of three transistors

$$V_{DD} = V_{GS6} + V_{GS8} + V_{GS5} \tag{5.74}$$

two of the transistors are *n-channel* and one is *p-channel*, therefore

$$V_{DD} = V_{Th,p} + 2V_{Th,n} + \sqrt{\frac{2}{k'_n}\left(\frac{L}{W}\right)_8 I_8} + \left(\sqrt{\frac{2}{k'_p}\left(\frac{L}{W}\right)_5} + \sqrt{\frac{2}{k'_n}\left(\frac{L}{W}\right)_6}\right)\sqrt{I_6} \tag{5.75}$$

$$\sqrt{I_6} = \frac{V_{DD} - V_{Th,p} - 2V_{Th,n} - \sqrt{\frac{2}{k'_n}\left(\frac{L}{W}\right)_8 I_8}}{\sqrt{\frac{2}{k'_p}\left(\frac{L}{W}\right)_5} + \sqrt{\frac{2}{k'_n}\left(\frac{L}{W}\right)_6}} \tag{5.76}$$

A possible variation of the supply voltage (in a commercial power supply generator it can be ± *10%*) and the technological alterations of threshold voltages make pretty low the accuracy of the first three terms in the numerator of (5.76). Thus, when using an uncontrolled current mirror (as it is done if in Fig. 5.33 V_{B1}=V_{B2}) the current in M_6, may vary significantly. A possible way to fix the problem is to control the overdrive voltage of M_8 and compensate with it the variation of the first three terms in the numerator of (5.76).

> **TAKE HEED!**
>
> The mail limitation of the two stages AB class OTA in Fig. 5.33 derives from the uncertainty of the quiescent output current. Its control with a suitable feedback is essential.

Fig. 5.34 shows a possible bias network that we can use to drive the current source M_9. The part within the dashed lines matches half of the input stage. It replicates the voltage of the gate of M_8 and use it to drive M_{8b} (a replica of M_8) The current mirror M_{5b1}-M_{5b} copies the current of M_{6b} and compares it with the reference current. Any difference between the two currents affects V_{B2}. This voltage drives M_{9b} so that the connection closes the loop. Voltage V_{b2} is then used in the scheme in Fig. 5.33.

Observe that we have three gain stages around the feedback loop. The load of stages M_{6b}-M_{5b} and M_{9b}-M_{8b} is given by $1/g_{m,5b}$ and $1/g_{m,8b}$ respectively. Consequently, their gain is limited. In practice, the entire gain loop is ensured by M_{5b1} and the output resistance of the current reference generator, I_{ref}.

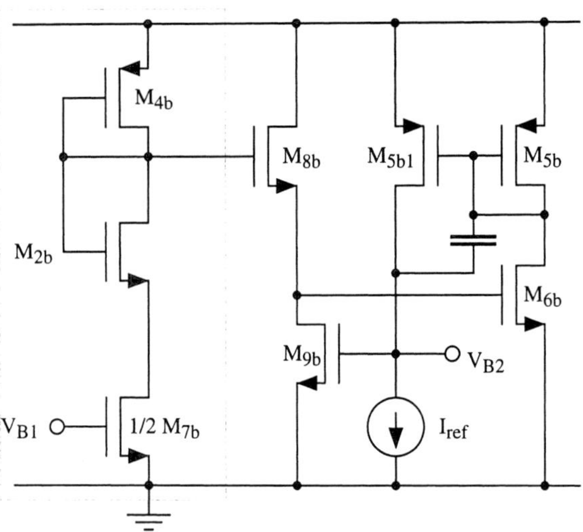

Fig. 5.34 - Possible bias network for driving transistor M_9 in Fig. 5.33

Because of the three stages around the loop, the circuit needs compensation. The capacitor across the input and output of M_{5b1} takes benefit of the Miller amplification. That is therefore the best position where to place the compensation capacitor.

5.9.2 Unfolded Differential Pair

Basically, the limit to slewing in a single stage amplifier occurs because of the current source used in the input differential pair. Even with a large differential signal the current that flows through one of the output nodes (see Fig. 5.35 a) can not be larger than the bias current, I_{bias}. In turn, we have a limit in the current that the circuit can deliver to the output nodes. It is possible to remove the above mentioned limitation by "unfolding" the input pair as shown in Fig. 5.35 b). The result is a stack of two complementary transistors that provide output currents from the two drains. The driving of the pair requires a *dc* difference between the control of the *n-channel* gate and the *p-channel* gate, $V_{AB,0}$. This difference controls the quiescent current, I_Q

$$V_{AB,0} = V_{th,n} + V_{th,p} + \sqrt{\frac{2I_Q}{C_{ox}}}\left(\sqrt{\frac{L_n}{\mu_n W_n}} + \sqrt{\frac{L_p}{\mu_p W_p}}\right) \tag{5.77}$$

moreover, the circuit allows an extensive increase ΔI of the current if V_{AB} augments to demand slewing. Equation (5.77) permits us to calculate the current-

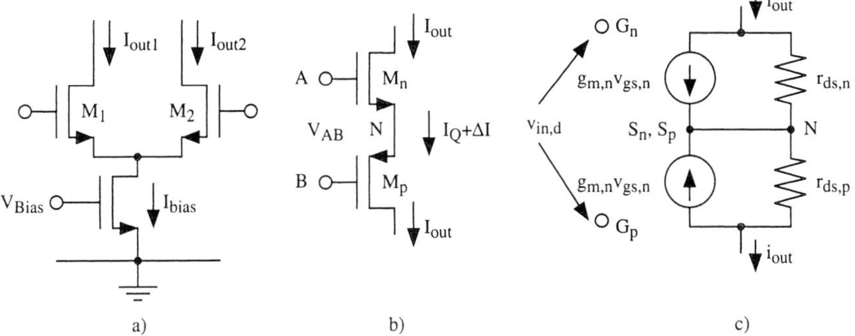

Fig. 5.35 - The "unfolded" differential pair: it is used in a class AB single stage amplifier

voltage relationship for any possible value of V_{AB}. When V_{AB} becomes lower than $V_{th,n} + V_{th,p}$ the current goes to zero. For voltages higher than such a condition we have

$$\frac{2I_{out}}{C_{ox}\mu_n\left(1+\frac{\mu_p W_p L_n}{\mu_n W_n L_p}\right)} = \left(\sqrt{\frac{2I_Q}{C_{ox}\mu_n\left(1+\frac{\mu_p W_p L_n}{\mu_n W_n L_p}\right)}} + \Delta V_{AB}\right)^2 \tag{5.78}$$

that is a non-linear relationship between the input (signal) voltage and the output current.

Observe that any increase of V_{AB} is shared between the overdrive of the *n-channel* and the *p-channel* transistor. We can write

$$\Delta V_{AB} = \Delta V_{ov,n} + \Delta V_{ov,p} \quad ; \quad \frac{\Delta V_{ov,n}}{\Delta V_{ov,p}} = \sqrt{\frac{\mu_p W_p L_n}{\mu_n W_n L_p}} \tag{5.79}$$

therefore, if we want to have equally split the variation of V_{AB} we have to use transistor aspect ratios inversely proportional to the mobility ratio. This will keep constant the voltage of the intermediate node *N* for a symmetrical change of node *A* and *B* voltages.

Fig. 5.35 *c)* shows the low frequency small-signal equivalent circuit of the unfolded pair. Neglecting the drain resistances permits to achieve an approximated (but effective) result. Without the drain resistances the small signal output current becomes equal to the one of two current generators

$$i_{out} = g_{m,n} v_{gs,n} = -g_{m,p} v_{gs,p} \tag{5.80}$$

moreover, $v_{gs,n}$ and $v_{gs,p}$ are linked to the input differential voltage by

$$v_{in,d} = v_{gs,n} - v_{gs,p} \tag{5.81}$$

using (5.80) and (5.81) one obtains

$$i_{out} = \frac{g_{m,n} \cdot g_{m,p}}{g_{m,n} + g_{m,p}} v_{in,d} = g_{eq} v_{in,d} \tag{5.82}$$

that defines the equivalent transconductance gain g_{eq} of the unfolded pair. Observe that the equivalent transconductance equals half of the transconductance of each single transistor if $g_{m,n} = g_{m,p}$; moreover, the signal voltage is the entire differential input. Therefore, if $g_{m,n} = g_{m,p}$ the unfolded pair operates exactly as the normal (folded) counterpart.

COMMENT

For small signals, an unfolded differential pair operates like the folded counterpart. The equivalent transconductance is given by

$$g_{eq} = \frac{g_{m,n} \cdot g_{m,p}}{g_{m,n} + g_{m,p}}$$

A more precise analysis that includes the effect of drain resistances is not done here. The result will show some slight difference with respect to the simple result given above. The reader can do such an analysis as an exercise.

In addition to the equivalent transconductance estimation we should determine the equivalent output resistances of the unfolded pair. That is done by inspection of the circuit in Fig. 5.35 c). Nulling the input signal, we immediately estimate the total equivalent resistance: it is the series connection of $r_{ds,n}$ and $r_{ds,p}$. For input differential signal half of that series value depicts the resistance between each output node and analog ground.

5.9.3 Single Stage AB-class OTA

The circuit in Fig. 5.36 uses two unfolded pairs. Their connections are crossed so that the current of the pair M_1-M_2 is accessible on the right-bottom side and the one of the pair M_3-M_4 is at hand on the right-top side. These currents are mirrored by Wilson schemes and summed up at the output node. The same input voltage controls transistors M_1 and M_4. By contrast, two level shifts, ΔV, (represented in the figure with two batteries) disjoint the positive input from the gates of M_3 and M_2. Alternatively, we can use level shifts to disjoint the voltages of the gate of M_1-M_3 from the differential input applied throughout M_2 and M_4. However, the latter option is not optimum: parasitic capacitances will surely affect the level shifts; it is better to use them on the positive input since it is typically connected to the analog ground.

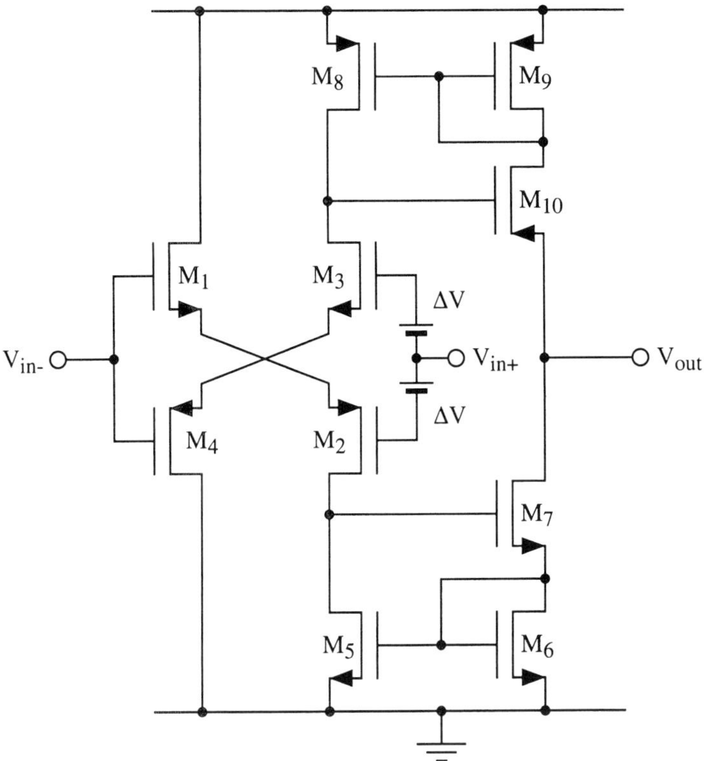

Fig. 5.36 - Single stage AB class op-amp

If, for instance, the input umbalancement brings the positive input higher than the negative one, the current in the pair M_1-M_2 diminishes (and, possibly goes to zero) while the current in the pair M_3-M_4 goes up. Therefore, thanks to the mirroring, the circuit accomplishes a push pull operation.

In the quiescent conditions the level shifts ΔV controls the current in the output stage. In fact, if the current from the output node is zero, the following relationships hold

$$I_7 = k_{56} I_1 = I_{10} = k_{89} I_3 \tag{5.83}$$

$$k_{56} = \frac{(W/L)_6}{(W/L)_5}; \quad k_{89} = \frac{(W/L)_9}{(W/L)_8} \tag{5.84}$$

where k_{56} and k_{89} are the current mirroring factors, assumed equal. The above equations assume that all the transistors are in the saturation region. Moreover, matched transistors and matched mirroring factors avoid a systematic offset in the stage. In addition we have

$$\Delta V = V_{th,n} + V_{th,p} + \sqrt{\frac{2I_1}{C_{ox}}}\left(\sqrt{\frac{L_1}{\mu_n W_1}} + \sqrt{\frac{L_2}{\mu_p W_2}}\right) \quad (5.85)$$

Therefore, a given ΔV leads to a well defined quiescent current.

A possible output current results from the difference

$$I_{out} = k_{89}I_3 - k_{56}I_1 = k_{56}(I_3 - I_1) \quad (5.86)$$

remembering that

$$V_B + V_{in} = V_{GS2} + V_{GS4} = \quad (5.87)$$

$$= V_{Th,n} + V_{Th,p} + \left(\sqrt{\frac{2}{k'_n}\left(\frac{W}{L}\right)_2} + \sqrt{\frac{2}{k'_p}\left(\frac{W}{L}\right)_4}\right)\sqrt{I_2}$$

$$V_B - V_{in} = V_{GS1} + V_{GS3} = \quad (5.88)$$

$$= V_{Th,n} + V_{Th,p} + \left(\sqrt{\frac{2}{k'_n}\left(\frac{W}{L}\right)_3} + \sqrt{\frac{2}{k'_p}\left(\frac{W}{L}\right)_1}\right)\sqrt{I_1}$$

it results

$$I_{out} = k_{56}(I_1 - I_2) = \alpha k_{56} V_B V_{in} \quad (5.89)$$

CALL UP

The time allowed by a sampled data system to charge or discharge a capacitance is partaken between the slewing phase and the settling of the output voltage. The longer is the slewing the shorter is the time for settling.

where α is a proper coefficient that results from the above equations. It depends on the technology and the transistor sizing. Observe that the compensation of quadratic terms makes the output current linearly proportional to the large signal input voltage. Equation (5.89) holds until some current flows on both the unfolded pairs. When the current in one of them goes to zero the output current increases in a non-linear way as described by equation (5.78). Fig. 5.37 sketches the relationship between output current and input differential voltage. In a given interval, as predicted by (5.89) the relationship is linear. Outside the linear region the current increases faster than in the linear region. Such a feature is effective for enhancing the slew-rate, but, on the other hand, leads to harmonic distortion.

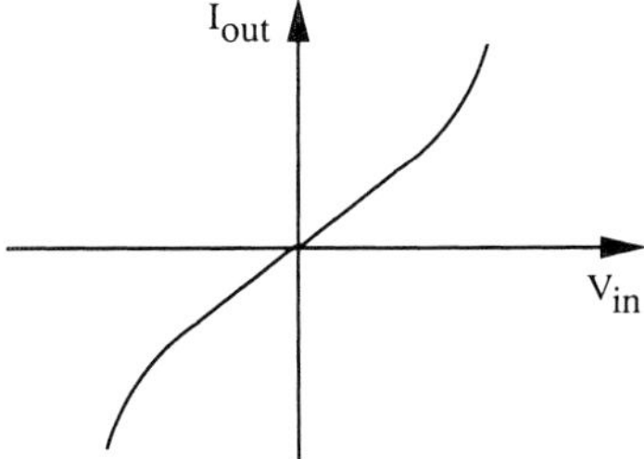

Fig. 5.37 - Output current as a function of the input voltage in the class AB stage of Fig. 5.36

Designer can derive the small signal performances of the circuit in Fig. 5.36 by inspection of the circuit. The unfolded stages generate small signal currents that, after being multiplied by the mirror factor k_{56} (= k_{89}) are summed up at the output node. The combined signal current flows into the output resistance and leads to the voltage gain

$$A_v = 2k_{56}g_{eq}r_{out} \tag{5.90}$$

The output resistance is the parallel connection of the resistances of the two Wilson current mirror. Remembering the study done in the previous chapter, we have

$$r_{out,p} = \frac{g_{m8}}{g_{m9}} \cdot \frac{r_{ds10}g_{m10}}{g_{ds3}+g_{ds8}} = \frac{1}{k_{89}} \cdot \frac{r_{ds10}g_{m10}}{r_{ds3}+r_{ds8}} \tag{5.91}$$

$$r_{out,n} = \frac{g_{m5}}{g_{m6}} \cdot \frac{r_{ds7}g_{m7}}{g_{ds2}+g_{ds5}} = \frac{1}{k_{67}} \cdot \frac{r_{ds7}g_{m7}}{g_{ds2}+g_{ds5}} \tag{5.92}$$

Equation (5.82) establishes the equivalent transconductance of the unfolded pair. Thus, the voltage gain is

$$A_v = 2k_{56}\frac{g_{m1}g_{m2}}{g_{m1}+g_{m2}}\left[\frac{1}{k_{56}}\left(\frac{r_{ds7}g_{m7}}{g_{ds2}+g_{ds5}}\right) // \frac{1}{k_{89}}\left(\frac{r_{ds10}g_{m10}}{g_{ds3}+g_{ds8}}\right)\right] \tag{5.93}$$

that, using the condition $k_{56} = k_{89}$, becomes

$$A_v = 2\frac{g_{m1}g_{m2}}{g_{m1}+g_{m2}}\left[\frac{r_{ds7}g_{m7}}{g_{ds2}+g_{ds5}} // \frac{r_{ds10}g_{m10}}{g_{ds3}+g_{ds8}}\right] \approx \gamma(g_m r_{ds})^2 \tag{5.94}$$

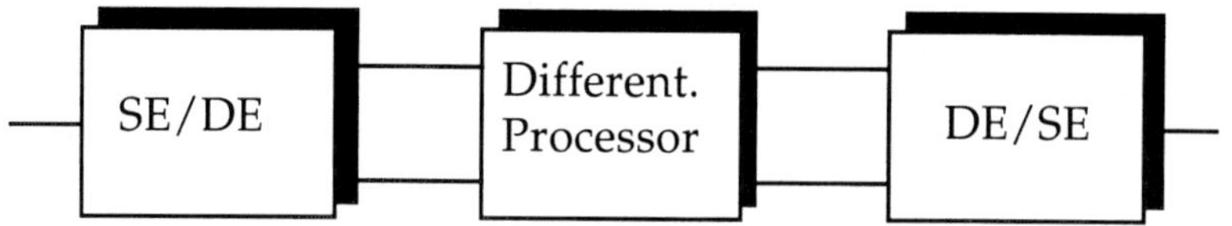

Fig. 5.38 - Fully differential signal processing chain.

Therefore, the single stage *AB* class *OTA* achieves a *dc* gain equivalent to the one of a two stage or a conventional single stage *OTA* studied so far. Observe that a possible benefit achieved with a gain in the current mirror vanishes because of an equivalent reduction in the output resistance. Thus, an amplifying factor in the current mirrors doesn't produce any benefits as far as the dc gain is concerned. Instead, higher quiescent currents in the output branch (that, keep note, produces higher power consumption) hand over higher output current for the (capacitive) load.

5.10 FULLY DIFFERENTIAL OP-AMPS

Mixed signal circuits extensively use fully-differential signal processing. We have seen that a fully-differential solution is beneficial for the clock feedthrough cancellation. We acquire similar advantages in the rejection of spur affecting the power supplies. In general, a fully differential solution rejects all the common mode components that we have at the input. Therefore, assuming that disturbing signals affect in the same extent the differential path, when considering the differential effect they are cancelled out.

In many processing systems a single-ended input and not a differential one is usually available. Moreover, numerous applications require a single ended output. Therefore, in addition to the differential processor we need a single-ended to differential and a differential to single-ended converter, as shown on Fig. 5.38. The additional blocks consume silicon area and power and, more importantly, add noise. Therefore, at the system level, it is essential to carefully estimate the global costs and the benefits of a fully-differential implementation.

5.10.1 Circuit Schematics

All the single ended schemes studied so far can be easily transformed into a fully-differential version. Often, it is just required to remove the differential to single ended converter and, possibly, use two second gain stages.

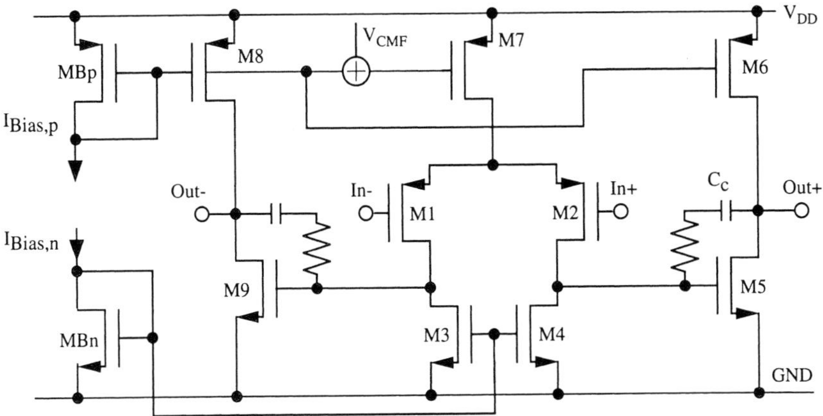

Fig. 5.39 - Fully differential two stages OTA.

Fig. 5.39 shows the fully-differential version of the two stages *OTA*. The differential to single ended conversion, achieved by connecting M_4 in the diode configuration has been removed. M_3 and M_4 both operate as active load and the dc gain of the first stage becomes

$$A_1 = \frac{1}{2}\frac{g_{m1}}{g_{ds1} + g_{ds4}}. \tag{5.95}$$

Two second stages with zero nulling compensation networks complete the scheme. Observe that the circuit in Fig. 5.39 requires one additional bias line to control the gate of M_3 and M_4. Therefore, the designer must generate that bias voltage and, more importantly, bring around another connection line.

An important function required by a fully differential scheme is the common mode feedback. We discuss this necessity by considering the circuit in Fig. 5.39. A proper operation of the network requires to balance the quiescent current of a *p-channel* transistor with the current of the equivalent *n-channel* transistor while both remain in saturation. The scheme in Fig. 5.39 has three of such a necessities: M_6 and M_8 need to match the current of M_5 and M_9 respectively; the current in M_8 demands to be equal to the sum of the currents in M_3 and M_4. Unavoidable mismatches between current references and mirror factors prevent to achieve the fittings. It is therefore necessary to use a specific circuit that takes care of the current matching issue: this circuit is called the *common mode feedback*. It detects the average value of the output voltages, compares it with a desired level and produces a controls signal that, in feedback, regulates one of the currents that we want to match.

Fig. 5.39 shows a possible input of the common mode control, V_{CMF}. Assume that the currents drained by the p-channel transistors of the second stages are less than what the n-channel counterparts sink to ground. The common mode output voltage would drop down and the circuit doesn't operate properly. A positive V_{CMF} control solves the problem. It diminishes the overdrive of M_7 thus reducing the bias current in the input stage. Therefore, the quiescent currents of M_5 and M_9 diminish until they match the current of M_6 and M_8. A specific circuit capable to generate V_{CMF} is not discussed here. It is achieved by an inverting amplification of the common mode output voltage.

Fig. 5.40 and Fig. 5.41 show the fully differential version of the mirrored cascode and the folded cascode *OTAs*. The two schemes follow directly from the single ended versions: the connections of the cascode current mirrors have been taken away to provide the fully differential outputs. In both circuits we need to properly bias the transistor pairs M_3-M_4, M_5-M_6 and M_7-M_8. Therefore, with respect to the single ended version, the circuit requires additional bias voltages. The designer can possibly limit the design burden by using the same voltage used to control M_9 for the biasing of M_7-M_8.

The schematics in Fig. 5.40 and Fig. 5.41 highlight the points where the designer can introduce the common-mode feedback. We can achieve the nec-

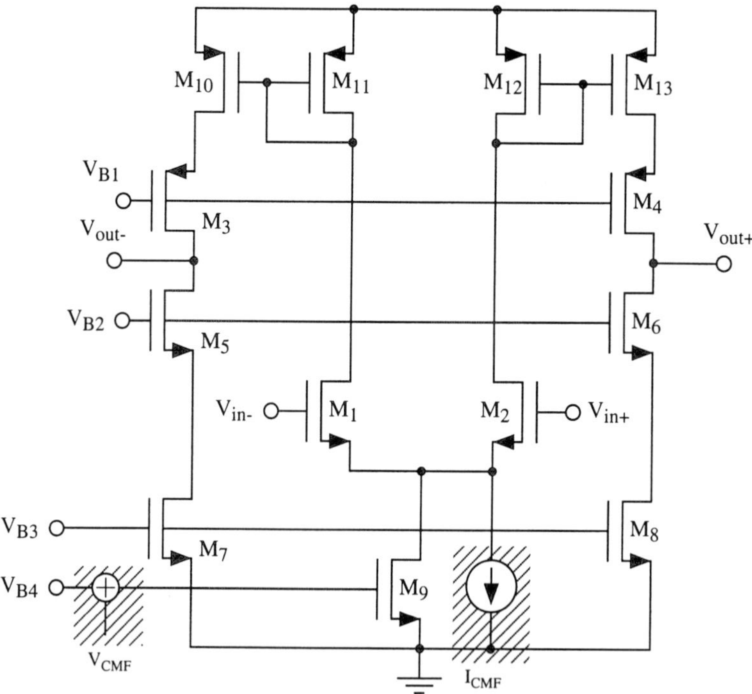

Fig. 5.40 - Fully differential version of the mirrored cascode OTA. The circuit shows two options for the common-mode feedback input: one for a voltage-mode the other for a current-mode implementation.

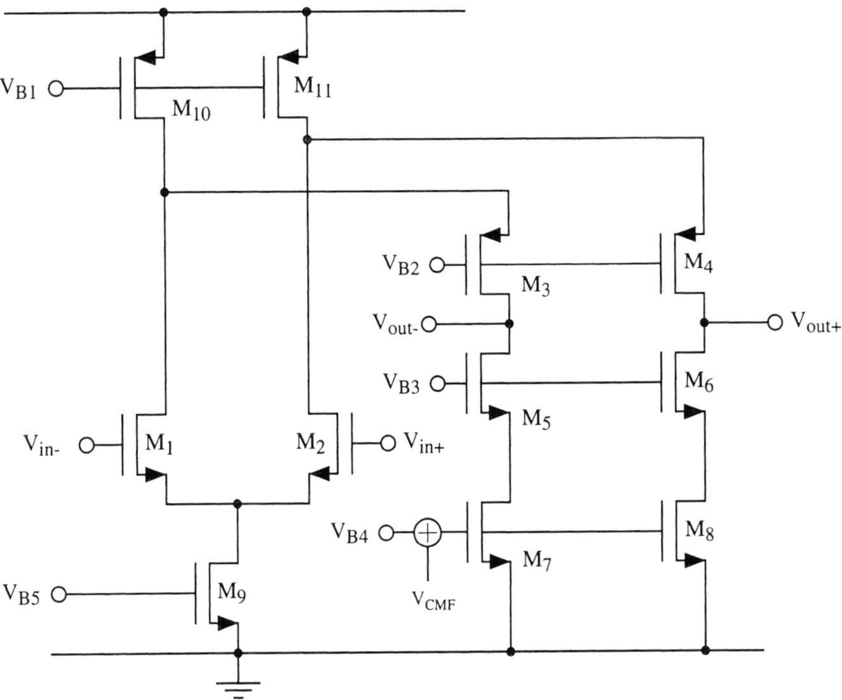

Fig. 5.41 - Fully differential version of the folded cascode OTA.

essary balancement of currents by adjusting in feedback the gate voltage of M_9 (Fig. 5.40) or the one of M_7-M_8 (Fig. 5.41). Moreover, we balance currents by a straight injection of a common-mode feedback current.

The output nodes, similarly to the single ended version, are the only high impedance nodes. All the other nodes are connected to the source of a transistor or to a diode connected element. Therefore, the pole splitting compensation or other compensation techniques are not necessary. The stability depends on the position of the dominant pole. Its location can be controlled by a possible increase of the capacitive loads at the two differential outputs.

5.10.2 Common Mode Feedback

Before studying circuit implementations of common mode feedback let us discuss again, from a different perspective, why a fully differential op-amp needs the common mode feedback. Fig. 5.42 shows a single ended application of an op-amp (or an *OTA*) and the corresponding fully differential version. In the single-ended circuit the non-inverting input is connected to the analog ground while the feedback network links the inverting terminal to the output. If

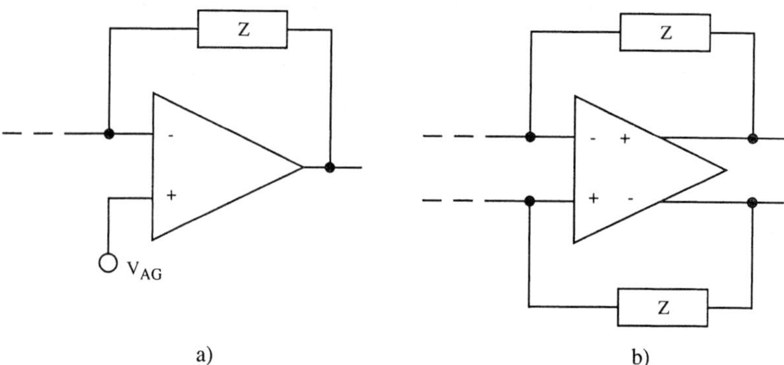

Fig. 5.42 - a) single ended use of an op-amp (or an OTA); b) fully differential version.

the voltage gain is large the differential input is low, at the limit zero. Therefore, as we normally say, the inverting terminal is virtually connected to the analog ground. In turn, since the feedback binds output and inverting terminal, even the common-mode output equals the analog ground. Therefore, the biasing of the non-inverting terminal fixes either the common-mode input and the common-mode output.

> **OBSERVATION**
>
> All the high impedance nodes need some control to secure their dc voltage. When the differential feedback loose its effectiveness an additional specific control loop is mandatory.

By contrast, the differential version utilizes the two input terminals to establish the feedback in both paths. Consequently, the circuit doesn't have any bind to analog ground. The *dc* gain ensures that the differential input is small or, at the limit, zero but there is no condition that fixes the quiescent voltage of input and output terminals.

Fig. 5.43 shows the basic functions of the

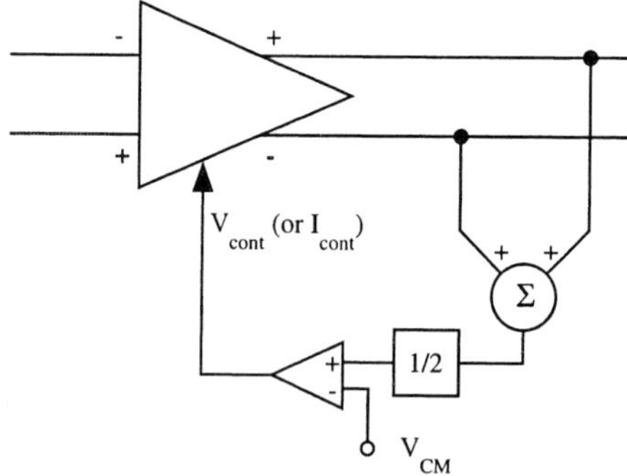

Fig. 5.43 - Basic functions performed by the common-mode feedback.

common mode feedback. The adder and the 1/2 amplifier determine the common mode output; the other block compares the result with the desired common mode level; the possible error (if necessary amplified) controls in feedback a suitable node of the fully-differential architecture. We will see in the next subsections that either a continuous-time or a sampled data approach achieve the above basic functions.

5.10.3 Continuous-time Common-mode Feedback

Fig. 5.44 a) shows a simple circuit sensitive to the average value of differential outputs. It is made by a matched pair of source-coupled transistors. The output is a shifted-down replica (by $V_{GS1} = V_{GS2}$) of equal inputs; small differential changes of the inputs doesn't modify the output. The result drives the current generator, M_I to produce a current signal. Fig. 5.44 a) doesn't incorporate the block that compares the common-mode output with a desired value. Nevertheless, the comparison results implicitly from the circuit

$$V_{CM} = V_{GS,I} + V_{GS1} \tag{5.96}$$

If the differential swing is larger than the overdrive of M_1-M_2 one of the transistors turns-off and the output follows the higher input voltage. It turns out that the circuit in Fig. 5.44 a) operates properly only for limited differential swings. Nevertheless, the solution is valuable in non-demanding cases because of its simplicity and the relative low power consumption. The circuit in Fig. 5.44 b) uses two diode connected transistors M_4 and M_5 to degenerate the pair M_1 and M_2. The range of operation of the common-mode feedback is extended at the expense of an increased V_{CM}.

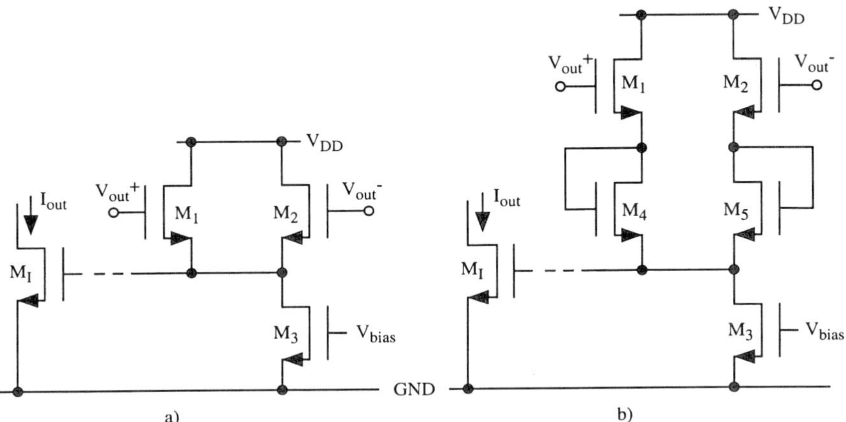

Fig. 5.44 - a) Simple source coupled common-mode feedback. b) Source coupled pair with degeneration (resistors can, possibly, replace the diode connected transistors).

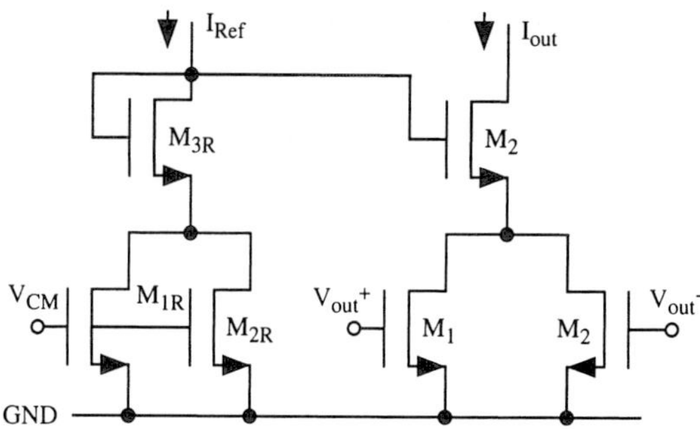

Fig. 5.45 - Use of a current mirror with adjustable mirror factor in the common-mode feedback.

Fig. 5.45 shows another solution that achieves a simple but effective common mode feedback. It is a modified version of the current mirror with adjustable mirror factor studied in Chapter 4. Two transistors, instead of one, degenerate M_{3R} and M_3. The gate of M_1 and M_2 are controlled by the output differential signals which common mode voltage must be kept equal to V_{CM}.

The transistors M_1 and M_2 are equal. Assuming them in the triode condition, we have

$$I_1 = \frac{1}{2}k\left(\frac{W}{L}\right)_1 [2(V_{out+} - V_{Th})V_{DS} - V_{DS}^2] \quad (5.97)$$

$$I_1 = \frac{1}{2}k\left(\frac{W}{L}\right)_1 [2(V_{out-} - V_{Th})V_{DS} - V_{DS}^2] \quad (5.98)$$

similarly, for the reference branch, being M_{1R} and M_{2R} equal and matched with M_1

$$I_{1R} = I_{2R} = \frac{1}{2}k\left(\frac{W}{L}\right)_1 [2(V_{CM} - V_{Th})V_{DS,R} - V_{DS,R}^2] \quad (5.99)$$

Transistors M_3 and M_{3R} are matched; then, if the currents I_{Ref} and I_{out} are not much different the V_{DS} of the pair M_{1R}-M_{2R} equals the one of M_1-M_2. Thus, using $V_{DS,R}=V_{DS}$ in (5.99), from (5.97) and (5.98) it results

$$I_{out} = I_1 + I_2 = I_{Ref} + 2k\left(\frac{W}{L}\right)_1 V_{DS}\left(\frac{V_{out+} + V_{out-}}{2} - V_{cm}\right) \quad (5.100)$$

that corresponds to the required control. The achieved transconductance gain is $2k'(W/L)_1 \cdot V_{DS}$.

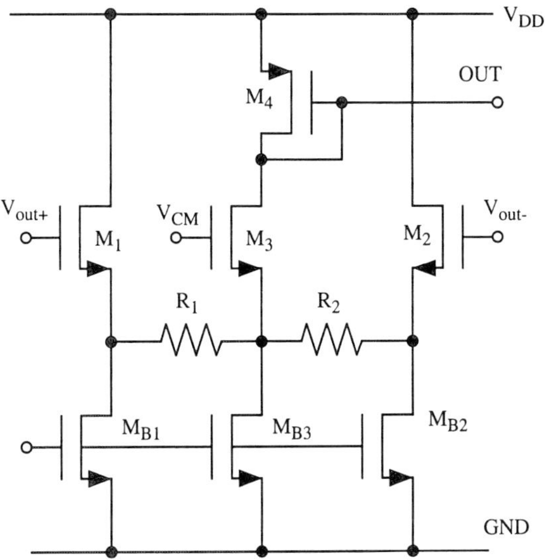

Fig. 5.46 - Common-mode feedback based on source followers.

Similarly to the solution in Fig. 5.44, the circuit in Fig. 5.45 doesn't accomplish a direct comparison of common-mode output with a desired value. The operation results implicitly because of the symmetrical action of V_{out+}/V_{out-} and V_{CM}.

Fig. 5.46 shows a more complex implementation of the common mode feedback. It uses two source follower to generate a shifted down replica of the differential outputs. Moreover, a third source follower shifts down the desired common-mode voltage, V_{CM}. The resistors R_1 and R_2, nominally equal connect the sources of transistors M_1, M_2 and M_3. If the voltages at the three gates are equal no current flows on the resistors. A common mode change of V_{out+} and V_{out-}, for example in the positive direction, causes two equal currents flowing out of the sources of M_1 and M_2. These currents will diminish by a corresponding extent the current in M_3 (and M_4) being M_{B3} a current source. Assuming one the gain of the source follower and neglecting the effect of the finite resistance of the follower, the current variation in the diode connected transistor M_4 is simply given by

$$\Delta I_{M4} = \frac{2(V_{out+} + V_{out-} - 2V_{CM})}{R_1} \tag{5.101}$$

By contrast a differential signal induces a balanced swing at the sources of M_1 and M_2. Therefore, a current proportional to the differential signal flows from the source of M_1 into the source of M_2 without affecting the branch in the middle.

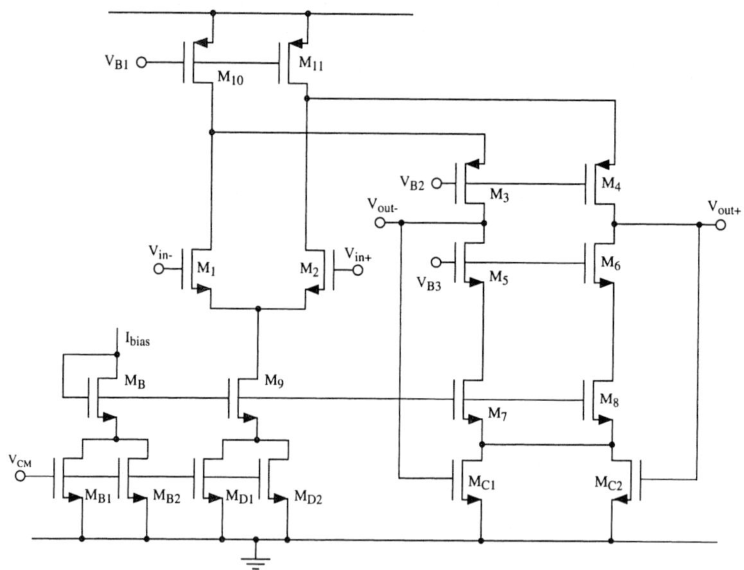

Fig. 5.47 - Fully differential folded cascode with the common-mode feedback of Fig. 5.45.

Fig. 5.47 incorporates the common mode feedback in Fig. 5.45 in the fully differential folded cascode configuration. The circuit includes the two additional transistors M_{C1} and M_{C2} in the output branch and the matching transistors M_{B1}-M_{B2} and M_{D1}-M_{D2} to degenerate the reference current and the source of the input stage. Moreover, he circuit uses the same control for both current transistors M_6 and M_7.

The circuit operates properly until M_{C1} or M_{C2} are in the triode region. Large output differential signals can bring one of them in the saturation or even in the off state. The entire current flows in the other transistor and the

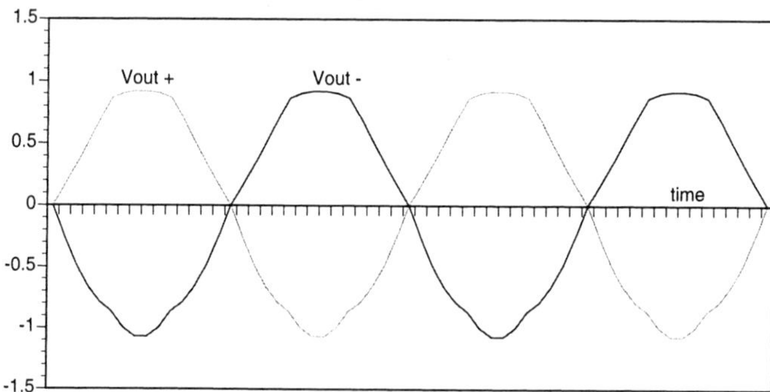

Fig. 5.48 - Typical distortion of the output differential voltages due to a poor common-mode control.

common mode control ceases working properly. The negative swing of the output doesn't compensate the positive one any more and the common mode output drops down. Fig. 5.48 shows a typical plot of the output voltage for large output differential signals. The result looks poor; however, what is important to us is the difference and not the single components shown in the figure. Actually, after the differential to single-ended conversion the signal becomes an accurate sine-wave. The limit that comes out from the response in Fig. 5.48 anyway regards the dynamic range. The output signal must remain in the region where the gain is large enough. Since a poor common mode control broadens downward the necessary dynamic range, the maximum output swing diminishes accordingly.

5.10.4 Sampled-data Common-mode Feedback

Fig. 5.49 shows how capacitors and switches can implement the basic functions required by the common mode feedback. Two complementary non-overlapped phases control the circuits. The network in Fig. 5.49 a) determines the average of the output voltages. It works as follow: during phase *1* the voltages V_{out+} and V_{out-} pre-charge the two nominally equal capacitors C. Then, during phase 2, the capacitors are connected in parallel. They share their charges and generate the voltage

$$V_{parall} = \frac{Q_T}{2C} = \frac{CV_{out+} + CV_{out-}}{2C} = \frac{1}{2}(V_{out+} + V_{out-}) \tag{5.102}$$

The circuit in Fig. 5.49 b) subtracts the two voltages V_2 and V_1. During phase *1* V_2 pre-charges the capacitor. During the next phase 2 the top plate is floating and V_1 biases the bottom plate. Therefore, the top plate pops to V_2-V_1.

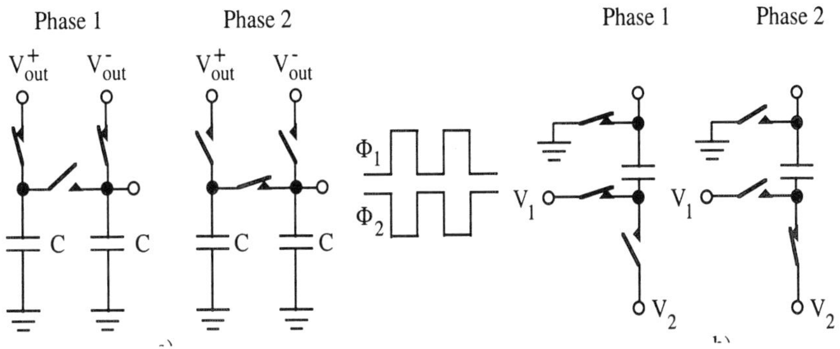

Fig. 5.49 - a) Switched capacitor network for the calculation of the average of two voltages. b) Circuit useful for the subtraction of two voltages,

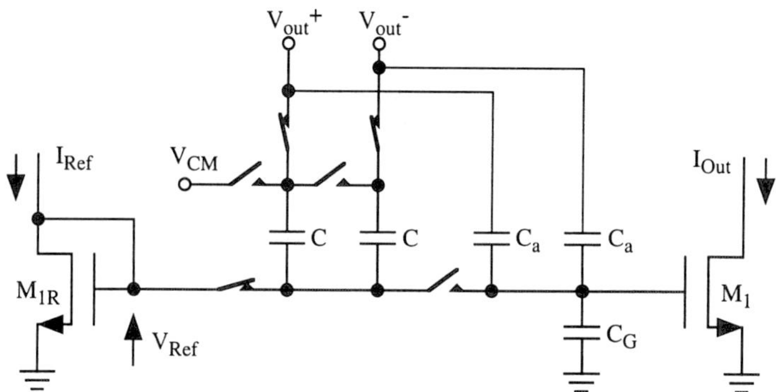

Fig. 5.50 - Sampled-data common-mode feedback. A common mode output larger than V_{CM} leads to an output current lower than the reference.

The circuit in Fig. 5.50 incorporates the two basic operation discussed above thus implementing a sampled-data common-mode feedback operation.

The reference current I_{Ref} injected in the diode connected transistor M_{1R} generates a gate reference voltage V_{Ref}. The output current would be equal to I_{Ref} if the voltage V_{Ref} were straight applied to the gate of M_1. The switched capacitor network in between the gate of M_{1R} and M_1 possibly modify this condition. During phase *1* V_{out+} and V_{out-} minus V_{Ref} charge the two equal capacitors C. During phase *2* (complementary to the one illustrated in the figure), a switch connects the capacitors in parallel while the top plate is connected to V_{CM}. If the average of V_{out+} and V_{out-} equals V_{CM} the voltage developed at the bottom plate becomes V_{Ref}. Now, if the average of V_{out+} and V_{out-} is larger than V_{CM} the voltage at the bottom plate becomes lower than V_{Ref}, and vice versa. Therefore, a common mode output voltage larger than what was expected diminishes the output current.

Actually during ϕ_2 the two capacitors C are placed in parallel with C_G. Capacitance C_G represents the gate capacitance of M_1 and an actual capacitor that is likely used to sustain the gate voltage of M_1 during the phase *1*. Capacitor C_G also smooths the changes induced every clock cycle. (The circuit acts like a switched-capacitor network whose function is equivalent to an *RC* low-pass filter).

The discussed common mode feedback measures the output voltage during one phase and controls the gate voltage during the successive phase, therefore, it doesn't react to instantaneous changes of the output voltages. Capacitors C_a, through a capacitive couplings between V_{out+} and V_{out-} and the gate of M_1, ensure a high speed path for the common mode feedback circuit.

The circuit in Fig. 5.50 diminishes the output current as response to a com-

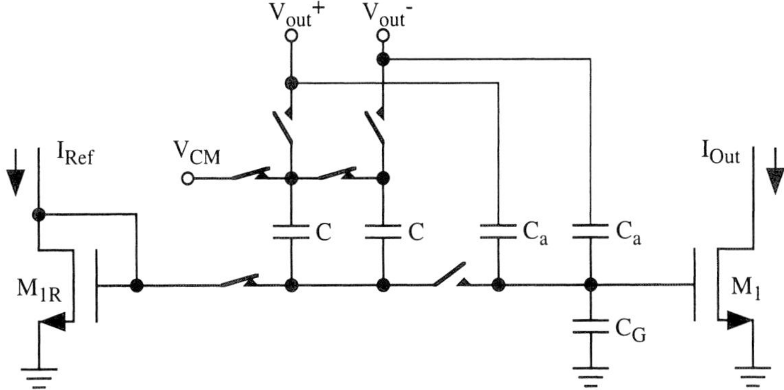

Fig. 5.51 - Sampled-data common-mode feedback with effect opposite to the one of the circuit in Fig. 5.50. A common mode output larger than V_{CM} leads to an output current higher than the reference.

mon mode higher than V_{CM}. In some cases the fully differential circuit demands for an opposite behaviour. Inverting the phases of the switches controlling the upper plate of capacitors *C* increases of the current. Fig. 5.51 shows the circuit configuration.

5.11 MICRO-POWER *OTA*'S

Many applications require very low power consumption. Portable apparatuses are powered by small batteries with a limited power capacity. Thus, to ensure the largest autonomy it is required to minimize the current (consequently, the power) in the basic blocks used. When the bias current in a *MOS* transistor becomes pretty low, the region of operation is no more the saturation but transistor enters in the sub-threshold region. Here as we have learned in Chapter 1, the current-voltage relationship is exponential and the transconductance becomes

$$g_m = \frac{I_D}{nV_T} \tag{5.103}$$

Moreover, the gain of a simple inverter with active load reaches a maximum value

$$A_v = -1/\left[n\frac{kT}{q}(\lambda_n + \lambda p)\right] \tag{5.104}$$

At room temperature, it can be around *60 dB*. Therefore, by using low bias current it is easy to obtain pretty high *dc* gains. Thus, a single stage or a two-stage amplifier are more than enough for satisfying usual gain requirements.

The key design problem results from the very small current available (typically in the ten of micro-ampere range or less). Small currents determine limited bandwidths and this can be accepted for low frequency signal processing. More important, small currents lead to small slew-rates; and this is, normally, the most severe design limitation. Therefore, when designing a micro-power *OTA* it is necessary to use specific techniques to entrance the slew-rate. The designer can use two methods:

- dynamic biasing of the current tail
- dynamic voltage biasing in push-pull stages

Below we will study both methods.

5.11.1 Dynamic-biasing of the Tail Current

The basic concept behind dynamic biasing is quite simple; to provide more current than the quiescent level when slewing needs it. To achieve the result it is necessary to use a slew-rate detector and to have a current bias boost.

In the slew-rate region the current in one branch of the input differential stage equals the tail current while the current in the other branch goes to zero. Thus, we detect the slew-rate condition by measuring the full current umbalancement in the input differential stage.

The circuit inside the dashed lines in Fig. 5.52 takes the difference between I_2 and I_2 that, if positive, flows into the sensing transistor $M_{8,1}$. The similar circuit (made by $M_{3,2}$-$M_{8,2}$) permits us to sense (I_1-I_2) (if positive). The com-

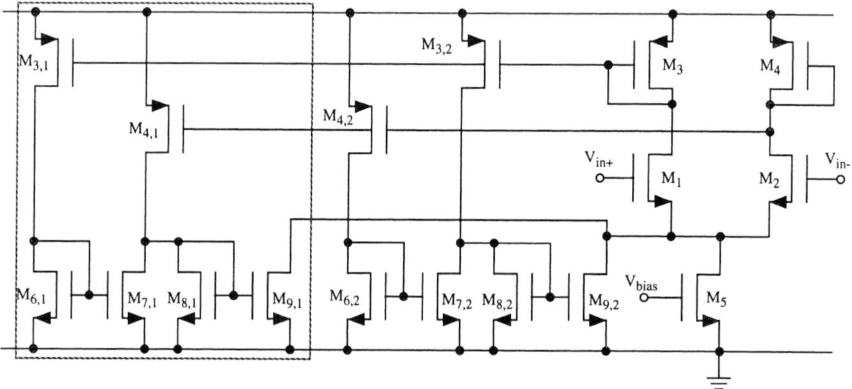

Fig. 5.52 - Dynamic biasing used in a simple gain stage.

bination of the two results, formed by the mirroring elements $M_{9,1}$ and $M_{9,2}$, provide $k|I_1\text{-}I_2|$. Where k comes from the mirror factors used.

The right part of the circuit is the input stage of the *OTA*. Possibly, a mirrored output stage completes the scheme. Assuming M_1 and M_2 in the subthreshold region the differential current is given by

$$|I_1 - I_2| = (I_D + I_{B1} + I_{B2}) atan \frac{|v_{in}|}{nV_T} \tag{5.105}$$

but

$$I_{B1} + I_{B2} = k|I_1 - I_2| \tag{5.106}$$

combining (5.105) and (5.106) it results

$$|I_1 - I_2| = \frac{I_D atan \frac{|v_{in}|}{nV_T}}{1 - k atan \frac{|v_{in}|}{nV_T}} \tag{5.107}$$

that denotes, as results from inspecting the circuit, a positive feedback that brakes into a regenerative grown when

$$k \cdot atan \frac{|v_{in}|}{nV_T} = 1 \tag{5.108}$$

Therefore, in order to have a significant increase of the current, k must be larger than 1 when the input differential voltage is comparable to nV_T. Note that the external feedback network used in the circuit controls the input differential voltage. At a given time, after slewing, the input umbalancement goes down and it becomes low enough to extinguish the regenerative feedback. Then, the current bias boosting turns off.

5.11.2 Dynamic Voltage Biasing in Push-pull Stages

A push-pull scheme is fairly adequate for low-power applications. It allows the use of very low currents and works with relatively low supply voltages.The *AB* class operation also provides large current when the op-amp is required to sustain the slewing phase.

Fig. 5.53 shows a fully-differential scheme based on the same approach used in Fig. 5.36. Two cross-coupled unfolded differential stages provide signal cur-

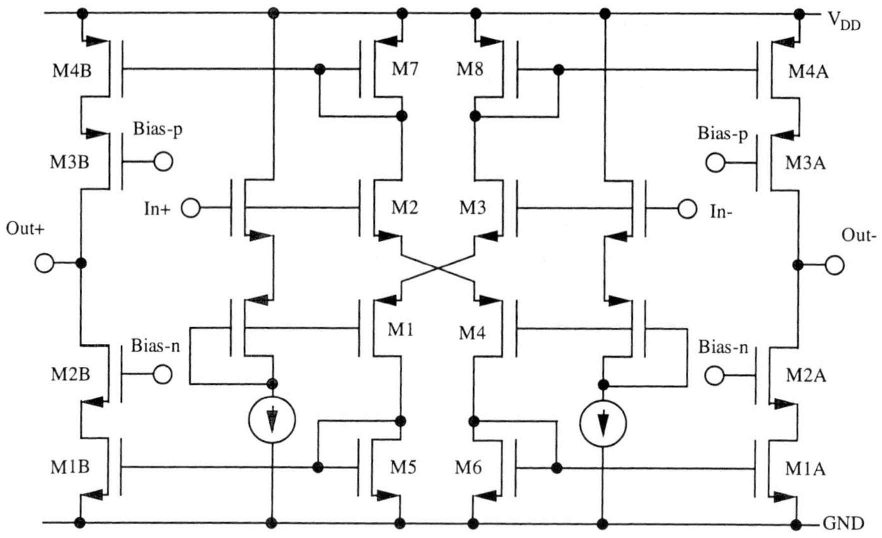

Fig. 5.53 - Fully-differential class AB

rents that, after mirroring, control the cascode output stages. The input section determines the minimum supply voltage. Assuming that the V_{DS} of the transistors of the unfolded pairs is kept a its minimum, we have

$$V_{DD,min} = V_{GS,n} + V_{GS,p} + V_{sat,n} + V_{sat,p} \quad (5.109)$$

that, for typical technologies, can be as low as $2\ V$ or so.

Observe that, in order to ensure low voltage, the circuit uses a simple diode connected transistor to detect the signal current in the input stage. Moreover, suitable bias voltages control the gates of the output cascodes. The input differential signal controls the n-channel input directly. Two level shifters shift down the input voltages by $V_{GS,n}+V_{GS,p}$ to provide the control of the p-channel inputs. For an optimum operation the crossing point in the input stage must be at $V_{DD}/2$. This means that the common mode input must be V_{GS2} higher than $V_{DD}/2$. By contrast the optimum common mode output is at $V_{DD}/2$. The above discrepancy is a minor problem and typically is solved with proper level shifters at the system level.

The differential gain is given by the transconductance of the input pair multiplied by the mirror factor and the output resistance

$$A_v = \frac{(\dot{W}/L)_{4A}}{(\dot{W}/L)_{8A}} \cdot \frac{g_{m1} g_{m3}}{g_{m1} + g_{m3}} \cdot \frac{g_{m3A}}{g_{ds3A} \cdot g_{ds4A}} // \frac{g_{m2A}}{g_{ds2A} \cdot g_{ds1A}} \quad (5.110)$$

that is pretty large especially with micro-currents.

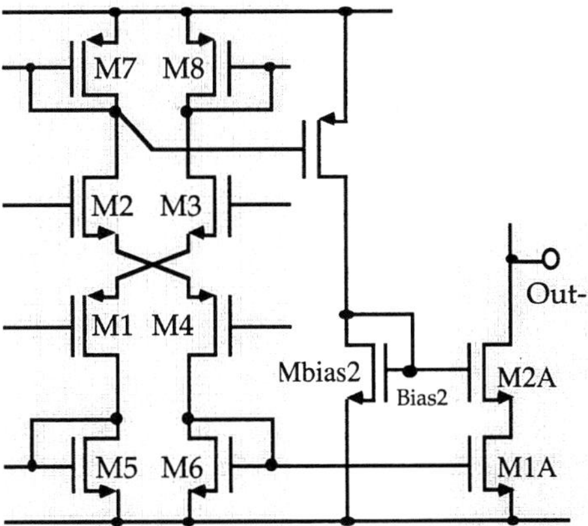

Fig. 5.54 - Dynamic control of the bias voltages in micro-power push-pull OTA.

The main design problem concerns the bias of the cascoding transistors. In order to maximize the output swing the bias voltages $V_{bias\text{-}p}$ and $V_{bias\text{-}n}$ must approach as close as possible to the rail voltages. However, during the slewing conditions the boost of the overdrive of transistors in the output stage doesn't lead to a substantial increase of the slew current. The restriction to V_{DS} caused by the cascoding elements leads the boosted transistors into the triode region. A solution to the problem comes from the use of a dynamic control of the bias voltages. During the slewing phase the dynamic control must increase the V_{DS} voltages and keep the boosted transistors into saturation.

Fig. 5.54 shows a possible implementation of the dynamic control of bias voltages. It applies to the cascode M_{1A}-M_{2A}. The other three cascodes will use similar circuits. The bias voltage V_{Bias2} results from a diode connected transistor whose current is a replica of the current in M_7. A suitable choice of the aspect ratio of M_{bias2} will determine the right V_{Bias2}. During slewing the current in M_6 increases. Accordingly the current in M_7 growths. Therefore, the augmented bias current in M_{bias2} pulls up V_{Bias2}. This happens until the circuit remains in the slewing phase. In the normal conditions of operation I_{M7} returns to its quiescent level and V_{Bias2} goes back to the nominal value.

5.12 NOISE ANALYSIS

The noise performances of the OTAs studied in the previous sections are controlled by the noise of the single transistors used in the circuit. As already discussed the noise analysis assumes uncorrelated the noise generators of dif-

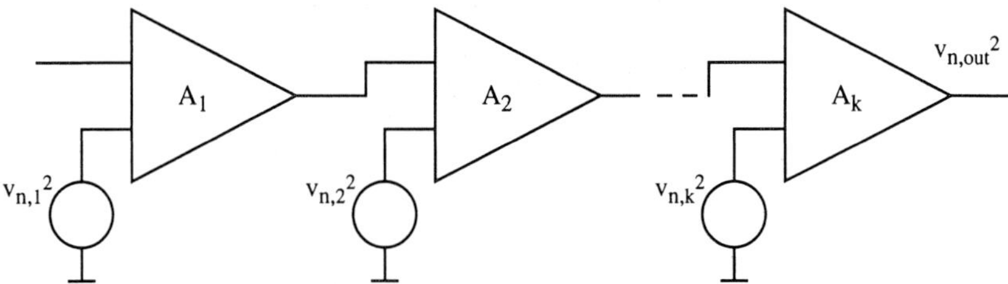

Fig. 5.55 - Noise representation in a cascade of gain stages.

ferent transistors. Therefore, their effect is superposed quadratically.

For the cascade of gain stages the most important contribution comes from the first stage. Assuming to represent the noise of each stage with an input referred generator as shown in Fig. 5.55, we have

$$v_{n,out}^2 = (A_1A_2\ldots A_k)^2 v_{n,1}^2 + (A_2\ldots A_k)^2 v_{n,2}^2 + \ldots + A_k^2 v_{n,k}^2 \qquad (5.111)$$

referred to the input of the chain, it results in

$$v_{n,in}^2 = v_{n,1}^2 + \frac{v_{n,2}^2}{A_1^2} + \ldots + \frac{v_{n,k}^2}{(A_1A_2\ldots A_{k-1})^2} \qquad (5.112)$$

that proves the above statement under the assumption that the gain in the first stage is large enough and expecting comparable input referred noise generators.

Fig. 5.56 shows the first stage of a two-stage amplifiers. It outlines the relevant input referred noise generators. The noise associated to M_5 is not included since it yields a minor effect on the output voltage. In fact, the noise generator of M_5 produces a noise current that, thanks to the differential input stage, flows in the two branches evenly. The pair M_3-M_4 mirrors half of the noise current and injects it into the output node. As a result the subtraction of two fully correlated terms following in the output node produces a null contribution. Therefore, the noise of M_5 can be neglected up to frequencies at which the current mirror operates properly.

The output noise caused by the generators in Fig. 5.56 is estimated by

$$v_{n,out}^2 = [(v_{n1}^2 + v_{n2}^2)g_{m1}^2 + (v_{n3}^2 + v_{n4}^2)g_{m3}^2]\left(\frac{1}{g_{ds2} + g_{ds4}}\right)^2 \qquad (5.113)$$

The input referred noise is derived by dividing $v^2_{n,out}$ by the square of the gain of the stage

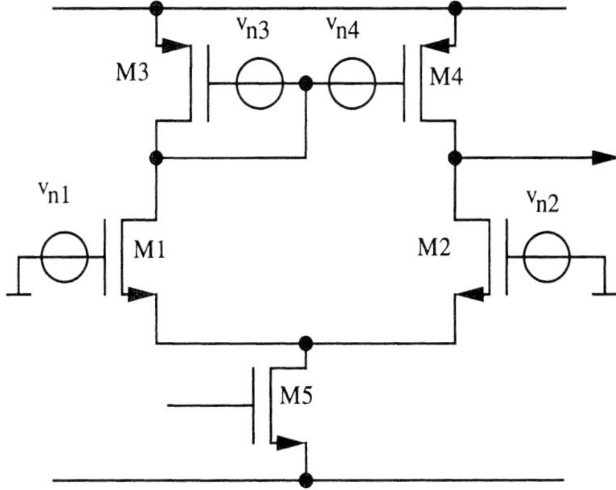

Fig. 5.56 - Schematic for the noise analysis of the first stage of a two-stages OTA.

$$v_{n,in}^2 = \frac{v_{n,out}^2}{g_{m1}^2}(g_{ds2} + g_{ds4})^2 = 2\left[v_{n1}^2 + \left(\frac{g_{m3}}{g_{m1}}\right)^2 v_{n3}^2\right] \tag{5.114}$$

Assuming the transistors in saturation, we can use for the transconductance the following relationship

$$g_m = \sqrt{2\mu C_{ox}\frac{W}{L}I} \tag{5.115}$$

Moreover, according to equations (1.69) and (1.70), the white and the *1/f* term of the spectrum of the noise generator are given by

$$v_n^2 = \left(\frac{8kT}{3g_m} + \frac{K_F}{2\mu C_{ox}}\frac{1}{WL}\cdot\frac{1}{f}\right)\Delta f \tag{5.116}$$

using the white component of the noise and (5.115) in equation (5.114) it results

$$v_{n,in,white}^2 = 2v_{n1}^2\left(1 + \frac{g_{m3}}{g_{m1}}\right) = 2v_{n1}^2\left(1 + \sqrt{\frac{\mu_3(W/L)_3}{\mu_1(W/L)_1}}\right) \tag{5.117}$$

indicating that the white component of the input referred noise, $v^2_{n,in,white}$, equals the noise of the input pair multiplied by a proper factor. Often it is required to have the input referred noise dominated by the input pair. Therefore, the multiplying factor should be kept at the minimum. By inspection of (5.117) we obtain the following design conditions:

- the mobility in the input pair should be larger than the one in the active load

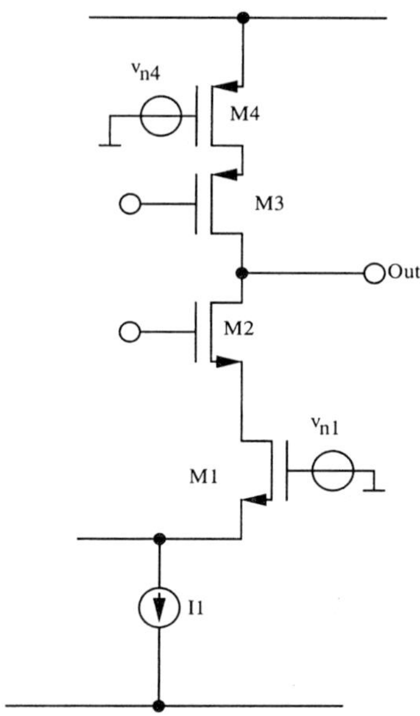

Fig. 5.57 - Schematic for the noise analysis of a telescopic cascode OTA

(that implies the use of *n-channel* transistors at the input);

- the aspect ratio of the input pair should be larger than the one of the active load (that satisfies the request to have a large transconductance gain).

The use of the expression of the *1/f* component of the noise leads to

$$v_{n,in,1/f}^2 = 2\frac{K_{F1}}{\mu_1 C_{ox} W_1 L_1} \cdot \frac{1}{f}\left(1 + \frac{K_{F3} L_1^2}{K_{F1} L_3^2}\right) \tag{5.118}$$

Again, the input referred noise is given by the noise of the input transistor enlarged by a multiplier factor. The designer keeps it at the minimum using a length of the active load larger than the one of the input pair.

Fig. 5.57 shows a schematic useful to estimate the noise performances of a telescopic cascode. The figure depicts half of the circuit only (the other half brings in the same amount of noise). Moreover, the figure shows the noise generators of M_1 and M_4 only. The noise of

NOISE OPTIMIZATION TIPS

- Use of an *n-channel* input pair to minimize the white noise.
- Employ a (W/L) of the input pair larger than one of the active load.
- Use an active load longer than the input pair (to minimize the 1/f noise).

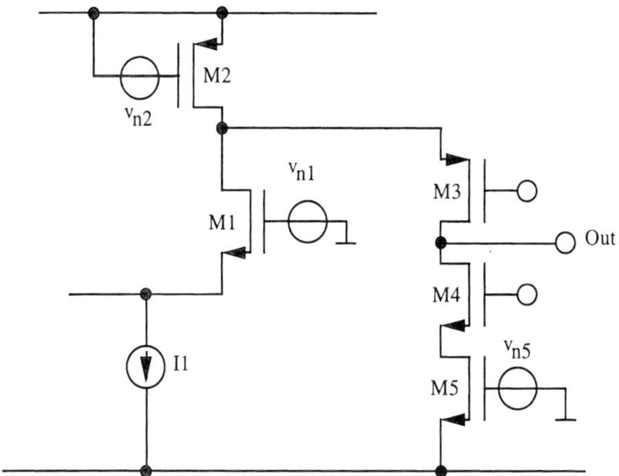

Fig. 5.58 - Schematic for the noise analysis of a folded cascode OTA

the current generator I_1 brings about a negligible effect for the same reasons discussed in the simple differential amplifier. The noise generators of M_2 and M_3 are not considered because the gain from their gates to the output is considerably lower than the gain from the gates of M_1 and M_4. In fact, by inspection of the circuit, one obtains

$$A_1 = g_{m1}R_{out}; \qquad A_4 = g_{m4}R_{out} \tag{5.119}$$

$$A_2 = g_{ds1}R_{out}; \qquad A_3 = g_{ds4}R_{out}. \tag{5.120}$$

The input referred noise becomes

$$v_{n,in}^2 = 2\left[v_{n1}^2 + \left(\frac{g_{m4}}{g_{m1}}\right)^2 v_{n4}^2\right] \tag{5.121}$$

that equals the equation that we obtained for the simple differential stage. Therefore, the designer should follow for input pair and the transistors connected to the positive supply voltage the same recommendation given for the input pair and the active load of Fig. 5.56.

In the folded cascode scheme (see Fig. 5.58) we have an additional transistor that contributes to the noise. It is the transistor used to achieve the current generator M_2. The analysis of the circuit leads to

$$v_{n,in}^2 = 2\left[v_{n1}^2 + \left(\frac{g_{m2}}{g_{m1}}\right)^2 v_{n2}^2 + \left(\frac{g_{m5}}{g_{m1}}\right)^2 v_{n5}^2\right] \tag{5.122}$$

therefore, it is necessary to follow consistent design recommendations for both

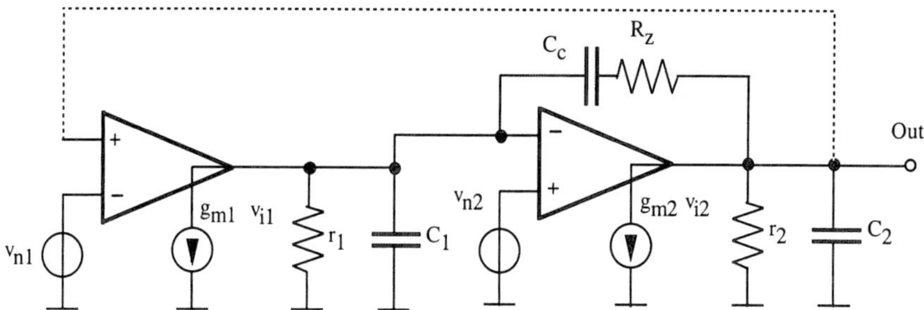

Fig. 5.59 - Equivalent circuit for estimations the noise transfer functions in a two-stages OTA.

transistors M_2 and M_5 in Fig. 5.58.

The previous analysis didn't accounted for the frequency behaviour of the gain stages and the effect of the feedback network. Fig. 5.59 shows the equivalent circuit of a two stages amplifier whose external feedback establishes (for noise calculation purposes) a unity gain configuration. The analysis of the network in Fig. 5.59 or a Spice simulation lead to the transfer functions for the two noise generators. Fig. 5.60 shows the results of a Spice simulation. The overall gain is 80 dB, the gain-bandwidth is 8 MHz and the phase margin is 60°. The transfer function from the first stage shows that the noise is conveyed at the output with a unity gain until the f_T of the OTA. Two poles, f_T and f_2 determine a roll-off of the transfer function. Since f_T and f_2 are pretty close one to the other the roll-off of the Bode diagram rapidly becomes *-40 dB* per

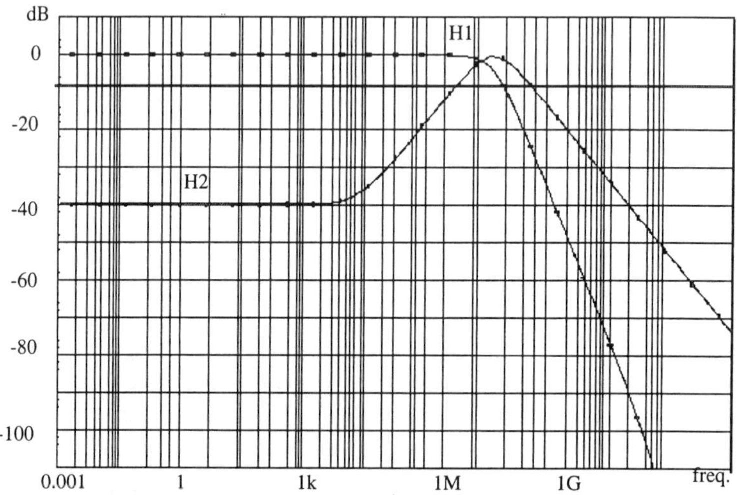

Fig. 5.60 - Noise transfer functions of a typical two-stages OTA.

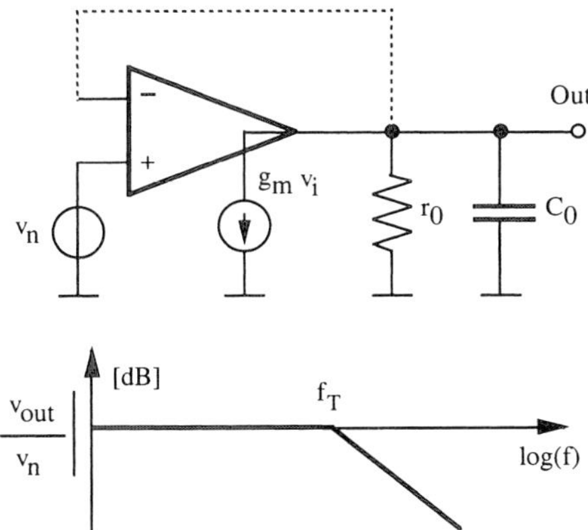

Fig. 5.61 - Equivalent circuit for estimating the noise transfer functions in a single-stages OTA.

decade. The noise transfer function form the input of the second stage has a different behaviour. As expected at low frequency the gain is $1/A_2$ (- *40 dB*). The Bode diagram shows a zero at a given frequency and two poles at f_T and f_2. The zero stays approximately at f_T/A_2. The result confirms that the noise performances are dominated by the first stage. However, at frequencies around the gain-bandwidth the noise provided by the second stage can become comparable with the one of the first stage.

Fig. 5.61 shows the equivalent circuit for estimating the noise transfer function in a single stage OTA. The model used outlines a single pole operation. Therefore, the input referred noise generator influences the output up to f_T. After f_T the noise transfer function rolls down by *20 dB* per decade.

Fig. 5.61 show a low pass filtering action of the op-amp (or OTA) over the input referred noise spectrum. Therefore, independent on the bandwidth of the load, the noise brings about a finite power. The maximum noise power results from the integral over the infinite interval of the output noise spectrum. Assuming the white term of the input spectrum represented by $v^2_{in} = 2(1+\alpha)(8/3)kT/g_m$ one obtains

$$\overline{V_{n0}^2} = \int_0^\infty v_n^2 \frac{df}{|1 + s/(2\pi f_T)|^2} = \tag{5.123}$$

$$= 2(1+\alpha)\frac{8}{3}kT\int_0^\infty \frac{1}{g_{m1}} \frac{df}{1 + (2\pi f C_0/g_{m1})^2} =$$

$$= 2(1+\alpha)\frac{8}{3}kT\frac{1}{2\pi C_0}\int_0^\infty \frac{dx}{1+x^2} = \frac{4}{3}2(1+\alpha)\frac{kT}{C_0}$$

where we assume $f_T = 2\pi C_O/g_{m1}$.

Equation (5.123) shows that the noise power depends on the load capacitor of the op-amp and the factor α, denoting the noise exceeding the input pair contribution. The fact that the result doesn't depend on the transconductance of the input pair can be confusing. To justify the result, observe that an increase of the transconductance leads to a lower noise but, at the same time, the bandwidth of the noise transfer function increases.

5.13 LAYOUT

The previous chapters examined the techniques for laying out a single transistor or a basic building block. This section considers the more comprehensive issue of laying-out a complete op-amp or an OTA. The performances of circuits studied so far critically depends on the electrical design. Additionally, any design presumes matching between transistors and symmetry between sections. Moreover, it is expected that parasitics and spur couplings produce negligible effects. Therefore, a properly executed layout is the key to comply the design targets.

In general, the layout should conform with the following guidelines:

- to procure the same symmetry that the circuit has at the electrical level;
- to ensure that transistors supposed to be equal or assumed to have a given aspect ratio match at the geometrical and technological level;
- to bring at minimum and/or to match the parasitic drops voltage across interconnections;
- to obtain balanced paths in signal interconnections;
- to avoid (or to make minimum when unavoidable) capacitive couplings especially the one with high impedance nodes.

Symmetry in circuits are important to minimize the random offset and to reject common-mode unwanted signals. Since the random offset comes from the mismatch of paired transistors, inter digitized or, even better, common centroid arrangements of critical components are necessary.

5.13.1 Parasitic Effects

Spurs affect circuits in different ways. They appear because of the interference between interconnection lines. They come from the substrate or can be the consequence of opening a switches. Therefore, the protection from spur signals involves both transistors and interconnections. The use of symmetrical

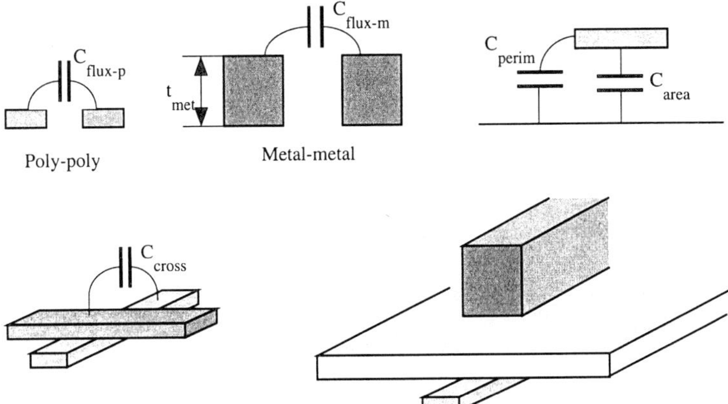

Fig. 5.62 - Possible parasitic coupling affecting metal and poly interconnections.

layouts made by two mirrored half-cells helps in obtaining balanced paths and symmetrical features. However, a mirrored layout ensures solely a symmetry axis; a gradient in direction orthogonal to that axis causes mismatch. Therefore, it is probably better to use small mirrored half-cells wired with balanced interconnections.

Fig. 5.62 shows the capacitive couplings that can occur in a typical layout. A flux capacitance affects two lines running in parallel. For polysilicon strips at $0.5\ \mu$ distance the value of the specific capacitance $C_{flux\text{-}p}$ is typically $50\ aF/\mu$. Therefore, two lines running in parallel at the minimum separation for $200\ \mu$ can be affected by a $10\ fF$ link. Moreover, the thickness of metal lines, $t_{met} \approx 0.5\text{-}1.5\ \mu$, is larger than the height of poly strips by a factor 2 to 3. Even if the minimum distance between metal is larger than the separation of poly the flux capacitance $C_{flux\text{-}m}$ can be larger than $C_{flux\text{-}p}$.

Since the metal-metal flux capacitance is large, some technologies exploit

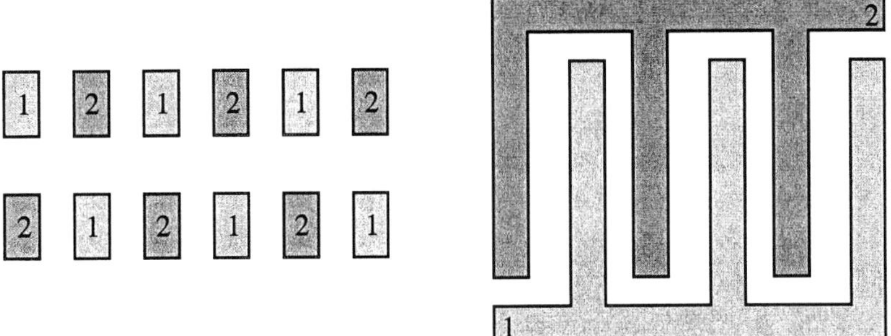

Fig. 5.63 - Cross section and layout of a flux capacitor using two metal layers.

the feature to obtain capacitors. The structures used connect in parallel the various metal layers available so that the flux capacitance enlarges. Fig. 5.63 shows the cross section and the layout of a possible flux capacitor. The structure uses two metal layers. An interleave in the horizontal and vertical directions of the two metal achieves plate 1and plate 2.

Table 5.2 provides the parasitic parameters for a typical *0.25 μ* CMOS technology. Observe that the parasitic in the vertical direction between substrate and connection lines or between different conductive layers depends on the area and, because or the fringing effects, on the perimeter of the considered geometry. Therefore, the estimation of the coupling due to the intersection of connecting wires must account for both crossing area and perimeter. Observe that a relatively narrow line (below *1 μ*) brings about fringing parasitic larger than the direct coupling. For example, a *0.5 μ metal-1* line running over the field oxide for *100 μ* leads to *2 fF* for the overlap and *6fF* for the fringing coupling. Moreover, the fringing term dominates the capacitive coupling in the crossing of a *1 μ metal-1* and a *1 μ metal-2*. The vertical coupling, *30 aF*, is a small fraction of the total, *230 aF.*

TABLE 5.2 - *Parasitic Capacitances*

Parameter	Value	Unit
Poly-poly flux	50	aF/μm^2
Metal1-metal2 flux	70	aF/μm^2
Metal2-metal2 flux	80	aF/μm^2
poly over field oxide area	40	aF/μm^2
poly over field oxide perimeter	30	aF/μm
metal1 over field oxide area	30	aF/μm^2
metal1 over field oxide perimeter	40	aF/μm
metal 2 over field oxide	15	aF/μm^2
metal2 over field oxide perimeter	30	aF/μm
metal1 to poly area	20	aF/μm^2
metal1 to poly perimeter	35	aF/μm
metal2 to metal1 area	30	aF/μm^2
metal2 to metal1 perimeter	50	aF/μm

Albeit the estimated values of parasitic capacitance are quite low, their effect is significant for analog applications. Example 5.8 shows how an improper layout induces significant spur on sensitive analog lines.

Example 5.8

A metal-2 line overlap for 50 μ a metal-1 line. The width of both lines is 0.5 μ. The first line carries a digital clock that switches from 0 to 3.3 V. The second line connects to the gate of an analog transistor which input capacitance is 0.2 pF. Calculate the disturbance caused by the parasitic coupling. Use the parasitic parameters given in Table 5.1.

Solution:
The coupling comes from the overlap area and the fringing contribution

$$C_p = (30 \cdot 25 + 50 \cdot 100)\text{aF} = 2.25\text{fF}$$

The digital clock, attenuated by the capacitive divider C_p-C_{in}, affects to the gate of the analog transistor by

$$V_{gate} = V_{clock}\frac{C_p}{C_{in} + C_p} = 36.7mV$$

which is a large number for any analog applications.

In any schematic an interconnection is intended to be a wire with zero resistance. Unfortunately, in real circuits that is not true. Metal lines achieve interconnections with a given series resistance and, relevant for high frequency operation, a parasitic capacitance. The specific resistance of metal depends on the technology: interconnections may use aluminium or copper; the thickness of metal lines can be a fraction or more than one micron. As a result, the specific resistance of metals varies over a pretty large range. A typical value is around *0.1 Ω/square*.

The width of the interconnections must account for the carried current. To avoid electromigration (that affects the reliability) it is necessary to ensure a given width per unity current. Table 5.3 furnishes typical parasitic parameters for metal interconnections. Observe that in addition to the metal connections it is necessary to account for the resistance of via when interconnections use different metal layers.

A wire of few ten of squares produces some *Ω* of parasitic resistance. When a relatively large currents flows, the drop voltage across the wire can be compa-

TABLE 5.3 - *Parasitic Resistances and Current Densities*

Parameter	Value	Unit
Poly-1 specific resistance	20	Ω/□
Metal-1 specific resistance	100	mΩ/□
Metal-2 specific resistance	70	mΩ/□
Copper Metal specific resistance	20	mΩ/□
Metal-1 poly contact resistance	10	Ω/contact
Metal-1 metal-2 via	1	Ω/via
Poly current density	0.4	mA/μm
Metal-1 current density	1	mA/μm
Metal-2 current density	1.5	mA/μm
Via current	1	mA/via

rable to the offset caused by the threshold mismatch in adjacent transistors. Widening metal lines reduces the series resistance. However, the parasitic coupling increases possibly limiting the frequency performances or augmenting the spur injection. Therefore, it is necessary to find the best trade-off between the parasitic resistance and the capacitance limitation.

Fig. 5.64 compares a layout that leads to mismatched interconnections and a more suitable implementation. The current toward the sources of M_1 and M_2 flows from the left side of the layout. Therefore, the voltage at the source of M_1 is $V_S - R_1(I_1+I_2)$. Moreover, the difference between the voltage of the sources of M_1 and M_2 is R_2I_2. Assuming M_1 and M_2 matched and in saturation

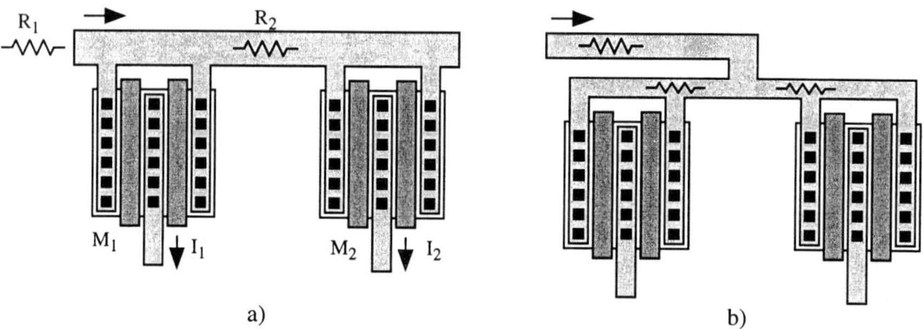

Fig. 5.64 - a) Interconnection of matched transistors causing an offset error.
b) Interconnection matching the drop voltage across the metal lines.

the current I_2 becomes

$$I_2 = k\left(\frac{W}{L}\right)(V_{GS1} - R_2 I_2 - V_{Th})^2 = I_1\left(1 - \frac{2R_2 I_2}{V_{GS1} - V_{Th}}\right) \tag{5.124}$$

If, for example, the drop voltage across R_2 is $1\ mV$ and the overdrive voltage is $200\ mV$ the current mismatch is 1%. Actually, what degrades the circuit performances is not the drop voltage $R_1(I_1+I_2)$ but $R_2 I_2$. Therefore, instead of using wider metals it can be more effective to match the parasitic drop voltages. Fig. 5.64 b) shows the modified the layout. The current flows through the common wire until a junction point; from that point the connections to the sources of M_1 and M_2 are symmetrical. The use of two metals leads to a same result.

5.13.2 Stacked Layout

Splitting transistors in a number of fingers favours a stack arrangement and improves the layout matching. For example Fig. 4.3 presented the layout of a simple current mirror. The four finger of each transistor were interleaved so that the centroid of the two transistors are one close to the other. The arrangement of the stack was *AABBAABB* (where *A* and *B* represent the fingers of M_1 and M_2 respectively). An alternative organization was *ABBAABBA* that lead to an identical common centroid. However, the boundary conditions are not symmetrical: two fingers of M_1 establish the two boundaries while M_2 has all the fingers inside the array.

The above layout strategy can be generalized for more complex cells. However, it is important to work on a design that favours the stacked approach. The transistors' finger that should be laid-out on the same stack must have the same width. This is often possible: all the circuits include transistors which sizes are not particularly critical. A small change of the widths don't modify the performances but permits a better layout.

We will discuss the stacked layout technique with examples. Let us consider the two stages OTA of Fig. 5.65. The scheme specifies the number of fingers and the (W/L) of each transistor. However, the numbers used are not real dimensions. A scaling factor *:2* is used. Therefore, the fingers of the sizes of input pair are $W/L = 10\mu/0.4\mu$. The input pair sums up *12* fingers. The transistor M_{N2} and M_{N3} have *12* fingers; the same is valid for the trio M_{P0}, M_{P1}, and M_{P2}.The above observation suggests to arrange the layout using three stacks

CCDDEEEEEEEE

XABBAABBAABBAX

FFFFGGGGGGGG

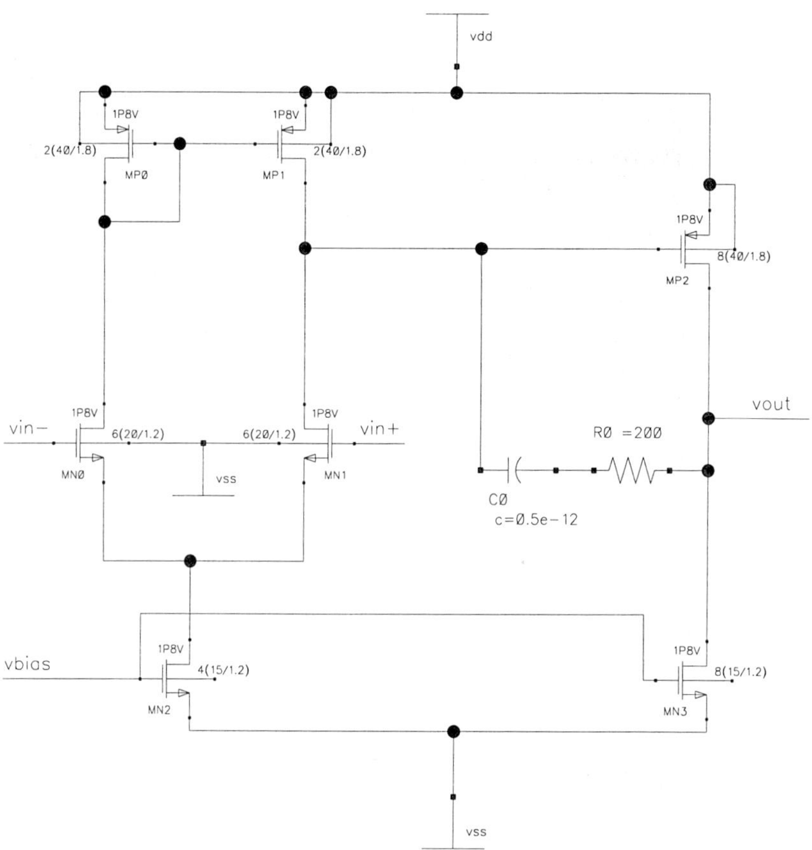

Fig. 5.65 - Circuit schematic of a two-stages OTA. The transistor sizes are convenient for a stacked layout. Use a shrink factor of 2 to have the real dimensions.

The top one arranges the fingers of the p-channel transistors. The letters *C*, *D*, and *E* represent the fingers of M_{P0}, M_{P1}, and M_{P2} respectively. The middle stack achieves the input pairs with the arrangement *ABBAABBAABBA* and two dummy fingers, *X*, at the endings. The third stack includes the tail current generator and the n-channel transistor of the second stage.

Note that the ending node of every pair of fingers is the same. Therefore, if the left terminal of the stack *CCDDEEEEEEEE* is V_{DD}, a connection to V_{DD} every second finger is required. Therefore, it is possible to interchange pair of fingers and obtain different arrangements. For example, the first stack can become *EEEECCDDEEEE* or any other convenient combination. Different finger sequences lead to a corresponding interconnection. Some routing are more problematic than other. This makes the choice of a stack arrangement a trade-off between transistor matching and interconnection balancement. For

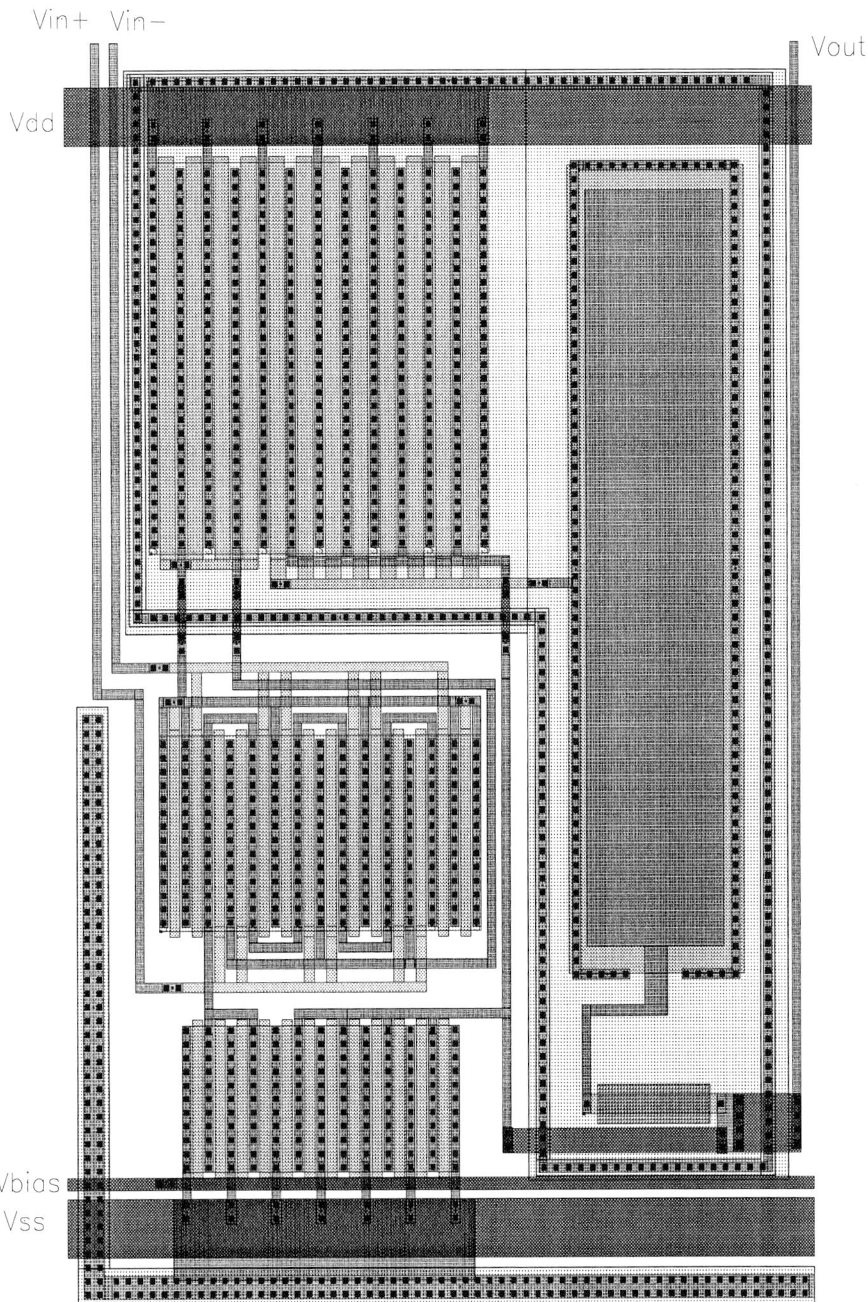

Fig. 5.66 - Layout of the two-stages OTA which schematic is shown in Fig. 5.65

low-power, medium-frequency circuits the matching between transistors is more important than the routing.

Fig. 5.66 shows the layout. Two metal layers favour the interconnections. Actually, metal-2 is used for V_{DD}, V_{SS} and V_{bias}. In addition, only few metal crossings use metal-2. Thus, metal-2 is largely available for the system level interconnections. Moreover, the use of metal-2 for V_{DD} enables the crossings with the vertical metal-1 lines leading out the inputs and the output. This layout organization favours the placement of various op-amps one side to the other with biasing running horizontally and external components (like capacitors and switches) placed on the top side.

The compensation network is on the right edge of the stacked structure. A poly1-poly2 arrangement obtains the capacitor while a poly strip achieves the zero nulling resistor. An n-well, biased all around the periphery, sits under the p-channel transistors and the compensation network. An *L*-shaped substrate bias closes the protection ring around the whole structure.

Fig. 5.67 shows the schematic of a folded cascode. The circuit is more

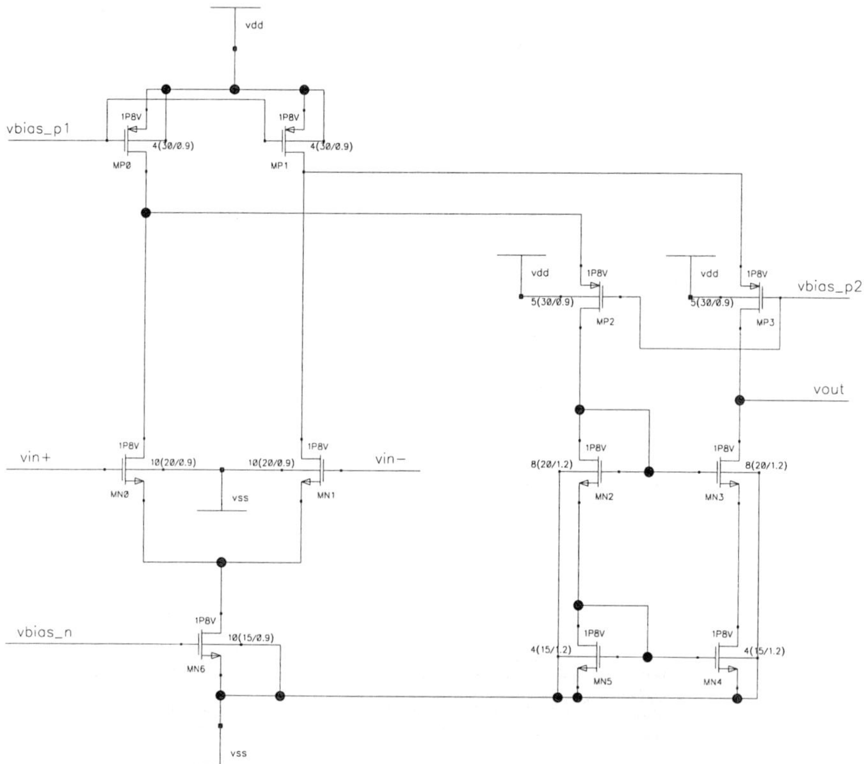

Fig. 5.67 - Circuit schematic of a a folded cascode. The transistor sizes are appropriate for a stacked layout. Use a shrink factor of 2 to have the real sizes.

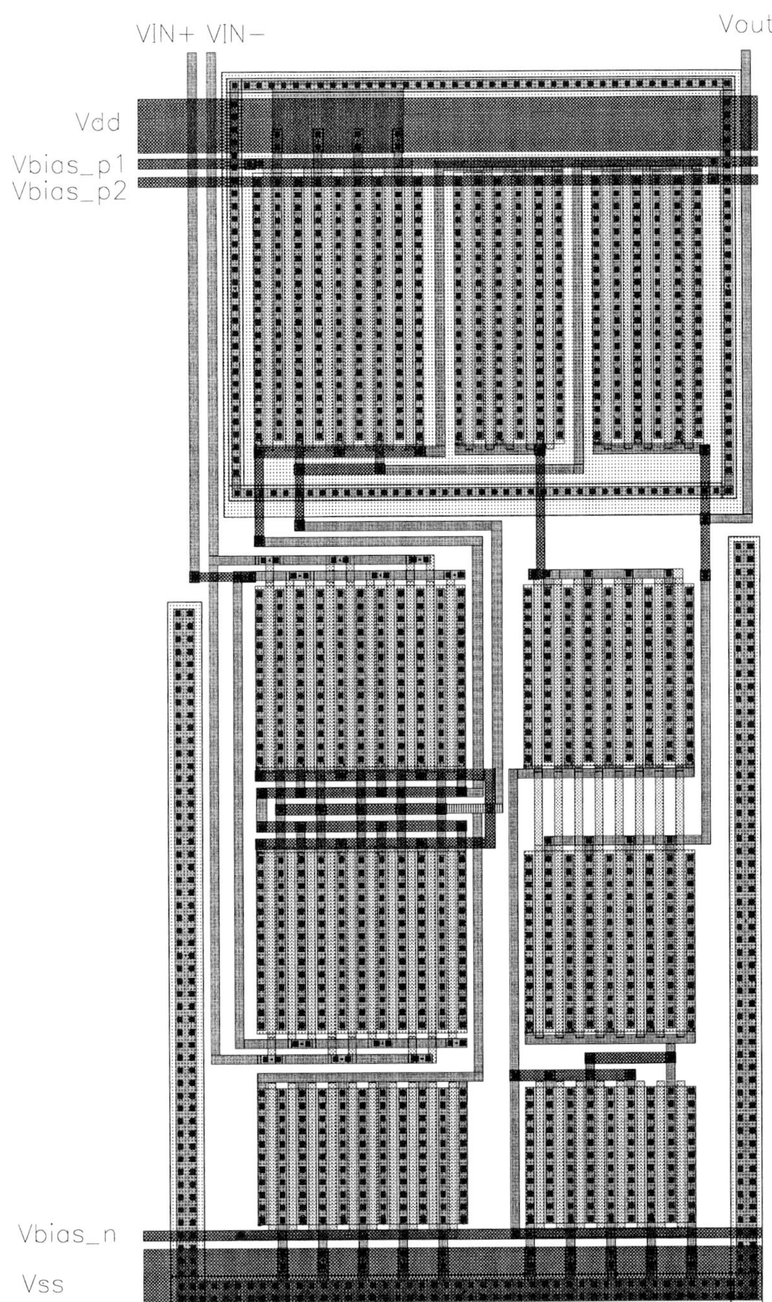

Fig. 5.68 - Layout of the folded cascode which circuit schematic is shown in Fig. 5.67

complex than the simple two-stages discussed above. However it is possible to identify a proper scheme for the stacked layout. The transistors of the input pair is made by 10 fingers. Such a large number advises the use of a common centroid layout. The below scheme shows a possible floor planning. It uses four rows of stacks divided in two vertical sections

CDDCCDDC FFFFF GGGGG
BAABBAABBA KKKKKKKK
ABBAABBAAB NNNNNNNN
EEEEEEEEEE PPQQPPQQ

The used letters denotes transistor fingers corresponding to

$$A \rightarrow M_{N0} \quad B \rightarrow M_{N1} \quad C \rightarrow M_{P0} \quad D \rightarrow M_{P1}$$
$$E \rightarrow M_{N6} \quad F \rightarrow M_{P2} \quad G \rightarrow M_{P3} \quad K \rightarrow M_{N2}$$
$$N \rightarrow M_{N3} \quad P \rightarrow M_{N5} \quad Q \rightarrow M_{N4}$$

Fig. 5.68 shows the obtained layout. The rectangular shape and the vertical metals leading out inputs and output resembles, once again, the approach used for a standard cell layout.

The transistors M_{P2}, M_{P3}, M_{N2} and M_{N3} are non critical elements: they enhance the output resistance of the OTA or improve the symmetry of the circuit. However a possible inaccuracy of transistors' sizes don't affect much the performances. Therefore, in the layout M_{P2}, M_{P3}, M_{N2} and M_{N3} are separate elements. Actually, the width of M_{P0}, M_{P1}, M_{P2} and M_{P3} would permit a single stack arrangement. It is

FFFFFCCDDCCDDGGGGG

Instead, the sizes of M_{P2}, M_{P3}, M_{N4} and M_{N5} don't offer the same possibility.

Observe that the left stack on the top row is *CDDCCDDC*. It ensures a good matching between transistors but it would require dummy elements at the ending. This option is not used in the layout of Fig. 5.68.

The input pair has a common centroid arrangement. The use of two metal layers facilitates the required interconnection crossing. The input pair doesn't use dummy elements. The reason is that the common centroid yields an *A* and *B* element at both endings. Metal wiring joining the gates at the top and the bottom of the common centroid structure minimize the gate resistance.

A well bias surrounds the well where the *p-channel* transistors sit. Additionally, a *U* shaped substrate bias encloses the *n-channel* region.

5.14 REFERENCES

- D. A. Johns and K. Martin, *Analog Integrated Circuit Design*, J. Wiley & Sons, New York, NY, 1997.
- P. R. Gray and R, Meyer, *Analysis and Design of Analog Integrated Circuits*, J. Wiley & Sons, New York, NY, 1993.
- P.R. Gray and R.G. Meyer. *MOS operational amplifier design-a tutorial overview*, IEEE Journal of Solid-State Circuits (1982) SC-17.6 (Dec. 1982, pp. 969-982.
- B.K. Ahuja, *An improved frequency compensation technique for CMOS operational amplifiers*, IEEE Journal of Solid-State Circuits, Vol. 28, 1993, pp. 629-633.
- K. Gulati and Hae-Seung Lee, *A High-Swing CMOS Telescopic Operational Amplifier*, IEEE J. Solid-State Circuits, vol. 33, 1998, pp. 2010–2019.
- J. N. Babanezad, *A low-output-impedance fully differential op-amp with large output swing and continuous-time common-mode feedback*, IEEE J. Solid-State Circuits, vol. 26, 1991, pp. 1825–1833.
- K. Bult, G, Geelen, *A fast-settling CMOS op-amp for SC circuits with 90-dB dc Gain*, IEEE Journal of Solid-State Circuits, Vol. 25, 1990, pp. 1379-1384.
- E. Sackinger and W. Guggenbuhl, *A high-swing, high-impedance MOS cascode circuits*, IEEE Journal of Solid-State Circuits vol. 25, 1990, pp. 289-298.
- R. Castello, P. R. Gray, *A high performance Micropower Switched-capacitor Filter*, IEEE Journal of Solid-State Circuits, Vol. 20, 1985, pp. 14122-1132.
- M. Banu, J. M. Khoury, and Y. Tsividis, *Fully-differential Operational amplifiers with Accurate Output Balancing*, IEEE Journal of Solid-State Circuits, Vol. 23, 1988, pp. 1410-1414.
-S. Mallya and J.H. Nevi, *Design procedures for a fully differential cascode CMOS operational amplifier*, IEEE Journal of Solid-State Circuits vol., 24, 1989, pp.1737-1740.
- L.G.A. Callewaert and W.M Sansen, *Class AB CMOS amplifiers with high efficiency*, IEEE Journal of Solid-State Circuits, vol. 25, 1990, pp. 684-691.
- R. Kline, B. J. Hosticka, and H. J. Pfleiderer, *A very-high slew rate CMOS operational amplifier*, IEEE J. Solid-State Circuits, vol. 24, 1989, pp. 744–746.
- J. F. Duque-Carrillo, *Control of the Common-mode Components in CMOS Continuous-time Fully-differential Signal Processing*, International Journal of Analog Integrated Circuits and Signal Processing, Kluwer Academic Publishers, 1993
- M. Degrauwe, J. Rijmenants, E. A. Vittoz, and D. Man, *Adaptive biasing*

CMOS amplifiers, IEEE J. Solid-State Circuits, vol. 17, 1982, pp. 522–528.
- R. Harjani, R. Heineke and Feng Wang, *An integrated low-voltage class AB CMOS OTA*, IEEE Journal of Solid-State Circuits, vol.34, 1999, pp. 134-142.

5.15 PROBLEMS

5.1 Calculate the response of the circuit in Fig. 5.2 assuming that Z_1 and Z_2 are capacitors, C_1 and C_2 respectively, and $Z_3 = Z_4 = 0$. Assume finite the gain and the bandwidth of the op-amp.

5.2 Simulate, using Spice, the time response of the circuit in Fig. 5.3. Use the following parameters $C_1 = 0.5\ pF$; $C = 2\ pF$; $C_0 = 3\ pF$; $g_m = 1\ mA/V$; $r_0 = 1\ M\Omega$. Compare the result with the case: $C_1 = 0$ and a charge injection by *1 pCoul* at the time $t = 0$.

5.3 Using Spice, determine the spur signal produced on the positive supply bias of the op-amp of Fig. 5.6. The bonding pad causes $8\ \mu H$. The resistance on the analog wire is $1\ \Omega$ and the current I_A is $2\ mA$. The current I_D is made by bipolar glitches with $1\ \mu sec$ periodicity and amplitude $20\ mA$. The duration of the glitches is $40\ psec$.

5.4 Simulate, using Spice and the models of Appendix B the circuit in Fig. 5.12. Use the following design parameters: $(W/L)_1 = (W/L)_6 = 100/1.5$; $(W/L)_5 = 200/2$. All the sizes are in μ. $C_c = 2pF$. Determine the differential gain, the common mode gain, the offset and the power supply rejections.

5.5 Repeat Problem 5.4 using the models of Appendix C. Shrink the geometry of all the transistors by a factor *2*. Use a suitable resistance in series with C_c and estimate the value that permits the circuit to ensure a 60° phase margin with a capacitive load of *2pF*.

5.6 Calculate the systematic offset of the two stages amplifier described in Problem 5.4 for the following values of $(W/L)_5$: *190/2; 210/2; 100:1*. Moreover, perform a simulation where two equal transistors connected in parallel ($W/L = 50/1.5$) replace M_6 and four parallel equal transistors ($W/L = 50/2$) replace M_5. Compare the obtained offset with the initial design and comment the result. All the sizes are in μ.

5.7 Determine the random offset produced by a *5%* mismatch in the *W/L* ratio of M_1-M_2 and M_3-M_4 of the two stages amplifier discussed in

Problem 5.5 (shrink geometers). Compare the obtained random offset with the one that comes out from a *5%* inaccuracy of the *W/L* of M_6.

5.8 Design a two stages *OTA* with an *n-channel* input pair able to fulfil the following specifications: *dc* gain *80 dB*; f_T = *50MHz*; phase margin *60°*; C_L = *3pF*. Use the transistor models of Appendix C.

5.9 Determine the power supply rejection as a function of the frequency of the *OTA* designed in Problem 5.8. Estimate the effect of a capacitive coupling between the supply voltages and the inputs. The coupling results from a capacitor *C* joining inputs and V_{DD} and another capacitor *2C* linking inputs and ground. Study the effect of a spur voltage superposed to V_{DD} or to ground and the effect of a noisy current affecting I_{ref}.

5.10 Simulate the effect of the parasitic coupling in the circuit of Fig. 5.19. Assume C_{p1+}= C_{p2+} = *20fF*; C_{p1-} =C_{p2-} = *40fF*; C_1 = *1pF*; C_2 = *2pF*. Use for the op-amp the small signal equivalent circuit of Fig. 5.20. The gain of each stage is *40 dB*. Moreover, C_1 = *100fF*; C_2= *2pF*, C_c = *3pF* and g_{m1} = g_{m2}= *1mA/V*.

5.11 Design a two stages OTA using the following design conditions: Input pair *(W/L)=200/1*; tail current *300 μA*; current in the second stage *600 μA*; C_L = *5pF*. Use the Spice models of Appendix B and V_{DD} = *3.3V*. Determine the transistor sizes that lead to A_1=*35 dB*; A_2 = *40dB*. Design the zero nulling compensation network that provides a phase margin better than *60°*.

5.12 Design a compensation network with unity gain buffer to be used with for the *OTA* designed in Problem 5.11. Identify the possible zero-pole doublet that comes out because of the limited transconductance of the buffer. Use for the buffer a current lower than 50 μA.

5.13 With Spice simulations, determine the equivalent resistance of the transistor pair M_n–M_p of Fig. 5.24. $(W/L)_n$ =*10/1*; $(W/L)_p$ =*10/1*; V_{DD} = *+1.65 V*; V_{SS} = *-1.65V*; V_1 = *0*. Plot the result for V_1 ranging from *+1 V* to *–1V*. Compare results given by the transistor models of Appendix B and C. All the sizes are in *μ*.

5.14 Repeat Example 5.1 but use a supply voltage of *3.3 V*. The key design target is to limit the power consumption below *0.2 μW*. The input pair should operate at the limit of the saturation/sub-threshold regions. Determine gain and bandwidth (C_L = *2pF*, phase margin *40°*).

5.15 Repeat Example 5.2. Use the design conditions given in Problem 5.14.

and employ real current generators.

5.16 Given the *OTA* designed in Example 5.1, determine with Spice simulations, the positive and negative slew-rate. Connect the op-amp in the unity gain arrangement and utilize an input step signal jumping between *0.5* and *1.3V* (and vice-versa).

5.17 Find the small-signal voltage gain of the telescopic cascode of Fig. 5.27. Account for the substrate transconductance of transistors M_3 and M_4. Determine the gain from the gate of M_3-M_4 to the output.

5.18 Repeat Example 5.4 but assume that the capacitive load is *5 pF*. What is the bias current that permits to achieve the same gain bandwidth product of Example 5.4. What is the *dc* gain?

5.19 Using the Spice simulation results of Example 5.4 find the position of the non-dominant poles. Estimate the effect of the source-substrate and drain-substrate parasitic capacitances on the location of non-dominant poles.

5.20 Consider the mirrored cascode of Fig 5.29. The transistor sizing is the following: $(W/L)_1 = (W/L)_2 = 150/0.4$; $(W/L)_{10} = (W/L)_{11} = (W/L)_{12} = (W/L)_{13} = 50/0.8$; $(W/L)_7 = (W/L)_8 = (50/0.6)$; $(W/L)_9 = 25/1$; $V_{DD} = 3.3V$. The current in M_8 is *250 μA*. Use the transition models of Appendix C. Find the value of V_{B1} and suitably modify the *n-channel* current mirror in order to obtain the widest and symmetrical output swing. Use for all the cascading transistors $(W/L) = 200/0.6$. All the sizes are in *μ*.

5.21 Use the simulation results of Problem 5.20 to estimate the location of the non-dominant poles. Find the load capacitance that establishes a phase margin of 60° and calculate the ratio between the non-dominant poles and the unity gain frequency.

5.22 Repeat Example 5.6 but use a scaled version of transistors. Use a scaling factor *:2* and the Spice models of Appendix C. The supply voltage is *2.2 V*, and the capacitive load is *2 pF*.

5.23 Consider the folded cascode of Example 5.6, where the transistors are shrink by a factor *:2.5* and the supply voltage is *2.2 V*. The reference currents are reduced by the factor *1.5*. Use the local feedback technique to boost the *dc* gain. For the two amplifiers required for the gain boosting use a differential single gain stage.

5.24 Repeat Example 5.7 and try to increase the dc gain to *120 dB*. Any transistor sizes and current level are accepted.

5.25 Consider the bias network of Fig 5.34. Use the following design conditions: $(W/L)_{7b} = 50/0.6$; $(W/L)_{2b} = 200/0.4$; $(W/L)_{4b} = 50/0.8$; $I_{ref} = 200\ \mu A$; $I_{M7b} = 200\ \mu A$; $V_{DD} = 3.3V$. Design the rest of the circuit using reasonable criteria. Find sensitivity of the current I_{M6b} for a $\pm 10\%$ variation of V_{DD}. All the sizes are in μ. Use the models of Appendix C.

5.26 Design the single stage *AB* class op-amp of Fig. 5.36. The required specifications are: $A_0 = 60\ dB$; $I_{tot} = 0.1\ \mu A$. Estimate the slew rate of the output for $4\ pF$ load and open loop conditions Use the technology described in Appendix C.

5.27 Estimate, with Spice simulations and $(W/L)_1 = (W/L)_2 = 50/2$; $(W/L)_3 = 20/0/6$; $I_{M3} = 100\ \mu A$, the range of operation of the common-mode feedback of Fig 5.44. Find the possible range extension achieved by two degeneration resistances of $2\ K\Omega$ connected in series with the source of M_1 and M_2. Use the technology of Appendix C.

5.28 Design and simulate the current mirror with adjustable mirror factor of Fig. 5.45. $V_{S3R} = V_{S2} = 0.1\ V$. The common mode output signal is $2.5\ V$. The nominal current is $0.2\ mA$ and must increase to $0.3\ mA$ when the control voltages rises to $3\ V$. Plot the variation of I_{out} versus the amplitude of the differential control.

5.29 Sketch the layout of the two stages op-amp of Fig. 5.65 but with input stage made by $8\ (20\mu/1.2\mu)$ fingers and the active load by $4\ (20\mu/1.8\mu)$ fingers. It is recommended a common centroid layout for the input stage and an interdigitized active loads. The currents in the wire connections are significant. Find a proper wiring (using two metal layers) that matches the parasitic drop voltages.

5.30 Sketch the layout of the folded cascode of Fig. 5.67. The width of the fingers of M_{N4}-M_{N5} is 20μ. Transistors M_{N2}, M_{N5} and M_{N6} are made by *5, 4, 18* fingers respectively. Find the arrangement that uses *4* stacks of transistors. They must include M_{p0}-M_{p1}-M_{p2}-M_{p3}; M_{N0}-M_{N1}; M_{N6}; and M_{N2}-M_{N3}-M_{N4}-M_{N5} respectively.

5.31 Sketch the layout of the two stages amplifier designed in Example 5.1. Use a common centroid layout for the input pair. Divide the transistors in fa suitable number of fingers and propose possible modifications of the sizes in order to achieve a more efficient layout.

5.32 Layout the telescopic cascode of Example 5.4. Change the transistor sizes or the number of finger to have the p-channel transistors in the same stack. The current tail generator and its control must have an interdigitized layout. Try to obtain a rectangular shape and ensure that the inputs and output leads are on the top side of the rectangle.

5.33 Identify two stacked arrangement for the mirrored cascode of Example 5.5. The first solution accommodates the transistors in 4 stacks, two for the p-channel transistors and two for the n-channel. The second arrangement uses three stacks, one for the input pair, one for the p-channel transistors and the third one for the remaining n-channel elements.

Chapter 6

CMOS COMPARATORS

In this chapter we shall deal with the design of CMOS comparators. A comparator is the basic component mainly used in analog-to-digital converters. Ideally, it generates an output logic signal as an instant response to the sign of an analog input (voltage or current). Obviously, a real circuit doesn't achieve the ideal function. The most important limits are the finite sensitivity, the offset and the finite speed. All the above limitations affect the performance of systems where comparators are used, especially when it is required to achieve high speed (or a high conversion rate) and high resolution.
Analog designer must know well how to properly face various design issues. This chapter will provide a number of suitable guidelines for that purpose.

6.1 INTRODUCTION

The electrical function of a comparator is to generate an output voltage which value is *high* or *low* depending on whether the sign of the input is positive or negative (Fig. 6.1). We can have two different types of input: voltage or current. In the former case the input voltage is measured with respect to a given reference level. Therefore, the comparator determines whether the amplitude of the input is higher or not than a reference. When the current is the input varia-

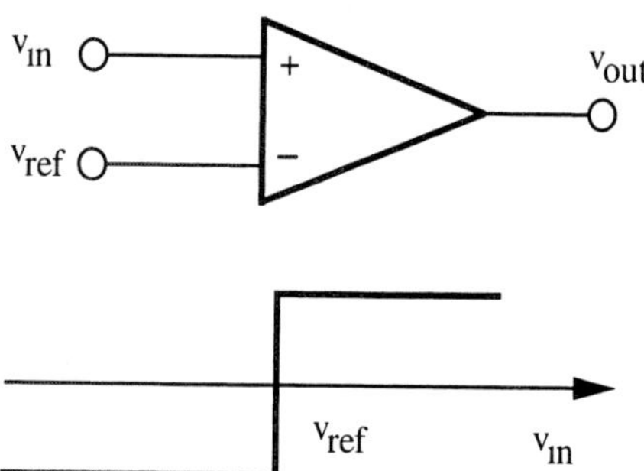

Fig. 6.1 - Symbol and ideal transfer characteristics of a comparator.

ble the comparator determines whether the input current is flowing in or out the input terminal.

A logic signal denotes the output. The amplitude of electrical representation of the *high* or the *low* state should match the convention used in the associated digital logic to clearly distinguish between a logic *1* and a logic *0*.

> **REMEMBER**
>
> A comparator can be "continuous time" or "sampled data". In the latter case the use of a clock to control the comparator operation permits us to achieve higher speed and to properly handle the offset issue.

When a comparator is used in a sampled data system a clock controls the action of the circuit. The comparator provides the output with a given periodicity synchronous with the clock. Therefore, a given time interval is available to achieve the result. Often, the fast variation of the input signal and the defined speed of the circuit used determine the need to separate the two functions inherent to the comparison process: to "catch" the value of the input signal and to generate the logic output. A sampled data system favour this disjunction: the clock period can be divided into two (or more) phases: one completes the sampling of the input and the other transforms the result into the logic signal. The latter non linear operation can take advantage of the use of a latch. A latch effects a regenerative amplification of the input and, thanks to its positive feedback, preserves the achieved output.

6.2 PERFORMANCE CHARACTERISTICS

Fig. 6.2 shows a quite general architecture of a sampled data comparator. It is the cascade connection of three blocks: a sample and hold, an amplifier and

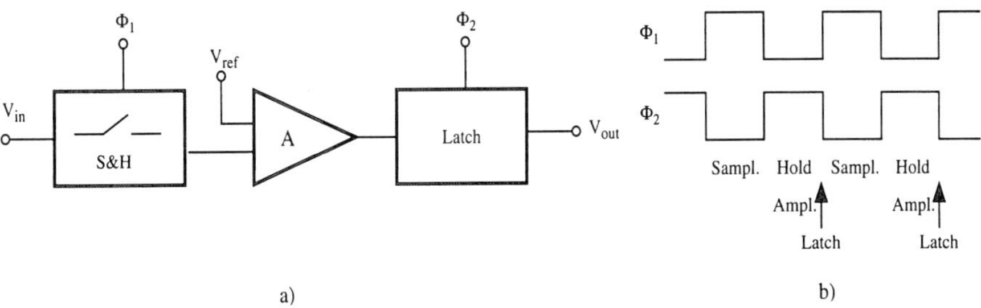

Fig. 6.2 - Block diagram of a clocked comparator and its phase control scheme.

a latch. The circuit samples the input during Φ_1. The gain stage amplifies the hold value during Φ_2, and at the end of the phase Φ_2 the regenerative action of the latch starts. In addition during Φ_1 or Φ_2 the circuit can take some action against limits like the offset, or the saturation caused by a previous overdrive.

Only few applications require a continuous-time operation. In such cases the architecture of the comparator becomes a simple amplifier which gain is large enough to achieve an output voltage representing the logic levels.

We discuss below the most important features of the comparator. We will mainly refer to the architecture of Fig. 6.2. However, the features that apply to the amplification block characterize continuous-time comparators as well.

Sensitivity: it is the minimum input voltage (or current) that produces a consistent output signal within the expected comparison time. For modern applications the sensitivity required is pretty low. For example, a *10 bit* data converter requires a sensitivity of *1/2 LSB* (less significant bit) that corresponds to *1/2048* the reference voltage (or current) used. Thus, for *10* bit and *1 V* reference we need *0.5 mV* sensitivity.

Input offset (V_{os}): It is the voltage that must be applied to the input to obtain the crossing point between low and high logic level. The feature is analogous to the offset of op-amps. Even the causes of the limitation are similar.

Amplifier response time (t_r): It is the minimum time-interval required to achieve the proper logic output as a response to a minimum input step. In the architecture of Fig. 6.2 the latch needs a given signal at its inputs to ensure the logic output. The pre-amplifier stage achieves this level in a time that depends on the input step amplitude: the step response of a gain stage is a ramp during the slewing that turns into an exponential in the linear region.

Overdrive recovery time (t_{rec}): When the input signal is pretty large the gain stage (or part of it) saturates to the positive or negative rails. If the input becomes small with the opposite sign, the gain stage takes some time to react and generates the voltage required to produce the output logic. The time required is higher than the response time and the extra time required is called overdrive recovery time. Often, the recovery from overdrive is much more time consuming than the amplification. Therefore, designer should put a special attention to this limitation.

Latching compatibility: The output of the gain stage should properly drive the latch. Therefore, depending on the specific scheme used it is necessary to provide the effective voltage levels at the input of the latch.

Power supply rejection: This is a feature equivalent to the one already discussed for op-amps. Spur signals affecting the power supply lines can modify the input sample and produce wrong outputs. The power supply rejection describes the ability of the circuit to avoid the limit.

Power Consumption: The clocked operation leads, in addition to the static power consumption, to a dynamic contribution. It depends, like in digital circuits, on the clock frequency and the capacitances that the preamplifier and the latch are required to charge and discharge.

Hysteresis: The comparison threshold for input signals changing from low to high can be different from the threshold to signals changing in the opposite direction. The difference between the two thresholds is the hysteresis. The effect can be a limit or a benefit, depending on the applications. For continuous-time applications the crossing of a noisy input signal through the reference may produce many transitions high and low of the output voltage. A hysteresis larger than the noise level avoids the effect and results are beneficial. When the comparator is used in data converters, hysteresis is a limitation: it can cause different output codes depending on the sign of the input signal derivative.

Example 6.1

A simple comparator is made by the cascade of four digital inverters. Perform simulations using a the digital inverter of a library available and estimate the response time and the delay due to the recovery from overdrive. Simulate the response time using a step input changing by ± 2 mV around the threshold. For the recovery from overdrive use a step input that bounces from - 30 mV below to 2 mV above the threshold.

Solution:
The reader that doesn't have the access to a digital library should design the inverter by herself. Use an n-channel transistor whose W/L is two times the minimum allowed. The p-channel transistor must bring the inverter threshold at 1.65 V. Use 3.3 V supply voltage.
The figure of the next page shows the schematic used for simula-

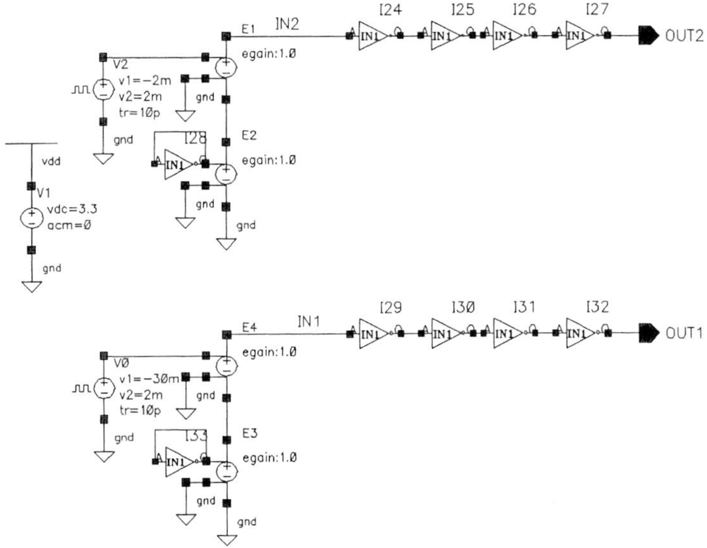

tions. A proper bias circuit drives the cascade of the four inverter. It includes an inverter whose output is connected to the input to provide the threshold voltage. A voltage-controlled-voltage-source copies the threshold and a second VCVS adds up the input. The schematic includes two equal versions of the same circuit. The network on the top measures the response time; the one on the bottom determines the overdrive delay.
The figure below shows the simulation results. OUT2 indicates that the response time is 1.7 nsec, a pretty low value. However, we have

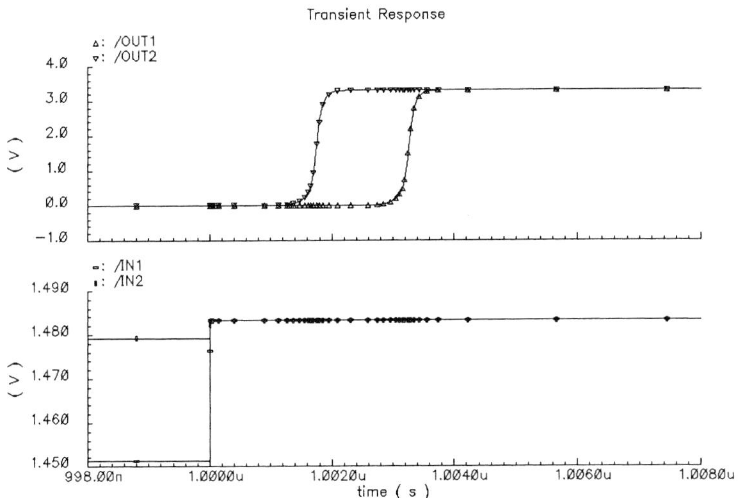

to remember that the input step is as large as 4 mV. The second trace (OUT1) marks the low to high transition at 3.15 nsec. Therefore, the recovery from overdrive is 1.45 nsec, a significant fraction of the response time.
Additional simulations show that as the overdrive becomes more negative the recovery time increases. When the overdrive reaches about -300 mV the recovery time saturates at 2.2 nsec. Below that input level the output of the first inverter becomes V_{DD}.
The input waveform exhibits an inverter threshold of 1.481 V. A bit lower than $V_{DD}/2$.

6.3 GENERAL DESIGN ISSUES

Many applications ask for a comparator sensitivity equal or below *1 mV*. However, the offset of a typical CMOS amplifier is larger than *1 mV* (the systematic and the random contributions come to 5-*10 mV* or so). Therefore, offset compensation is one of the most important design problems. We will study appropriate techniques that alleviate the problem shortly.

An important limitation to the speed is the time to recover from overdrive. When a gain stage goes into saturation it requires current and time to charge (or discharge) nodes that have been pushed close to V_{DD} or ground. A frequent remedy consists in the clipping of voltage nodes. The clipping limits the swing of nodes and prevents saturation. Moreover, when the timing of the system consents a dedicated time-slot the problem can be solved by resetting the critical nodes before every comparison.

Continuous-time comparators are difficult to design because they can not take advantage of the regenerative action of a latch. The gain must be large enough to generate the logic signals with the minimum input. By contrast, a sampled-data comparator require less effort: using a latch the gain stage of Fig. 6.2 must generate relatively low outputs (often differential). Typically a differential signal of ± *100 mV* is large enough to drive the latch and to account for any possible mismatch. Accordingly, the amplification of the gain stage can be relaxed by *20 - 30 dB* with respect to the continuous-time case.

For high frequency applications the static gain is not the key parameter. The time required to obtain a proper output level is normally much less than the time constant of the gain stage. Therefore, just the initial part of the step response is effective. What is important is not the asymptotic value but the voltage amplitude that the output reaches in the time-slot available.

Noise over the power supply (or the substrate) and the noise of electronic components affect the output. An op-amp is required to control the noise over the entire band of the signal. Instead, a clocked comparator must minimize the noise when the latch starts up. Therefore, it is important to use a proper clock timing. Namely, the clock used by the latch must be suitably apart from the clock controlling digital sections.

CALL UP

Very high speed comparators exploit just the initial part of the gain stages time response. Successive stages and a latch take care of further signal amplification.

6.3.1 Architecture of the Gain Stage

Two different approaches permit us to obtain the required gain: to employ a single complex amplifier or to use the cascade of many simple stages. In the design of op-amps (or *OTA*'s) stability requirements limit the number of stages that we can use. By contrast a comparator operates in open loop conditions. Therefore, it is possible to use a cascade with any number of stages.

The small signal equivalent circuit describes the speed limitation of a linear network. A comparator mainly handles large signals. However, to study the speed performance of comparators we assume the signals small enough to justify a small signal analysis.

Let us consider the cascade of n equal gain stages. The one-pole equivalent circuit of Fig. 6.3 represents the small-signal behaviour of each stage. The response to an input step is an exponential with time constant R_oC_o

$$V_o(t) = V_i g_m R_o(1 - e^{-t/R_oC_o}) \tag{6.1}$$

For high-speed applications it is not possible to wait for a long time and, typically, only the initial part of the response is used. Therefore, (6.1) can be approximated by

$$V_o(t) \cong V_i t \frac{g_m}{C_o} \quad for \quad t \ll R_oC_o \tag{6.2}$$

showing that the output voltage changes linearly with slope g_m/C_o. Observe that the output resistance (and the *DC* gain) does not influence the transient response in the initial part.

The output for two equal stages is the convolution of an exponential with an exponential or, for times lower than the time constant, the convolution of a

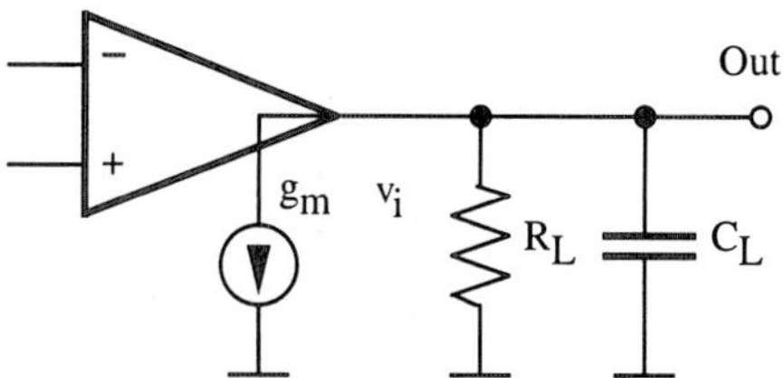

Fig. 6.3 - One-pole equivalent circuit of a single gain stage.

ramp with a ramp

$$V_{o,2}(t) = V_i \frac{(g_m/C_o)^2}{2} t^2 \tag{6.3}$$

in general, for *n* equal stages the output voltage is

$$V_{o,n}(t) = V_i \frac{(g_m/C_o)^n}{n!} t^n \tag{6.4}$$

Fig. 6.4 represents equation (6.1) for *1*, *2*, *3*, and *4* stages. At the very beginning the output of the first stage is higher than the others. At $t = 2C_o/g_m$ the output of the second stage becomes the higher one; later, at $t = 3C_o/g_m$, the output of the third stage takes over and so on. It turns out that there is an optimum number of stages that achieves a given required gain in the minimum time.

The use of equation (6.4) leads to the optimum transient performance for *n* cascaded stages

$$V_{o,n,opt} = V_i \frac{(n+1)^n}{n!} \tag{6.5}$$

that occurs at the time

$$t_{n,opt} = (n+1)\frac{C_L}{g_m} \tag{6.6}$$

showing that, for example, a small-signal gain larger than *10.6* and lower than *26* is optimal achieved with the cascade of *4* stages.

The above analysis provides just a design hint: the equations used imply a small signal operation. Since the sensitivity of typical applications is in the order of fractions of *mV*, the above results are useful for gains up to few tens.

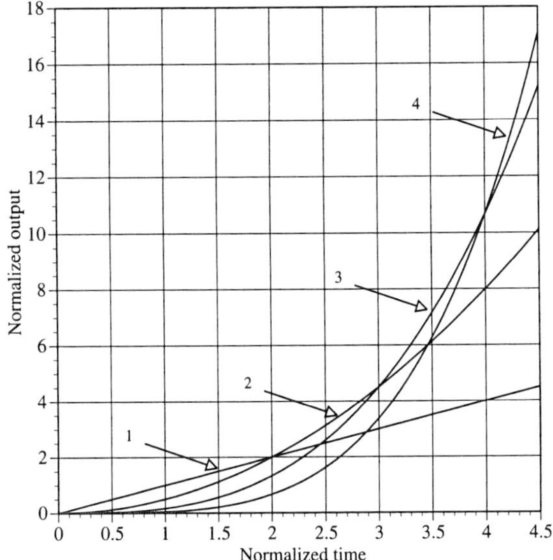

Fig. 6.4 - Step response of the cascade of n equal stages. A single pole model describes each stage. The normalized input is tC_0/g_m; the normalized output is V_o/V_{in}.

6.4 OFFSET COMPENSATION

The auto-zero technique is the basis of all the schemes used to compensate the offset. Fig. 6.5 depicts the underlying concept. The approach is appropriate for sampled-data operations being the scheme controlled by two non-overlapped phases. During the phase *1* the sample-and-hold reads the offset, V_{os}. During phase 2 the input signal, V_{in}, and the stored offset are summed up and applied to the inverting terminal of the gain stage. Therefore the differential input becomes

$$V_d = V_{os} - (V_{in} + V_{os}) = -V_{in} \tag{6.7}$$

showing an inverting operation and, more important, the cancellation of the offset contributions.

Equation (6.7) assumes that the offset at the sampling time and the one during phase 2 are equal. A more precise analysis leads to

$$V_d\left(nT + \frac{1}{2}T\right) = V_{os}\left(nT + \frac{T}{2}\right) - V_{in}\left(nT + \frac{T}{2}\right) - V_{os}(nT) \tag{6.8}$$

Showing a subtraction of the actual offset and its delayed version. The

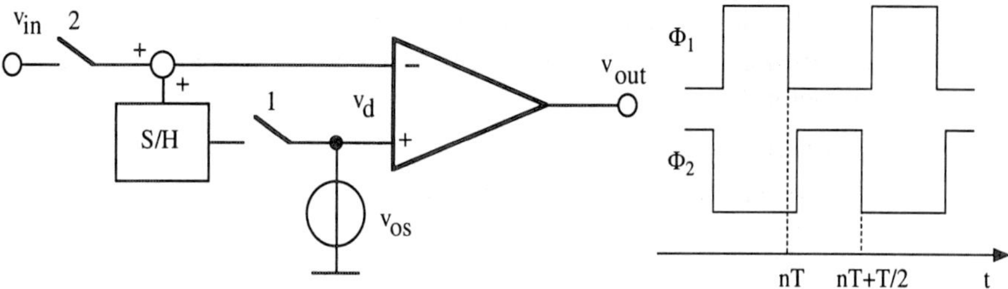

Fig. 6.5 - Conceptual scheme of the offset cancellation method.

method is effective anyhow: the offset is a *dc* or a very slow varying signal. It mainly comes from static errors (geometrical mismatch or technological disuniformity) and temperature drifts. However, if we incorporate the effect of the input referred noise a given frequency dependency will result. The delay between the two offset terms in (6.8) leads to the following offset transfer function

$$H_{os}(\omega) = 1 - e^{-sT/2} \tag{6.9}$$

or, using the *z*-transform

$$H_{os}(z) = 1 - z^{1/2} \tag{6.10}$$

that, using $z = e^{j\omega T}$, leads to

$$F_{os}(\omega) = \frac{e^{-j\omega T/4}}{2j} sin\left(\frac{\omega T}{4}\right) \tag{6.11}$$

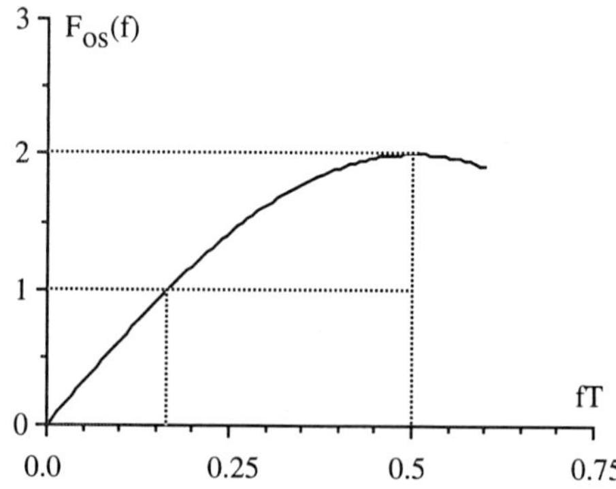

Fig. 6.6 - Offset transfer function of the circuit in Fig. 6.5.

Equation (6.11) denotes a high pass transfer function: the *dc* component vanishes while the low frequency terms are significantly attenuated. By contrast, at $f=\pi/T$ the noise transfer function (depicted in Fig. 6.6) shows an amplification by a factor 2.

The described method is often named *correlated double sampling technique*. As a matter of fact, the circuits sample the offset two times: one by the sample-and-hold and a second time by the input terminal of the gain stage. The correlated part of the two samples is cancelled out. The other name, *auto-zero* technique, describes the capability of the method to measure and zeroing the offset without any external help.

TAKE NOTE

The correlated double sampling technique (or auto-zero) procures a high pass transfer function. The technique is beneficial only for the low frequency components of the input referred disturbances.

6.4.1 Implementation of the Auto-zero Technique

Fig. 6.7 shows a possible circuit capable to implement the auto-zero technique. The circuit uses a gain stage, a capacitor C_A and three switches controlled by two non-overlapped phases. During phase *1* the switch S_1 connects the gain stage in the unity gain configuration. Assuming the gain large enough, the voltage of the inverting terminal equals the offset, V_{os}. During phase *2* the gain stage goes in the open loop configuration and the switch S_3 connects the left terminal of C_A to the input voltage, V_{in}. We assume that at the first approximation the capacitor operates like a level shifter. Therefore, the inverting terminal of the gain stage is a shifted replica by V_{os} of the input voltage, as requested.

Strictly speaking the voltage at the inverting terminal during phase *1* is not the offset but

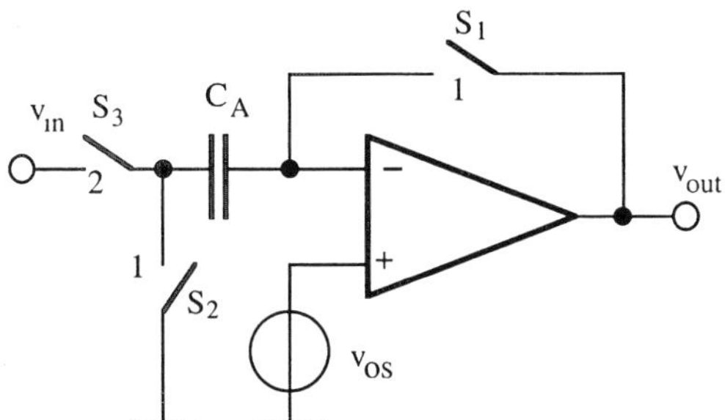

Fig. 6.7 - Block diagram used to implement the offset cancellation approach

$$V_- = \frac{V_{os}(A_0 - 1)}{A_0} \qquad (6.12)$$

therefore the circuit doesn't perform a complete compensation of the offset. The residual offset is

$$V_{os,res} = \frac{V_{os}}{A_0} \qquad (6.13)$$

That is negligible, if the gain of the stage is large enough.

The circuit in Fig. 6.7 has two drawbacks

- During phase 1 the stage is connected in the unity gain configuration: this may require to compensate the stage.
- The opening of the switch S_1 causes an injection of charge on C_A because of the clock feed through effect.

The first limit can be detrimental to the speed of the circuit. During the phase *1* the switch S_1 connects the circuit as a unity gain buffer, calling for a compensation capacitor. During the phase 2 the stage works as a comparator. Thus, every clock period the output node swings from analog ground to a large positive or negative level causing a periodic charging and discharging of the compensation capacitance.

Fig. 6.8 shows a possible remedy to the problem. Assume that the gain stage is a single stage architecture. The capacitor C_C loading the output node ensures stability. Since during the phase *2* the gain stage is in the open loop condition no compensation is required. The switch S_4 disconnects C_C, thus avoids charging and discharging of C_C. The same solution used for a two stages amplifier requires to disconnect during the phase *2* the pole-splitting capacitance.

The second limit affecting the circuit of Fig. 6.7 concerns the clock feedthrough. The critical node of the network is the inverting terminal. When

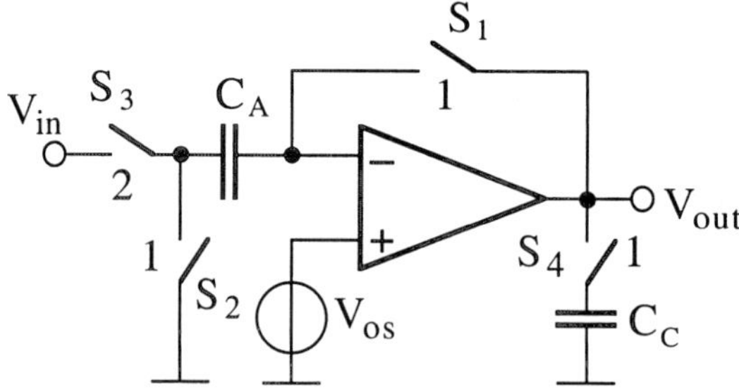

Fig. 6.8 - Disconnecting the compensation capacitance during Φ_2 speeds up the circuit,

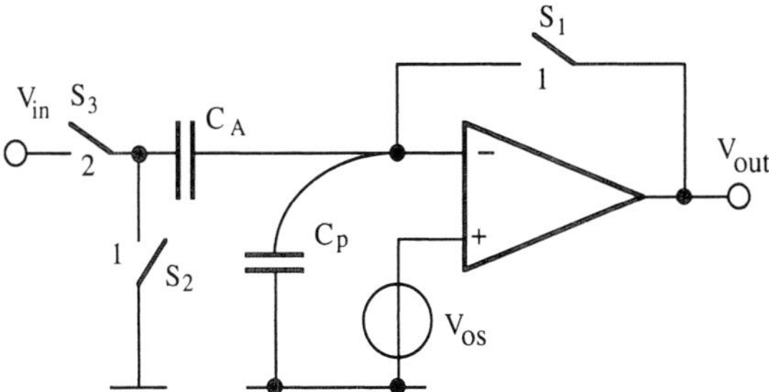

Fig. 6.9 - Circuit for estimating the residual offset caused by the clock feedthrough from S_1.

the switch S_1 opens it injects a charge that is trapped in the inverting terminal node and reduces the effectiveness of the offset cancellation. By contrast, the charge injected on the other terminal of C_A by the opening of S_2 doesn't affect the operation of the circuit. After a short period of time S_3 is closed and the control of the voltage of the right side of C_C is taken by the low-impedance input generator V_{in}.

The amount of charge injected by S_1 depends on the fall time of the clock phase controlling S_1 and the boundary conditions on the two sides of the switch. Namely, the boundary conditions at the virtual ground side depends whether S_2 is closed or opened. Since S_1 and S_2 are controlled by the same phase, what may happen is that when S_1 opens S_2 is in an undefined condition. It is therefore convenient to use for the control of S_1 and S_2 slightly delayed phases. The two switches do not open at the same time and the boundary conditions for S_1 are firmly established. It is recommended to open S_1 before. The left plate of C_A will be secured to ground by S_2 closed. Otherwise, the left plate of S_2 can be at a voltage that is not so well controlled.

The parallel of capacitance C_A and the parasitic load C_p (see Fig. 6.9) receive the charge, Q_{inj}, injected by S_1. The residual offset is

$$V_{os,\,res,\,inj} = \frac{Q_{inj}}{(C_A + C_p)} \tag{6.14}$$

Moreover, the parasitic loading the non inverting input attenuates the input signal by the factor $C_A/(C_A+C_p)$.

Example 6.2

Determine, with Spice simulations, the residual offset caused by the clock feedthrough. Use the scheme of Fig. 6.8 and C_A = 1 pF; C_C = 2 pF. Moreover, use the op-amp of Example 5.4 with a supply voltage of 3.3 V. Introduce an artificial offset equal to 4 mV by a proper

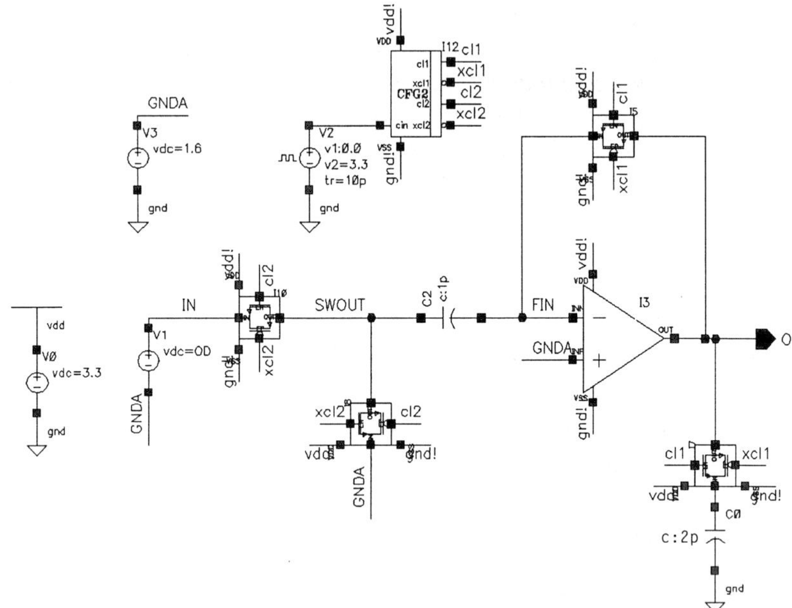

mismatch of the active load transistors. Use the switches considering the following three cases: complementary transistors which aspect ratio is $(W/L)_n = 5\mu/0.3\mu$; $(W/L)_p = 5\mu/0.3\mu$; only n-channel transistor; only p-channel transistor.

Solution:

The op-amp of Example 5.4 requires a compensation of 3 pF. The proposed solution provides the same output load: 1pF given by C_A and 2 pF by C_C. The control of the switches require no-overlapped complementary phases. The reader can achieve them by a set of pulse generators or by a non-overlapped phases generator driven by a master clock. The above figure shows the circuit diagram. The switch constitute a sub circuit. The choice facilitates the replacement of complementary transistors with an n-channel or a p-channel as required.

A number of trial simulations determines the mismatch that brings the offset to 4 mV: the width of one of the active load must be reduced to 334 μ.

The charge injected by each transistor of the switch is estimated by

$$Q_{inj} = \frac{1}{2}C_{ox}WL \cdot V_{DD} = \frac{2.1 \cdot 5 \cdot 0.3 \cdot 3.3}{2}10^{-15} = 5.2 \cdot 10^{-15}\text{Coul}$$

where the gate oxide specific capacitance is 2.1 fF/m^2. The charge Q_{inj} integrated over C_A = 1 pF will produce an offset of 5.2 mV. The use of an n-channel element leads to a negative residual offset, the use of a p-channel transistor causes a positive offset, the use of complementary transistors leads, in first approximation to a compensation of the charge injections.
The simulations give the following results: only p-MOS 5.63 m V; only n-MOS - 5.71 mV; complementary transistors -0.14 mV.

6.4.2 Auto-zero in Multi-stages Architectures

The amplifier of the block diagram in Fig. 6.2 can be a single amplifier whose gain is large enough or a cascade of gain stages with a relatively low gain. In the latter case the auto-zero technique is not very effective because, according to equation (6.13), the residual offset is inversely proportional to the gain. Moreover the offset of the second stage is referred to the input divided by A_1. Therefore, the attenuation of $V_{os,2}$ can be insufficient. When the value of residual offset is non acceptable it is necessary to use the auto-zero both in the first and in the second stage.

A straight use of the auto-zero technique would lead to the schematic of Fig. 6.10 a). An auto-zero network made by the auto-zero capacitor and three switches operate on the second amplifier. However, one can observe that the left plate of capacitor C_2 is connected to the output of A_1 during phase *1* and to

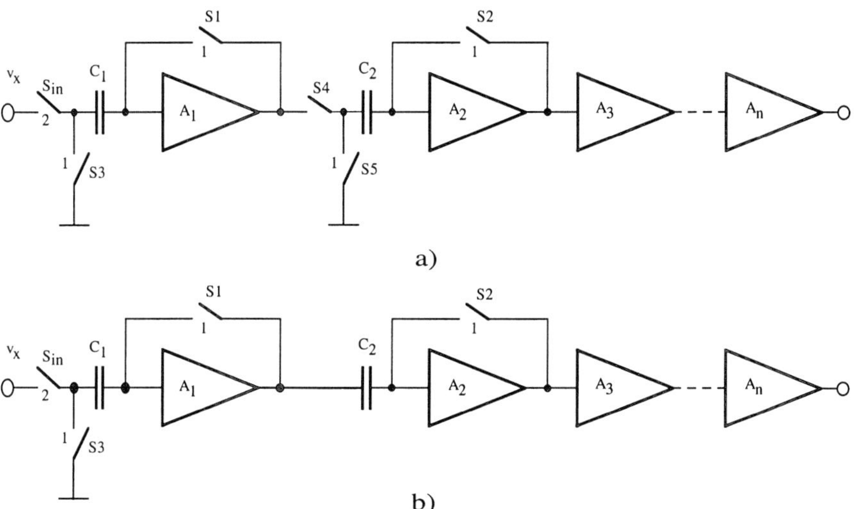

Fig. 6.10 - a) Direct use of the auto-zero in a cascade of gain stages. b) improved solution.

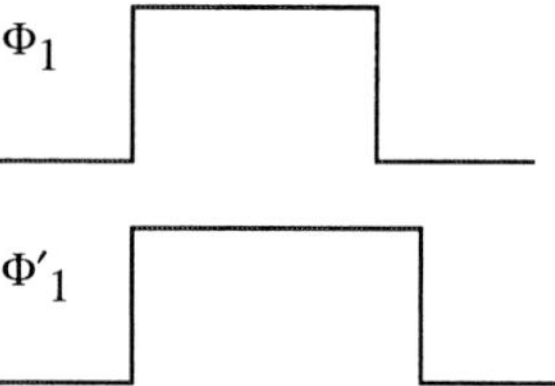

Fig. 6.11 - Phase Φ_1 and Φ'_1 useful for the control of the switches S_1 and S_2 of Fig. 6.10 b).

the analog ground during phase 2. These two voltages are the ones that the output node of A_1 develops, assuming negligible the effect of $V_{os,1}$. Therefore, it is convenient to remove the two switches in the second auto-zero network to obtain the schematic of Fig. 6.10 b).

The circuit of Fig. 6.10 b), is not only less complex than the one in Fig. 6.10 a), it provides an additional benefit. In reality, during the phase *1* the capacitor C_2 is charged at the difference between the offset of the second stage and the offset of the first stages. Therefore, it samples at the same time the offset of the second stage and the consequence of $V_{os,res,1}$. It turns out that $V_{os,res,1}$ is auto-zeroed by C_2 and the total the residual offset becomes

$$V_{os,res} = \frac{V_{os,2}}{A_1 A_2} \tag{6.15}$$

The opening of the switches S_1 and S_2 produces a clock feedthrough injection into the auto-zero capacitors C1 and C2. The injected charges cause an additional input referred residual offset

$$V_{os,res,ck} = \frac{Q_{inj,1}}{C_1} + \frac{Q_{inj,2}}{A_1 C_2} \tag{6.16}$$

COMMENT ON FIG. 6.10 B

The residual offset of the first stage is auto zeroed only if its effect is stored on the auto-zero capacitor of the second stage. This, in turn, impose the use of the phase scheme of Fig. 6.11.

The already discussed benefit of the circuit of Fig. 6.10 b) can be exploited to cancel the first term in (6.16). For this it is just necessary to use the clock phases Φ_1 and Φ'_1 shown in Fig. 6.11 to drive S_1 and S_2 respectively. The following operation results: S_1 opens before and injects its clock feedthrough charge, $Q_{inj,1}$, into C_1. The residual offset $V_{res,inj,1} = Q_{inj,1}/C_1$ amplified by A_1 comes out at the output of the A_1. Assuming that residual offset and gain A_1 are not too large, the output of A_1 remains out of the region where the gain drops. Moreover, since S_2 is still

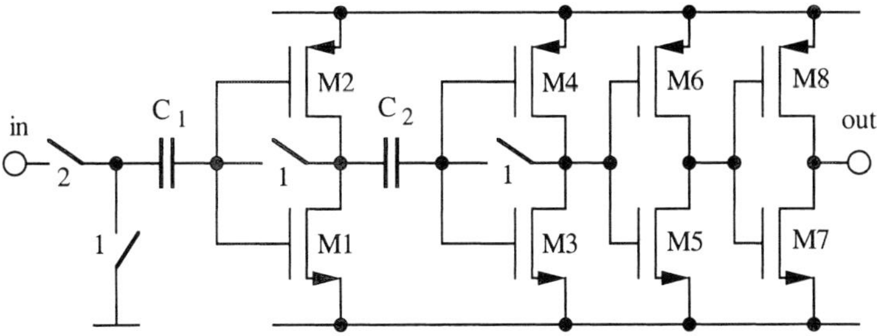

Fig. 6.12 - Possible circuit implementation of the auto zeroed comparator of Fig. 6.10 b).

closed the auto-zero capacitor C_2 samples $A_1 V_{res,inj,1}$. Therefore, the auto-zero operation cancels the residual offset of the first stage, the clock feed through of the first stage and, partially the offset of the second stage. Accounting for all the discussed effects, one obtains an input referred offset given by

$$V_{os,res} = \frac{V_{os,2}}{A_1 A_2} + \frac{Q_{inj,2}}{A_1 C_2} \tag{6.17}$$

Fig. 6.12 shows a possible circuit implementation of the scheme of Fig. 6.10 b). Simple inverters realize the gain stages. They have only one input (not a differential input) as outlined in Fig. 6.10. Actually, a differential input is not necessary, the feedback connections established by S_1 and S_2 bring about the offsets at the input terminals anyway. Moreover, a reference connection to the virtual ground is not necessary. The auto-zero capacitor provides a possible level shift between the input of the gain stage and the node at which the right plate is connected.

The gain of the inverters used in Fig. 6.12 is typically small (around *10*). Moreover, in order to keep low the power consumption, the designer must use pretty long transistors. Another limit of the circuit in Fig. 6.12 is the poor power supply rejection. On the other hand, the circuit doesn't need any bias voltage. Therefore, the circuit in Fig. 6.12 is suitable for architectures that use a large number of comparators like, for example, flash converters but don't require a very high accuracy.

6.4.3 Fully Differential Implementation

A fully differential architecture increases the defence of op-amp against common mode disturbances. The same technique can be conveniently used with comparators. Fig. 6.13 shows the fully differential version of the auto-zero pre-amplifier. It uses a fully differential gain stage and duplicates the auto-zero

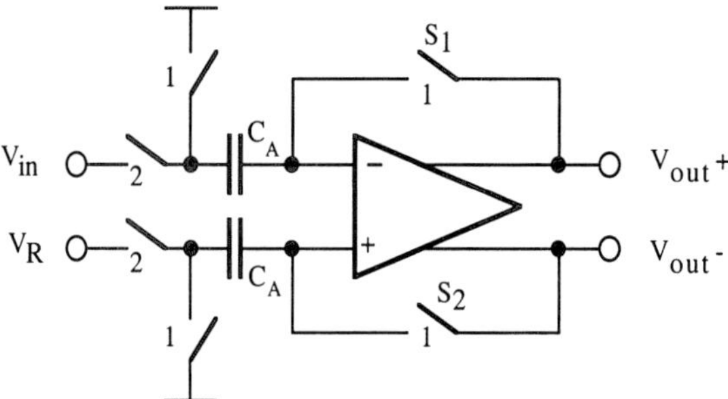

Fig. 6.13 - Fully-differential auto-zeroed gain comparator.

network. The injection of charge at the opening of S_1 and S_2 causes a common mode signal at the input that *CMRR* rejects. Only a possible mismatch between the switches S_1 and S_2 may affect the circuit. Different injected charges produce a differential signal that generates an offset. Assuming that the matching accuracy of minimum size transistors is in the order of *5%*, the fully differential architecture improves by a factor *20* the residual noise generated by the clock-feedthrough.

The use of a fully differential architecture provides four input terminals. In Fig. 6.13 two of them are connected to the analog ground and the remaining two are used for the inputs. However, the designer can use the four input to compare the combination of four voltages: the difference of the voltages applied to the positive input compared to the difference of the voltages applied to the negative input.

An important design issue concerns the choice of the gain-stage architecture. It can be made by the cascade of low-gain stages or by the use of stages with moderate-high gain. Fig. 6.14 a) shows a simple gain stage. Its gain is

$$A_v = \frac{g_{m1}}{g_{m3}} = \sqrt{\frac{\mu_n (W/L)_1}{\mu_p (W/L)_3}} \tag{6.18}$$

A proper choice of the transistors' aspect ratio easily leads to gain in the order of *10*.

An interesting feature of the circuit in Fig. 6.14 a) is that the output impedance is relatively low. The diode connected elements M_3 and M_4 control the output quiescent voltage; therefore, the common mode feedback is not necessary. We know that for high-speed applications, it is not the gain that is the most important parameter but the g_m/C_L ratio. The load capacitance of the stage in Fig. 6.14 a) mainly comes from the loading capacitors C_l and the gate capacitance of M_3-M_4

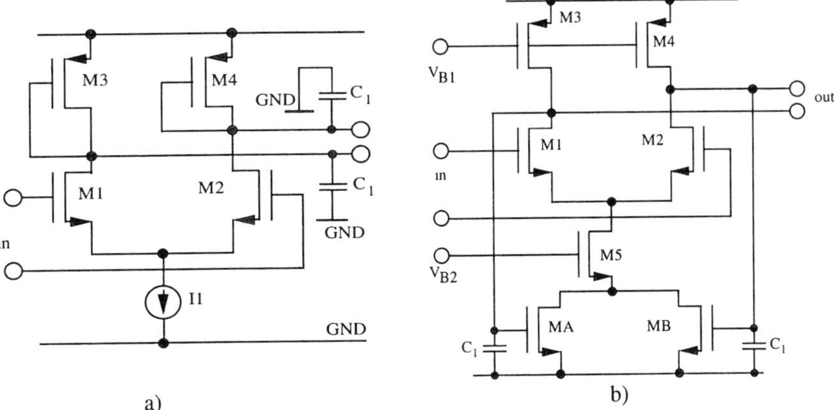

Fig. 6.14 -a) Fully differential stage with low gain b) Fully differential stage with moderate gain and common mode feedback.

$$C_L = C_{ox}(WL)_3 + C_{db1} + C_1 \tag{6.19}$$

The gate area of M_3 can be large, thus limiting the speed performances. The use of the circuit of Fig. 6.14 b) partially solves the problem. Fig. 6.14 b) is a single stage differential amplifier with active load. As it is known the gain is given by

$$A_v = \frac{g_{m1}}{g_{ds3}} \tag{6.20}$$

there is no need any more to use long active loads. Moreover the parasitic contribution of transistor M_3 to C_L is just from $C_{db,3}$. However, since the impedance of the output node is pretty high a common node feedback is necessary. The circuit of Fig. 6.14 b) uses one of the continuous-time solution discussed in the previous chapter. Transistors M_A and M_B degenerate the tail current generator, M_5. A differential output doesn't affect the current, while a common mode output changes the bias conditions.

> **OBSERVATION**
>
> In order to speed up the operation of a comparator it is essential to bring at the minimum the capacitances that integrate the signal current. Loads made by small transistors and a careful layout help in achieving the target.

Observe that the common mode feedback loads the output thus affecting C_L. Being M_A and M_B in the triode region, we have

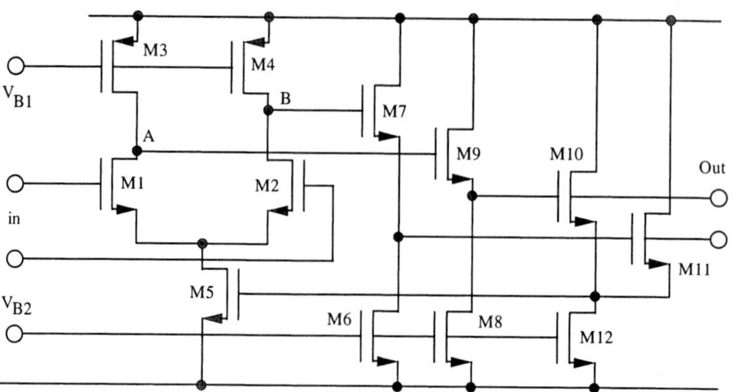

Fig. 6.15 - Fully differential gain stage with common mode feedback after decouple buffers.

$$C_L = C_{db1} + C_{db3} + \frac{1}{2}C_{ox}(WL)_A + C_1 \tag{6.21}$$

That is likely smaller than the load capacitance expressed by (6.19). In fact, the drain to bulk capacitances are normally negligible and the area of transistor M_A can be smaller than the one of M_3 in Fig 6.13 a).

Fig. 6.15 shows another realization. It uses a differential amplifier with active load like the solution of Fig. 6.14 b).However, it employs a different common mode solution. Two-source follower replicates the outputs. The obtained signals are use as input of the source-coupled pair M_9-M_{10} which source, in turn, controls the gate of M_5. The circuit of Fig. 6.15 enjoys two benefits. The common mode feedback doesn't load any more the output nodes of the gain stage (nodes A and B). Moreover, the use of the source followers leads to minimum capacitances loading the nodes A and B. Up to its unity gain frequency the followers provide a good replica of the input. Therefore, a bootstrap on C_{gs7} and C_{gs9} results. The capacitors, seen from nodes A and B, becomes

$$\begin{aligned} C_{gs7,boot} &= C_{gs7}(1 - A_B) \\ C_{gs9,boot} &= C_{gs9}(1 - A_B) \end{aligned} \tag{6.22}$$

Where A_B is the gain of the source follower. The capacitance loading nodes A and B becomes

$$C_L = C_{db1} + C_{db3} + C_{gs7}(1 - A_B) \tag{6.23}$$

Remember that the common mode feedback used in Fig. 6.15 has quite a limited region of linear operation. A relatively small differential signal completely imbalances the stage. One transistor turns off and the common mode source follows the gate of the other. This malfunctioning can cause a significant change of the tail current.

Example 6.3

Design the fully differential gain stage shown in Fig. 6.15. It is required to obtain an output signal 50 times bigger than the input in 50 nsec. Estimate the effect of a capacitance loading the nodes A and B. Use the Spice models of Appendix C.

Solution:

The transient response of the gain stage depends on the transconductance of the input pair and the capacitive loads on the nodes A and B (Fig. 6.15). The time available is pretty small; therefore, we expect to use only the initial part of the exponential transient. The use of equation (6.2)

$$V_o(t) \cong V_i t \frac{g_m}{C_o} \qquad \text{for} \qquad t \ll R_o C_o$$

permits us to determine the required value of $g_m/C_L = 10^9$.

The transconductance of the input pair increases with the square root of the input pair aspect ratio. However, the load capacitance increases because of the C_{db} contribution. Therefore, there is a trade-off between the transconductance increase and the worsening of the load capacitance.

The above considerations are not (on purpose) accounted for in the used design. The reader is asked to work at the schematic and

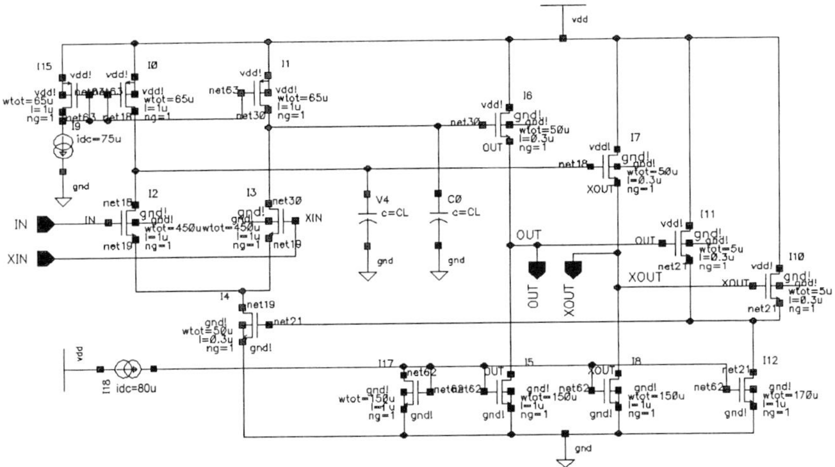

improve the performances of the circuit.

The below figure shows the schematic. Observe that the aspect ratio of the input pair is as largeas (450μ/1μ). That size and a tail current of 200 μA lead to g_m *= 1.7 mA/V.*

The use of small transistors in the common mode feedback network determines a large overdrive of the source coupled transistors. This, in turn, produces a reasonably large range of operation of the common mode feedback.

The below figure shows the transient response for a ± 1 mV step input. The curves refer to different capacitive loads at nodes A and B. The results indicate that in order to reach ± 50 mV the circuit needs 30 nsec with zero load; 55 nsec with 0.5 pF; 80 nsec with 1 pF, and so forth. The above results permit us to estimate the value of C_L*. It is approximately 0.6 pF. The relative large value of* C_L *suggests us that, as already mentioned, there is some room for improvement.*

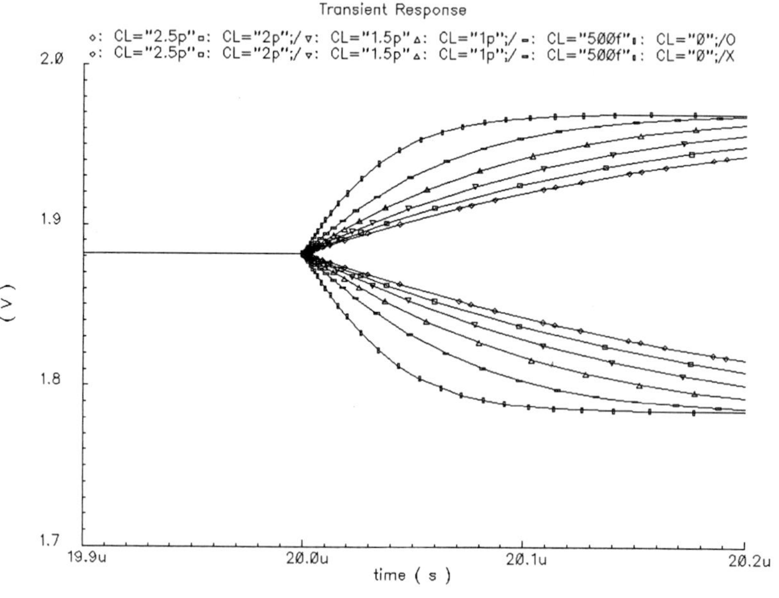

6.4.4 Use of an Auxiliary Stage

The auto-zero technique studied in the previous sub-section is based on the measurement of the offset at the input of the gain stage. An alternative possi-

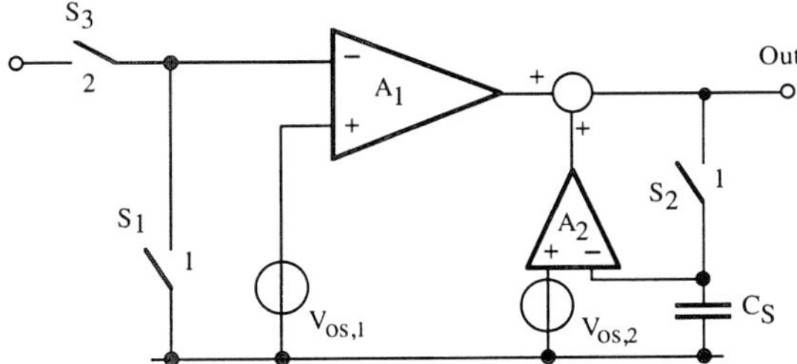

Fig. 6.16 - Conceptual scheme of the offset cancellation technique with auxiliary stage.

bility is to short the inputs and to measure the effect at the output. The gain stage amplifies the offset and this makes the compensation easier. The only limit is that the value of the offset and the gain must be small enough to avoid pushing the output voltage near to the supply voltages limits.

Fig. 6.16 shows the conceptual scheme of the idea. The switch S_1 connects the inverting terminal to the non-inverting one (assumed tied to the virtual ground). At the output of A_1 a feedback loop processes the signal. Assuming S_2 closed, the gain stage A_2 equals its two inputs. Therefore, at first approximation the output voltage becomes $V_{os,2}$. That means that A_2 generates a voltage capable to almost compensate $V_{os,1}$ amplified by A_1. A more accurate analysis of the circuit leads to the following balance equation

$$A_1 V_{os,1} + A_2(V_{os,2} - V_o) = V_o \tag{6.24}$$

That results in

$$V_o = \frac{A_1}{1 + A_2} V_{os,1} + \frac{A_2}{1 + A_2} V_{os,2} \tag{6.25}$$

The role of the two inputs can be interchanged: the negative terminal can be connected to the analog ground and the positive input to the input signal. Moreover, the switch S_3 connects the input generator to the input of the comparator. The impedance of the input generator is finite; therefore, a possible charge injected when S_1 opens is discharged through S_3. Therefore, the clock feedthrough produced by S_1 leads just to a glitch in the response. To procure this result it is necessary to open S_1 after S_2; otherwise the effect of the charge injection of S_1 is captured by the feedback loop and stored on C_S. The opening of the switch S_2 causes a charge injection on the storing element C_S. The injected charge, Q_{inj}, is integrated over C_S and amplified by A_2. when it is referred to the input, it determines a residual offset given by

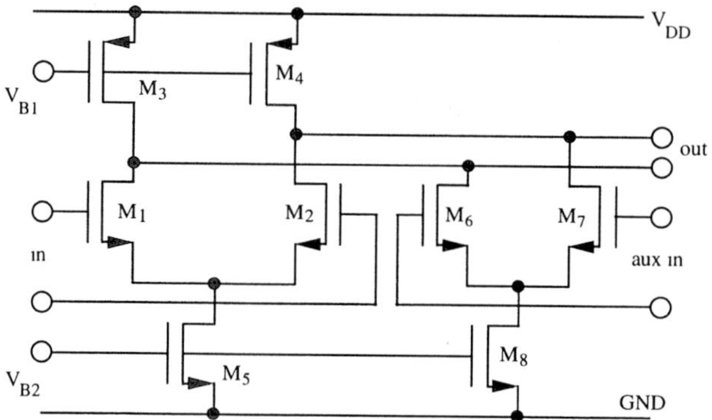

Fig. 6.17 - Four input OTA suitable for cancelling the offset cancellation by the technique that uses auxiliary stage.

$$V_{os,res} = \frac{Q_{inj}}{C_S}\left(\frac{A_2}{A_1}\right) \tag{6.26}$$

The designer minimizes the residual offset by using a pretty large capacitor C_S and designing the gain A_1 larger than the one of the auxiliary amplifier, A_2. When the gain stage A_2 is a single stage amplifier, capacitor C_S operates as its compensation element during the unity gain connection.

NOTICE

The best benefit brought by the use of auxiliary stages in the auto-zero consists in the disconnection between the storing element and the input node: we can use a direct connection of the input signal and we can use pretty large storing capacitors to minimize the residual offset (assuming the speed of the auxiliary amplifier large enough).

The technique discussed above can be implemented with a one-to-one translation of the basic scheme of Fig. 6.16. However, it is possible to identify solutions that achieve a more efficient implementation. Fig. 6.17 shows a fully differential solution that implements the two amplifiers of the scheme in Fig. 6.16. The two differential pairs realize the input stage of the main amplifier and the one of the auxiliary amplifier. The output current of the two stages are combined together and transformed into voltage thanks to the high impedance resistance of the two output modes. Therefore, the circuit, instead of generating voltages that must be summed up afterwards, combines the output currents and then it transforms the result into a voltage. This strategy simplifies significantly the circuit schematic. The used transistor sizes and the tail currents in the two differential stages control the two gains. Therefore, the designer can easily fulfil the

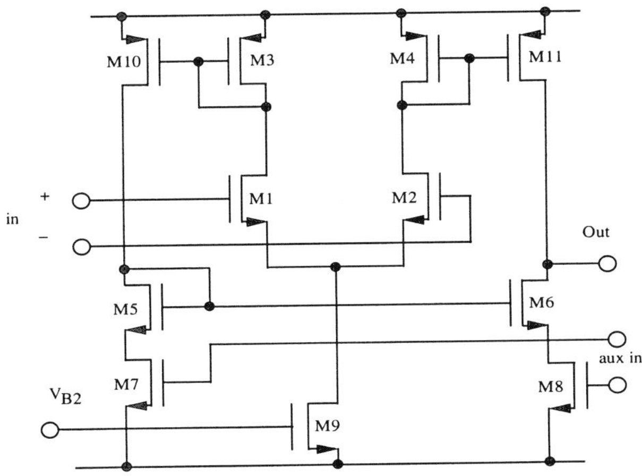

Fig. 6.18 - Four input OTA based on the current mirror degeneration.

request to have $A_1 > A_2$.

Fig. 6.18 shows a second circuit solution. It is based on a mirrored differential gain stage. The circuit includes two additional transistors $M_7 - M_8$ degenerate the current mirror $M_5 - M_6$. A possible umbalancement of the auxiliary inputs changes the mirror factor, thus modifying the output voltage.

The small signal differential gain of the main amplifier depends on the transconductance of the input pair, the mirror factor of $M_4 - M_{11}$ and the output resistance

$$A_1 = g_{m1} \frac{(W/L)_{11}}{(W/L)_4} r_{out} \tag{6.27}$$

The small signal differential gain of the auxiliary stage depends on the transconductance of the additional transistors $M_7 - M_8$.

$$A_2 = g_{m8} \, r_{out} \tag{6.28}$$

Since M_7 and M_8 operate in the triode region, their transconductance is typically lower than the one of the input pair. Therefore, as expected, $A_1 > A_2$.

6.5 LATCHES

The last block of the architecture in Fig. 6.2 is the latch. It furnishes supplementary gain to generate the logic output. In addition, it provides a stable output synchronous with the clock that brings into operation the regenerative loop.

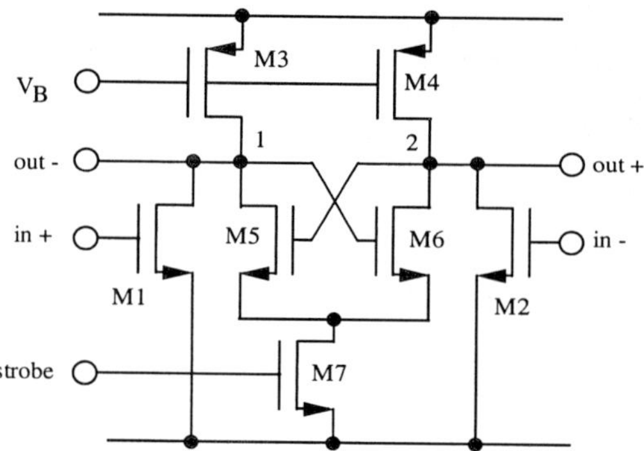

Fig. 6.19 - Simple latch configuration.

Fig. 6.19 shows a simple latch implementation. Assume that the strobe signal is low. The section $M_5 - M_6 - M_7$ of the circuit doesn't operate and the transistor pairs $M_1 - M_3$ and $M_2 - M_4$ form two inverters with active load. The voltage of node *1* and *2* are close to V_{DD} if the input voltages induce currents in M_1 and M_2 lower than the one drained by M_3 and M_4. When the common mode input exceeds a given level the currents in M_1 and M_2 become large and the voltage of node *1* and *2* drops down to ground.

Let assume that the common mode input is below the above-mentioned level and that the two inputs differ by a given extent. When the strobe control goes up the transistors M_5 and M_6 become active and they start the regenerative operation. Since the voltages of node *1* and *2* are pretty high the action of M_5 and M_6 will be significant. However, one of the voltages is more effective than the other and becomes more and more dominant with respect to the other.

If the input voltages are such that the node *1* and *2* are at a low level, at the limit, below the threshold of M_5 and M_6, when the strobe control is active the cross coupled pair M_5 and M_6 doesn't react properly or, at the limit remain in the sub-threshold region of operation. Therefore, the circuit in Fig. 6.19 works properly for a given range of the common-node input.

The nominal currents in M_3 and M_4 control the power consumption of the circuit.

Example 6.4

Design the latch of Fig. 6.20. The nominal current in M_3 and M_4 is 100 μA; the supply voltage is 3.3 V. The input signals differ by 100 mV. The circuit must operate properly with a common mode input voltage ranging from 0.95 V and 1.25 V. Determine the time required to get to the logic level low (0.3 V) and the logic level

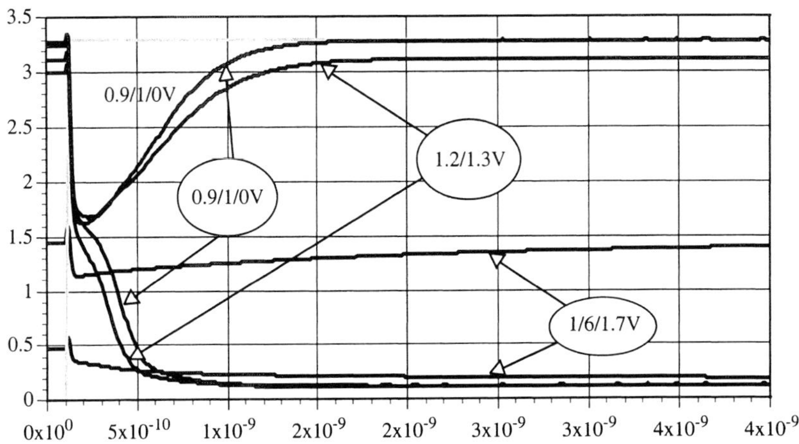

high. (3 V). Use the Spice models of Appendix A.

Solution:

The description of the behaviour given above recommends a high voltage level for nodes 1 and 2 in the preset phase. This requires to use small input transistors relatively. The highest input voltage must generate less than the nominal current of M_3 and M_4. Some Spice simulations determine $(W/L)_1 = (W/L)_2 = 4\mu/1\mu$. Transistor x are chosen larger than x to ensure a solid regenerative action. The resulting Spice list is shown below.

```
LATCH
M1   1 3 0 0  MODN W=4U   L=1U
M2   2 4 0 0  MODN W=4U   L=1U
M3   1 7 6 6  MODP W=4U   L=1U
M4   2 7 6 6  MODP W=4U   L=1U
M5   1 2 5 0  MODN W=8U   L=1U
M6   2 1 5 0  MODN W=8U   L=1U
M7   5 8 0 0  MODN W=8U   L=1U
MBP  7 7 6 6  MODP W=4U   L=1U

IBP 7 0 0.1M
VDD 6 0 3.3
V8 8 0 PULSE(0 3.3 0.1n 0.02n 0.1n 50n 100n)

VINp 3 0 0.9
VINn 4 0 1.0

.MODEL ......

.OP
.TRAN 0.01n 4n
.PRINT TRAN V(1) V(2) V(8)
.END
```

The figure displays the output responses for three different cases. They evidence the proper operation in the two limits of the required

range of operation. The time required to achieve the high level is between 1 nsec and 2 nsec. The swing in the downward direction is faster, between 0.9 nsec and 1.4nsec. Observe that when the strobe goes on the voltages of node 1 and 2 both drops down but after a while the stronger one takes over and goes to V_{DD}.
The third pair of curves shows the response for input voltages higher than the required range. It corresponds to a critical point of the transition region. The voltages of nodes 1 and 2 during the preset phase are down to 0.5 V and 1.5 V. This is not enough for a quick establishment of the logic level. The higher voltage is still below 1.5 V at 4 nsec.

Fig. 6.20 shows another latch configuration. It uses a p-channel transistors for the regenerative loop and n-channel transistors for preset phase. The circuit works as follow. When the strobe is high (that corresponds to the preset phase: the strobe controls the gate of a p-channel transistor) the differential pair M_1 and M_2 discharges the output nodes. The current generator M_5 determines the bias current in the circuit. A possible unbalancement of the input voltages ties down to ground one node better than the other. This determines the unbalancement that, when the strobe goes down, makes different the initial value of the outputs, thus driving the positive feedback in the right direction. In the latch phase the current of M_5 controls the power consumption.

In the circuit of Fig. 6.21 the input signal discharges the drains of M_2 and M_3 during the preset phase. Also, the drain of M_8 and M_9 are brought to V_{DD} by

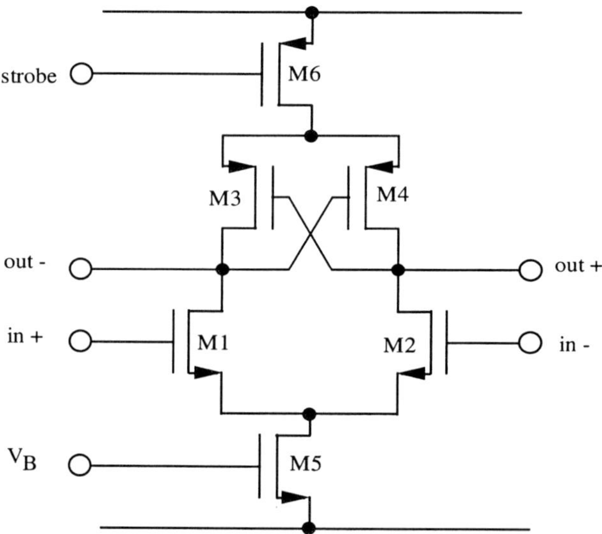

Fig. 6.20 - Simple latch configuration with the control of the tail current.

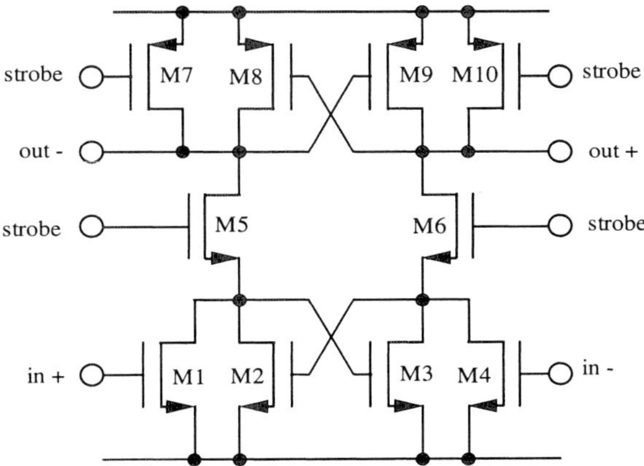

Fig. 6.21 - Latch with a double positive feedback.

the action of M_7 and M_{10} which gates controlled by the strobe signal. During the latch phase two regenerative feedback loop enforce one each other to obtain the output voltage.

The circuit of Fig. 6.22 is a combination of a gain stage and latch. It achieves its gain during the preset phase thanks to the transconductance of the pair $M_1 - M_2$ and the resistance at the output nodes and performs the latch function during the next phase by the use of $M_6 - M_7$. Transistors M_5 and M_6 switch the tail current established by M_9 toward the input pair $M_1 - M_2$ or toward the cross-coupled pair $M_6 - M_7$. Transistor M_9 controls the power consumption in the two phases of operation. In order to secure a proper gain in the preset phase the circuit requires the use of a common mode feedback (not shown in the circuit).

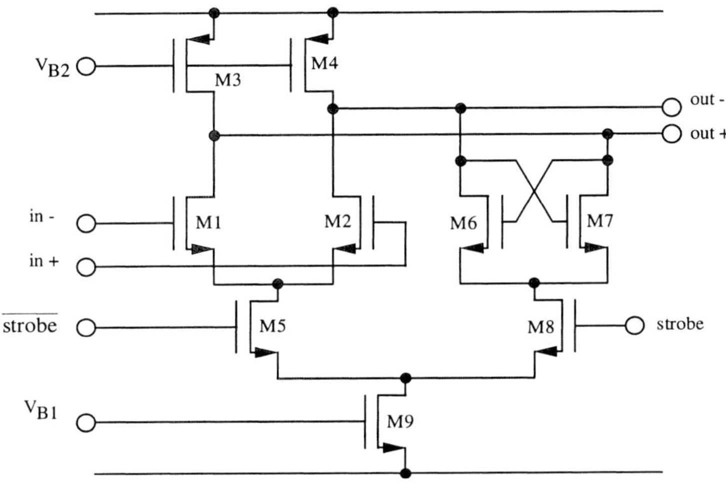

Fig. 6.22 - Combination of a gain stage and a latch.

Example 6.5

Study, with Spice simulations, the cascade of a pre-amplifier and a latch shown in the figure. Verify that the transistor sizes determine a pre-amplifier gain of 16. Determine the output response for 1 mV differential input. Estimate the kick-back on the output of the pre-amplifier that the latch activation causes.

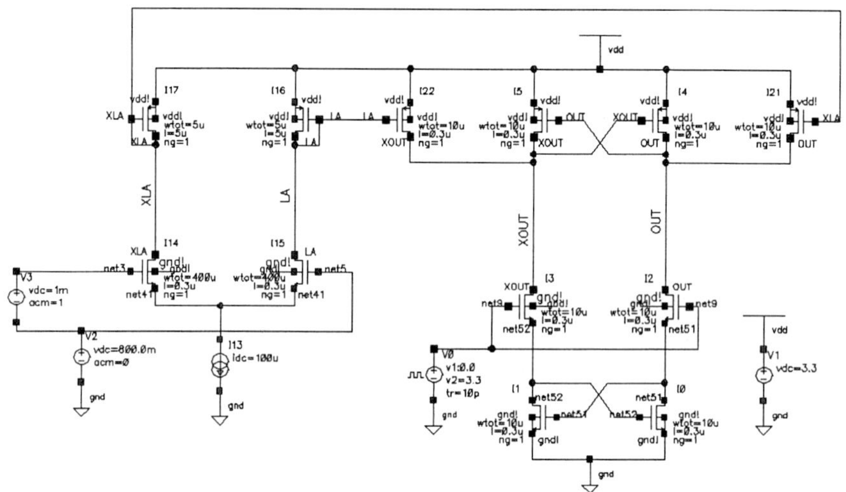

Solution:
Equation (6.18) provides the gain of the pre-amplifier

$$A_v = \frac{g_{m1}}{g_{m3}} = \sqrt{\frac{\mu_n(W/L)_1}{\mu_p(W/L)_3}}$$

where M_1 and M_3 are the input pair and the diode connected active load respectively. Assuming that the electron mobility is two times the one of holes, the above equation leads to a gain of 53. The simulation reveals a pretty lower result (A_1 =16) showing the limit of the approximate equation for pretty large transistors.
All the transistors of the latch have the same sizes (10μ/0.3μ). This choice facilitates the layout of the circuit.
The transient simulation of the circuit determines the result shown in the next figure. The first trace displays the strobe signal. The second one shows the outputs of the pre-amplifier and the last trace plots the output of the latch. The results show that after waking up the latch the output voltages drop down. This brings down the outputs of the pre-amplifier from the initial levels (1.048 V,

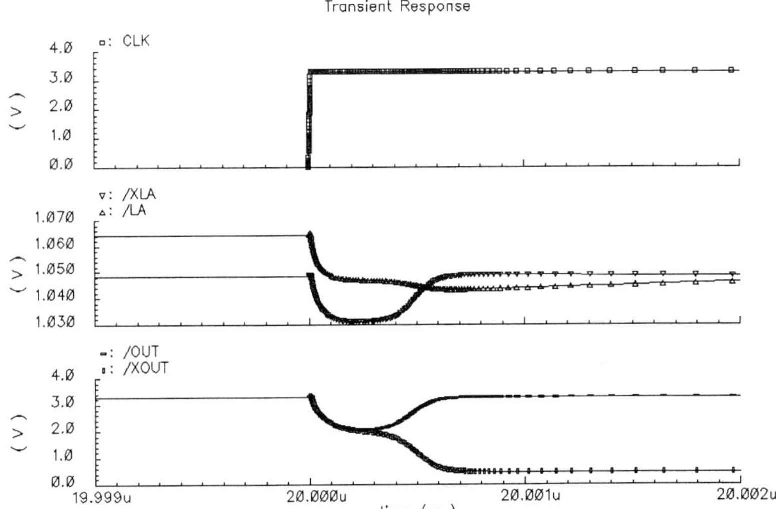

1.064 V) by 15 mV or so. After about 0.5 nsec the outputs of the latch reach the logic levels. By contrast the recovery of the outputs of the pre-amplifier is quite slow. Even, the curves cross one each other. The next diagram shows that the transient last for about 20 nsec. Therefore, the kick back produced by the latch disables the normal function of the pre-amplifier thus affecting the next comparison phase.

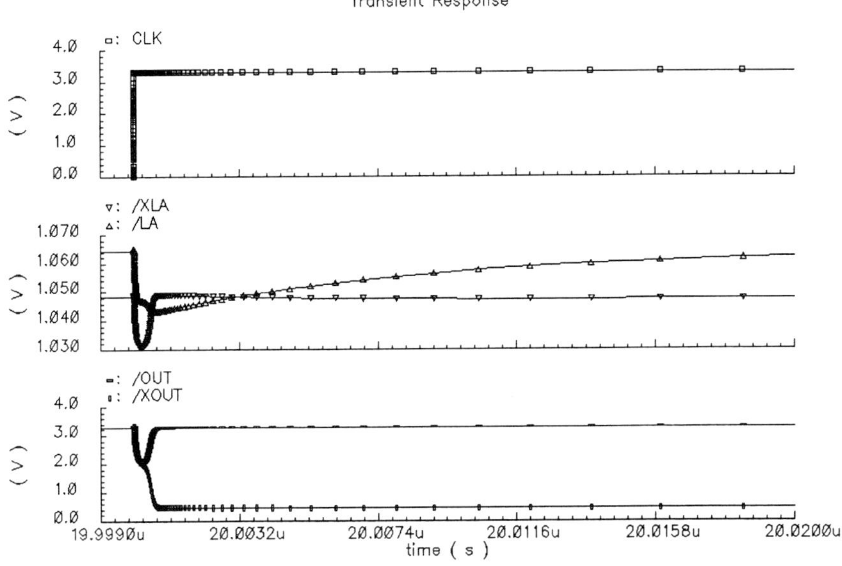

6.6 REFERENCES

- R. Gregorian, *Introduction to CMOS OP-AMPS and Comparators*, J. Wiley and Sons, New York, NY, 1999.
- B. Razavi and B.A. Wooley, Design techniques for high-speed, high-resolution comparators , IEEE Journal of Solid-State Circuits vol, 27, 1992, pp, 1916-1926.
- Jieh-Tsorng Wu and B.A. Wooley, *A 100-MHz pipelined CMOS comparator*, Journal Solid-State Circuits, Vol. 23, 1988, pp. 1379-1385.
- S. Tsukamoto, W. G. Schofield, and T. Endo, *A CMOS 6-b, 400-MSample/s ADC with Error Correction*, IEEE Journal of Solid-State Circuits vol, 33, 1998, pp, 1939-1947.
- M. P. Flynn, Member and B. Sheahan, *A 400-Msample/s, 6-b CMOS Folding and Interpolating ADC*, IEEE Journal of Solid-State Circuits vol, 33, 1998, pp, 1938-1938.
- I. Mehr, and D. Dalton, *A 500-MSample/s, 6-Bit Nyquist-Rate ADC for Disk-Drive Read-Channel Applications*,IEEE Journal of Solid-State Circuits vol, 34, 1999, pp, 912-920.

6.7 PROBLEMS

6.1 Repeat Example 6.1 but use an inverter with active load as basic gain stage. Use the following design parameters: $(W/L)_n = 20\mu/0.3\mu$; $(W/L)_p = 30\mu/0.4\mu$; $V_{DD} = 1.8\ V$. The n-channel element is the input device. Use the transistors models of Appendix C.

6.2 Simulate with Spice the cascode of three identical gain stages modelled by the equivalent circuit of Fig. 6.3. Determine the normalized response to a step input.

6.3 Repeat Example 6.2 but use the two stages op-amp designed in Example 5.2. Estimate the limitation to speed caused by the compensation network and show the benefit resulting from the switching-off of the compensation network during the comparator phase.

6.4 Simulate the circuit in Fig. 6.10 b). Use the gain stage of example 6.1 and $C_1 = C_2 = 1\ pF$. The switches are n-channel transistors whose sizes are $(W/L) = 3\mu/0.4\mu$. Verify that the residual offset (including the one caused by the clock feedthrough) of the first stage is cancelled by the use of phases like the ones shown in Fig. 6.11.

6.5 Design, using Spice and the models of Appendix C, the gain stage of Fig. 6.14 b). The gain must be larger than *25* and the output voltage must swing around *2.4 V*. V_{DD} = *3.3 V.*

6.6 Repeat Example 6.3. Modify the circuit design for an optimum speed. The current of the pre-amplifier can increase by a factor 2.

6.7 Design the latch of Fig. 6.20. The nominal current of M_5 is *150 mA*. The input unbalancement is *30 mV* around $V_{DD}/2$. Using the models of Appendix A and V_{DD} = *5 V* try to obtain a latching time below *0.7 nsec*.

6.8 Design the combination gain stage-latch of Fig. 6.22. The sizes of the transistors M_1–M_2 are $(W/L) = 40\mu/0.3\mu$ and the one of $M_6 - M_7$ are $(W/L) = 10\mu/0.3\mu$. The tail current is *50 μA*. Determine the sizes of M_3 and M_4 that leads to $A_1 = 50$. Design a proper common mode feedback and estimate the latch time.

APPENDIX A

Spice simulation parameters of an hypotetic 0.8 μm CMOS technology (suitable for level 2 Spice). Three cases, typical, fast and slow are given. Use only one for the pertinent simulation.

```
*                         TYPICAL CASE                        *
*
.MODEL MODN NMOS LEVEL=2
+ CGSO          =0.370E-09 CGDO  =0.370E-09 CGBO  =0.155E-09
+ CJ            =0.294E-03 MJ    =0.470E+00 CJSW  =0.294E-09 MJSW  =0.340E+00
+ JS            =0.011E-03 PB    =0.870E+00 RSH   =28.00E+00
+ TOX           =16.00E-09 XJ    =0.082E-06 LD    =-.010E-06
+ VTO           =0.840E+00 NSUB  =60.00E+15 NFS   =0.850E+12 NEFF  =11.00E+00
+ UO            =460.0E+00 UCRIT =36.00E+04 UEXP  =0.300E+00 UTRA  =0.000E+00
+ VMAX          =064.0E+03 DELTA =0.220E+00
*
.MODEL MODP PMOS LEVEL=2
+ CGSO          =0.370E-09 CGDO  =0.370E-09 CGBO  =0.110E-09
+ CJ            =0.340E-03 MJ    =0.500E+00 CJSW  =0.308E-09 MJSW  =0.230E+00
+ JS            =0.310E-03 PB    =0.870E+00 RSH   =32.00E+00
+ TOX           =16.00E-09 XJ    =0.068E-06 LD    =0.024E-06
+ VTO           =-.690E+00 NSUB  =16.00E+15 NFS   =0.550E+12 NEFF  =02.30E+00
+ UO            =181.0E+00 UCRIT =22.50E+04 UEXP  =0.260E+00 UTRA  =0.000E+00
+ VMAX          =046.0E+03 DELTA =0.820E+00
*
*                           FAST CASE                         *
*
.MODEL MODN NMOS LEVEL=2
+ CGSO          =0.353E-09 CGDO  =0.353E-09 CGBO  =0.140E-09
+ CJ            =0.282E-03 MJ    =0.470E+00 CJSW  =0.282E-09 MJSW  =0.340E+00
+ JS            =0.011E-03 PB    =0.870E+00 RSH   =27.00E+00
+ TOX           =14.00E-09 XJ    =0.082E-06 LD    =-.010E-06
+ VTO           =0.760E+00 NSUB  =60.00E+15 NFS   =0.850E+12 NEFF  =11.00E+00
+ UO            =470.0E+00 UCRIT =36.00E+04 UEXP  =0.300E+00 UTRA  =0.000E+00
+ VMAX          =064.0E+03 DELTA =0.220E+00
*
.MODEL MODP PMOS LEVEL=2
+ CGSO          =0.353E-09 CGDO  =0.370E-09 CGBO  =0.103E-09
+ CJ            =0.327E-03 MJ    =0.500E+00 CJSW  =0.327E-09 MJSW  =0.230E+00
+ JS            =0.310E-03 PB    =0.870E+00 RSH   =30.00E+00
+ TOX           =14.00E-09 XJ    =0.068E-06 LD    =0.024E-06
+ VTO           =-.600E+00 NSUB  =16.00E+15 NFS   =0.550E+12 NEFF  =02.30E+00
+ UO            =181.0E+00 UCRIT =22.50E+04 UEXP  =0.260E+00 UTRA  =0.000E+00
+ VMAX          =046.0E+03 DELTA =0.820E+00

*                           SLOW CASE                         *
*
.MODEL MODN NMOS LEVEL=2
+ CGSO          =0.400E-09 CGDO  =0.400E-09 CGBO  =0.170E-09
+ CJ            =0.312E-03 MJ    =0.470E+00 CJSW  =0.312E-09 MJSW  =0.340E+00
+ JS            =0.011E-03 PB    =0.870E+00 RSH   =29.00E+00
+ TOX           =18.00E-09 XJ    =0.082E-06 LD    =-.010E-06
+ VTO           =0.920E+00 NSUB  =60.00E+15 NFS   =0.850E+12 NEFF  =11.00E+00
+ UO            =450.0E+00 UCRIT =36.00E+04 UEXP  =0.300E+00 UTRA  =0.000E+00
+ VMAX          =064.0E+03 DELTA =0.220E+00
*
.MODEL MODP PMOS LEVEL=2
```

```
+ CGSO          =0.400E-09 CGDO  =0.400E-09 CGBO  =0.120E-09
+ CJ            =0.340E-03  MJ        =0.500E+00  CJSW     =0.340E-09  MJSW
=0.230E+00
+ JS            =0.310E-03 PB    =0.870E+00 RSH   =35.00E+00
+ TOX           =18.00E-09 XJ    =0.068E-06 LD    =0.024E-06
+ VTO           =-.770E+00  NSUB     =16.00E+15  NFS       =0.550E+12  NEFF
=02.30E+00
+ UO            =181.0E+00  UCRIT  =22.50E+04  UEXP      =0.260E+00  UTRA
=0.000E+00
+ VMAX          =046.0E+03 DELTA =0.820E+00
```

APPENDIX B

Spice simulation parameters of a 0.35 μm CMOS technology (suitable for Tanner Spice)

```
.MODEL MODN NMOS LEVEL=49
*
**** Common extrinsic model parameters ***
+ACM     =2
+RD      =0.000e+00 RS      =0.000e+00 RSH     =8.200e+01
+RDC     =0.000e+00 RSC     =0.000e+00
+LINT    =8.300e-09 WINT    =2.700e-08
+LDIF    =0.000e+00 HDIF    =6.000e-07 WMLT    =1.000e+00
+LMLT    =1.000e+00 XJ      =3.000e-07
+JS      =2.000e-05 JSW     =0.000e+00 IS      =0.000e+00
+N       =1.000e+00 NDS     =1000.     VNDS    =-1.000e+00
+CJ      =9.300e-04
+CJSW    =2.800e-10 FC      =0.000e+00
+MJ      =3.100e-01 MJSW    =1.900e-01 TT      =0.000e+00
+PB      =6.900e-01 PHP     =9.400e-01
*
**** Threshold voltage related model parameters ***
+NCH     =2.300e+17 VTH0    =4.600e-01
+VOFF    =-5.700e-02 DVT0    =2.200e+01 DVT1    =1.000e+00
+DVT2    =3.400e-03 KETA    =-6.200e-04
+K1      =6.000e-01
+K2      =2.900e-03 K3      =-1.700e+00 K3B     =6.300e-01
+PSCBE1 =2.700e+08 PSCBE2 =9.600e-06
+DVT0W   =0.000e+00 DVT1W   =0.000e+00 DVT2W   =0.000e+00
*
**** Parasitic resistance and capacitance related model parameters ***
+RDSW    =6.000e+02
+CDSC    =0.000e+00 CDSCB   =0.000e+00 CDSCD   =8.400e-05
+PRWB    =0.000e+00 PRWG    =0.000e+00 CIT     =1.000e-03
*
**** Flags ***
+MOBMOD =1.000e+00 CAPMOD =2.000e+00
*
**** Mobility related model parameters ***
+UA      =1.000e-12 UB      =1.700e-18 UC      =5.700e-11
+U0      =4.000e+02
*
**** Subthreshold related parameters ***
+DSUB    =5.000e-01 ETA0    =3.000e-02 ETAB    =-4.00e-02
+NFACTOR=1.100e-01
*
**** Temperature effect parameters ***
+AT      =3.300e+04 UTE     =-1.80e+00
+KT1     =-3.30e-01 KT2     =2.200e-02 KT1L    =0.000e+00
+UA1     =0.000e+00 UB1     =0.000e+00 UC1     =0.000e+00
+PRT     =0.000e+00
*
**** Overlap capacitance related and dynamic model parameters   ***
+CGDO    =2.100e-10 CGSO    =2.100e-10 CGBO    =1.100e-10
+CGDL    =0.000e+00 CGSL    =0.000e+00 CKAPPA =6.000e-01
+CF      =0.000e+00 ELM     =5.000e+00
+XPART   =1.000e+00 CLC     =1.000e-15 CLE     =6.000e-01
*
```

```
**** Saturation related parameters ***
+EM     =4.100e+07 PCLM   =6.800e-01
+PDIBLC1=1.100e-01 PDIBLC2=1.400e-03 DROUT  =5.000e-01
+A0     =2.200e+00 A1     =0.000e+00 A2     =1.000e+00
+PVAG   =0.000e+00 VSAT   =1.200e+05 AGS    =2.5000e-01
+B0     =-1.80e-08 B1     =0.000e+00 DELTA  =1.000e-02
+PDIBLCB=2.600e-01
*
**** Process and parameters extraction related model parameters ***
+TOX    =7.700e-09 NGATE  =0.000e+00
+NLX    =1.900e-07
+XL     =5.000e-08 XW     =0.000e+00
*
**** Substrate current related model parameters ***
+ALPHA0 =0.000e+00 BETA0  =3.000e+01
*
**** Geometry modulation related parameters ***
+LL     =0.000e+00 LW     =0.000e+00 LWL    =0.000e+00
+LLN    =1.000e+00 LWN    =1.000e+00 WL     =0.000e+00
+WW     =0.000e+00 WWL    =0.000e+00 WLN    =1.000e+00
+WWN    =1.000e+00
+W0     =1.200e-07 DLC    =8.300e-09
+DWC    =2.700e-08 DWB    =0.000e+00 DWG    =0.000e+00
*
**** Noise effect related model parameters ***
+AF     =1.400e+00 KF     =2.800e-27 EF     =1.000e+00
+NOIA   =1.000e+20 NOIB   =5.000e+04 NOIC   =-1.40e-12
+NLEV   =0
*
* END MODEL
*
*
.MODEL MODP PMOS LEVEL=49
*
**** Common extrinsic model parameters ***
+ACM    =2
+RD     =0.000e+00 RS     =0.000e+00 RSH    =6.000e+01
+RDC    =0.000e+00 RSC    =0.000e+00
+LINT   =9.900e-08 WINT   =3.900e-08
+LDIF   =0.000e+00 HDIF   =8.000e-07 WMLT   =1.000e+00
+LMLT   =1.000e+00 XJ     =3.000e-07
+JS     =2.000e-05 JSW    =0.000e+00 IS     =0.000e+00
+N      =1.000e+00 NDS    =1000.     VNDS   =-1.000e+00
+CJ     =6.000e-04
+CJSW   =3.300e-10 FC     =0.000e+00
+MJ     =4.400e-01 MJSW   =2.400e-01 TT     =0.000e+00
+PB     =8.400e-01 PHP    =9.400e-01
*
**** Threshold voltage related model parameters ***
+K1     =5.600e-01
+K2     =-1.600e-02 K3     =1.500e+01 K3B    =-1.40e+00
+NCH    =5.900e+16 VTH0   =-7.800e-01
+VOFF   =-1.100e-01 DVT0   =2.000e+00 DVT1   =5.000e-01
+DVT2   =-4.00e-02 KETA   =-7.700e-03
+PSCBE1 =5.000e+08 PSCBE2 =1.000e-10
+DVT0W  =0.000e+00 DVT1W  =0.000e+00 DVT2W  =0.000e+00
*
**** Parasitic resistance and capacitance related model parameters ***
+RDSW   =3.800e+03
+CDSC   =0.000e+00 CDSCB  =0.000e+00 CDSCD  =2.200e-04
+PRWB   =0.000e+00 PRWG   =0.000e+00 CIT    =3.200e-04
```

```
*
**** Flags ***
+MOBMOD =1.000e+00 CAPMOD =2.000e+00
*
**** Mobility related model parameters ***
+UA     =6.800e-11 UB     =1.000e-18 UC     =-1.200e-10
+U0     =1.100e+02
*
**** Subthreshold related parameters ***
+DSUB   =4.400e-01 ETA0   =4.800e-02 ETAB   =-3.50e-05
+NFACTOR=2.200e-01
*
**** Temperature effect parameters ***
+AT     =3.300e+04 UTE    =-1.40e+00
+KT1    =-5.70e-01 KT2    =2.200e-02 KT1L   =0.000e+00
+UA1    =0.000e+00 UB1    =0.000e+00 UC1    =0.000e+00
+PRT    =0.000e+00
*
**** Overlap capacitance related and dynamic model parameters   ***
+CGDO   =3.400e-10 CGSO   =3.400e-10 CGBO   =1.300e-10
+CGDL   =0.000e+00 CGSL   =0.000e+00 CKAPPA =6.000e-01
+CF     =0.000e+00 ELM    =5.000e+00
+XPART  =1.000e+00 CLC    =1.000e-15 CLE    =6.000e-01
*
**** Saturation related parameters ***
+EM     =4.100e+07 PCLM   =1.500e+00
+PDIBLC1=5.900e-03 PDIBLC2=3.400e-04 DROUT  =7.900e-02
+A0     =7.500e-01 A1     =0.000e+00 A2     =1.000e+00
+PVAG   =0.000e+00 VSAT   =9.500e+04 AGS    =1.700e-01
+B0     =3.400e-07 B1     =0.000e+00 DELTA  =1.000e-02
+PDIBLCB=-3.00e-01
*
**** Process and parameters extraction related model parameters ***
+TOX    =1.300e-08 NGATE  =0.000e+00
+NLX    =2.800e-07
+XL     =0.000e+00 XW     =0.000e+00
*
**** Substrate current related model parameters ***
+ALPHA0 =0.000e+00 BETA0  =3.000e+01
*
**** Geometry modulation related parameters ***
+W0     =7.300e-07 DLC    =9.900e-08
+DWC    =3.900e-08 DWB    =0.000e+00 DWG    =0.000e+00
+LL     =0.000e+00 LW     =0.000e+00 LWL    =0.000e+00
+LLN    =1.000e+00 LWN    =1.000e+00 WL     =0.000e+00
+WW     =0.000e+00 WWL    =0.000e+00 WLN    =1.000e+00
+WWN    =1.000e+00
*
**** Noise effect related model parameters ***
+AF     =1.800e+00 KF     =1.100e-26 EF     =1.000e+00
+NOIA   =1.000e+20 NOIB   =5.000e+04 NOIC   =-1.40e-12
+NLEV   =0

* END MODEL
```

APPENDIX C

Spice simulation parameters of a 0.25 μm CMOS technology (suitable for Cadence Spice).

P-type transistor

```
* ------------------------------------------------------------------
************************ SIMULATION PARAMETERS **********************
* ------------------------------------------------------------------
* format    : SpectreS
* model     : MOS BSIM3v3
* ------------------------------------------------------------------
*                    TYPICAL MEAN CONDITION
* ------------------------------------------------------------------
*
simulator lang=\spice
.model &1 bsim3v3 version=\1.0 type=\p   capmod=1.900e+00 &
  mobmod=1.000e+00   nqsmod=0.000e+00   noimod=1.000e+00 &
      k1=5.600e-01 &
      k2=-4.30e-02       k3=4.500e+00       k3b=-8.50e-01 &
     nch=1.000e+17     vth0=-4.60e-01 &
    voff=-1.10e-01     dvt0=1.400e+00      dvt1=3.800e-01 &
    dvt2=-1.10e-02     keta=-2.50e-02 &
  pscbe1=1.000e+09   pscbe2=1.000e-08 &
   dvt0w=0.000e+00    dvt1w=0.000e+00     dvt2w=0.000e+00 &
      ua=2.100e-10       ub=8.200e-19        uc=-5.20e-11 &
      u0=1.200e+02 &
    dsub=5.000e-01     eta0=2.200e-01      etab=-3.90e-03 &
 nfactor=8.200e-01 &
    pclm=2.900e+00 &
   drout=5.000e-01 &
      a0=1.400e+00       a1=0.000e+00        a2=1.000e+00 &
    pvag=0.000e+00     vsat=1.900e+05       ags=3.400e-01
    b0=2.700e-07         b1=0.000e+00 &
   delta=1.000e-02  pdiblcb=-1.70e-02 &
 pdiblc1=3.300e-02 &
 pdiblc2=1.000e-09 &
      w0=4.800e-08 &
    dlc=-5.60e-08 &
    dwc=3.800e-08       dwb=0.000e+00       dwg=0.000e+00 &
     ll=0.000e+00        lw=0.000e+00       lwl=0.000e+00 &
    lln=1.000e+00       lwn=1.000e+00        wl=0.000e+00 &
     ww=0.000e+00       wwl=0.000e+00       wln=1.000e+00 &
    wwn=1.000e+00 &
      at=3.200e+04      ute=-1.30e+00 &
     kt1=-5.60e-01      kt2=2.100e-02       kt1l=0.000e+00 &
     ua1=0.000e+00      ub1=0.000e+00        uc1=0.000e+00 &
     prt=0.000e+00 &
    cgdo=2.000e-10     cgso=2.000e-10       cgbo=1.000e-10 &
```

```
    cgdl=0.000e+00      cgsl=0.000e+00     ckappa=5.900e-01 &
      cf=0.000e+00       elm=5.000e+00 &
   xpart=1.000e+00       clc=0.900e-15        cle=5.800e-01 &
    rdsw=1.800e+03 &
    cdsc=6.900e-04     cdscb=2.900e-04      cdscd=1.900e-04 &
    prwb=0.000e+00      prwg=0.000e+00        cit=1.100e-04 &
     tox=7.500e-09     ngate=0.000e+00 &
     nlx=1.700e-07 &
      xl=5.000e-08        xw=0.000e+00 &
      af=1.200e+00        kf=1.000e-27         ef=1.000e+00 &
    noia=1.000e+20      noib=5.000e+04       noic=-1.40e-12 &
      rd=0.000e+00        rs=0.000e+00        rsh=1.560e+02 &
    minr=1.000e-03 &
     rdc=0.000e+00       rsc=0.000e+00       lint=-5.60e-08 &
    wint=3.800e-08      ldif=0.000e+00       hdif=6.000e-07 &
      xj=3.000e-07        js=2.000e-05 &
       n=1.000e+00 &
        dskip=\no             tlev=0             tlevc=0 &
      cj=1.300e-03      cjsw=3.700e-10 &
      fc=0.000e+00      fcsw=0.000e+00 &
      mj=5.500e-01      mjsw=3.800e-01 &
      pb=1.000e+00      pbsw=1.000e+00
* -----------------------------------------------------------------
```

N-type transistor

```
* -----------------------------------------------------------------
*********************** SIMULATION PARAMETERS **********************
* -----------------------------------------------------------------
* format    : SpectreS
* model     : MOS BSIM3v3
* -----------------------------------------------------------------
*                     TYPICAL MEAN CONDITION
* -----------------------------------------------------------------
*
simulator lang=\spice
.model &1 bsim3v3 version=\1.0 type=\n   capmod=1.900e+00 &
  mobmod=1.000e+00   nqsmod=0.000e+00   noimod=1.000e+00 &
      k1=6.000e-01 &
      k2=2.900e-03       k3=-1.70e+00       k3b=6.300e-01 &
     nch=2.300e+17     vth0=4.200e-01 &
    voff=-5.60e-02     dvt0=2.200e+01      dvt1=1.000e+00 &
    dvt2=3.300e-03     keta=-6.10e-04 &
  pscbe1=2.700e+08   pscbe2=9.600e-06 &
   dvt0w=0.000e+00    dvt1w=0.000e+00     dvt2w=0.000e+00 &
      ua=1.000e-12       ub=1.700e-18        uc=5.700e-11 &
      u0=4.000e+02 &
```

```
    dsub=5.000e-01      eta0=3.000e-02      etab=-3.90e-02 &
 nfactor=1.100e-01 &
    pclm=6.900e-01 &
   drout=5.000e-01 &
      a0=2.200e+00        a1=0.000e+00        a2=1.000e+00 &
    pvag=0.000e+00      vsat=1.100e+05       ags=2.400e-01
    b0=-1.70e-08          b1=0.000e+00 &
   delta=1.000e-02   pdiblcb=2.500e-01 &
 pdiblc1=1.000e-01 &
 pdiblc2=1.400e-03 &
      w0=1.100e-07 &
     dlc=8.100e-09 &
     dwc=2.600e-08       dwb=0.000e+00       dwg=0.000e+00 &
      ll=0.000e+00        lw=0.000e+00       lwl=0.000e+00 &
     lln=1.000e+00       lwn=1.000e+00        wl=0.000e+00 &
      ww=0.000e+00       wwl=0.000e+00       wln=1.000e+00 &
    wwn=1.000e+00 &
      at=3.300e+04       ute=-1.80e+00 &
     kt1=-3.30e-01       kt2=2.200e-02      kt1l=0.000e+00 &
     ua1=0.000e+00       ub1=0.000e+00       uc1=0.000e+00 &
     prt=0.000e+00 &
    cgdo=2.000e-10      cgso=2.000e-10      cgbo=1.000e-10 &
    cgdl=0.000e+00      cgsl=0.000e+00    ckappa=5.900e-01 &
      cf=0.000e+00       elm=5.000e+00 &
   xpart=1.000e+00       clc=0.900e-15       cle=5.800e-01 &
    rdsw=6.000e+02 &
    cdsc=0.000e+00     cdscb=0.000e+00     cdscd=8.400e-05 &
    prwb=0.000e+00      prwg=0.000e+00       cit=1.000e-03 &
     tox=7.500e-09     ngate=0.000e+00 &
     nlx=1.900e-07 &
      xl=5.000e-08        xw=0.000e+00 &
      af=1.400e+00        kf=2.800e-27        ef=1.000e+00 &
    noia=1.000e+20      noib=5.000e+04      noic=-1.40e-12 &
      rd=0.000e+00        rs=0.000e+00       rsh=8.200e+01 &
   minr=1.000e-03 &
     rdc=0.000e+00       rsc=0.000e+00      lint=8.200e-09 &
    wint=2.600e-08      ldif=0.000e+00      hdif=6.000e-07 &
      xj=3.000e-07        js=2.000e-05 &
       n=1.000e+00 &
         dskip=\no              tlev=0              tlevc=0 &
      cj=9.200e-04      cjsw=2.700e-10 &
      fc=0.000e+00      fcsw=0.000e+00 &
      mj=3.100e-01      mjsw=1.900e-01 &
      pb=6.900e-01      pbsw=6.900e-01
* ----------------------------------------------------------------------
```

Index

D

E

F

G

H

I

L

R

S